The Arts of the Microbial World

synthesis

A series in the history of chemistry, broadly construed, edited by Carin Berkowitz, Angela N. H. Creager, John E. Lesch, Lawrence M. Principe, Alan Rocke, and E.C. Spary, in partnership with the Science History Institute

The Arts of the Microbial World

Fermentation Science in   Twentieth-Century Japan

VICTORIA LEE

The University of Chicago Press

Chicago and London

The University of Chicago Press, Chicago 60637
The University of Chicago Press, Ltd., London
© 2021 by The University of Chicago
Published 2021
Printed in the United States of America

30 29 28 27 26 25 24   2 3 4 5

ISBN-13: 978-0-226-81274-8 (cloth)
ISBN-13: 978-0-226-81288-5 (e-book)
DOI: https://doi.org/10.7208/chicago/9780226812885.001.0001

Library of Congress Cataloging-in-Publication Data

Names: Lee, Victoria (Science historian), author.
Title: The arts of the microbial world : fermentation science in
    twentieth-century Japan / Victoria Lee.
Other titles: Synthesis (University of Chicago Press)
Description: Chicago : The University of Chicago Press, 2021. |
    Series: Synthesis | Includes bibliographical references
    and index.
Identifiers: LCCN 2021027282 | ISBN 9780226812748 (cloth) |
    ISBN 9780226812885 (ebook)
Subjects: LCSH: Fermentation. | Microbiology—Japan.
Classification: LCC QR151 .L35 2021 | DDC 572/.49—dc23
LC record available at https://lccn.loc.gov/2021027282

♾ This paper meets the requirements of ANSI/NISO Z39.48-1992
(Permanence of Paper).

"I merely borrowed the power of microbes."

ŌMURA SATOSHI

press conference, October 5, 2015

Contents

Microbe History

In twentieth-century Japan, scientists and skilled workers sought to use microbes' natural processes to create new products, from soy sauce mold starters to monosodium glutamate (MSG), from vitamins to statins. They emphasized their ability to deal with the enormous sensitivity and variety of the microbial world: to distinguish between the microbes, refine their qualities, and propagate them according to which was best for making the end products in mind. In the early twentieth century, in cities such as Kyoto and Osaka, small-scale, family-run *moyashi* makers, who made microbial starter in the form of dried mold spores for the brewing industries, relied mainly on sensory means to identify good spores. In color, yellow or yellow-green spores tended to darken to brown with time. From the smell, the maker could tell how dry the spores were and the method of production, with especially well-known regional differences between makers in Kyoto and Osaka. The makers could preserve and improve the chosen spores by cultivating them on special ash, or by new scientific techniques of pure culture to cultivate only one strain. They sold the best spores under their brand names to thousands of specialist *kōji* (the rice mold used in Japanese brewing) makers, or to sake, soy sauce, and *shōchū* (distilled liquor) houses.

Half a century later in the 1960s, scientists in the central laboratory of a large pharmaceutical company in Osaka deliberated whether to use gram-positive or gram-negative, *Actinomycetes* or *Bacillus*, known or wild microbes and their mutants to develop a product. Like *moyashi* makers, pharmaceutical companies had come to invest in the extensive knowledge and effort required to maintain microbial collections, which are alive and thus need to be constantly recultured. Often the researchers turned to microbes with established lineages, such as antibiotic producers, in the national strain collection housed in the laboratory's vicinity. One of the researchers instead isolated a new *Bacillus subtilis* microbe from a sample of *nattō* (fermented soybeans, a widely consumed food) purchased from a traditional Kyoto shop, which turned out to be a very high-potency strain. The researchers developed it to ferment nucleotide flavorings for coating MSG crystals. The pharmaceutical giant Takeda Yakuhin sold the flavoring products to instant ramen, miso, and *kamaboko* (boiled fish paste) producers, as well as to the rapidly expanding flavorings market among ordinary households.

This book tells the story of a society where scientists asked microbes for what they termed "gifts."[1] From the microbe's integration as a concept imported into political economy at the turn of the twentieth century to its application in recombinant DNA biotechnology since the 1980s, the ubiquity and significance of fermentation in Japan is hard to overstate: in daily life (as common foods like miso and *nattō*), in popular culture (the boy who can see microbes in the manga and anime *Moyashimon*), and in scientific achievements (the Nobel Prize–winning drug ivermectin). In areas including sake and soy sauce brewing, biofuels, antibiotics, and nutrition and flavors, our global history of biotechnology and the chemical industry is wanting without accounting for Japanese innovations. In turn, our understanding of Japanese industries, as well as of people's modern lives, is incomplete without a consideration of fermentation. In traditional brewing houses as well as in the food, fine chemical, and pharmaceutical industries across the country, scientists and skilled workers came to study microbial life and to tinker with life as fermentation phenomena.

Their view of life crystallized out of new institutions in agriculture and engineering in the late nineteenth century, particularly the discipline of agricultural chemistry (*nōgei kagaku*), which the government supported in the hope of improving industries that had existed in premodern Japan. Their efforts meant that indigenous techniques were a part of modern science as it took shape in Japan, contradicting earlier scholars' assumption that the intellectual and material culture of post-Meiji (1868–1912) Japan was an abrupt break from the country's Asian past.[2] They show how salient the question of cultural difference remains in twentieth-century science and technology—and how, when we look through the eyes of scientists and technicians in another society, we find that experts approach common problems through a distinctive set of categories.

Fermentation methods cut across promises to transform Japanese society using science and technology, which were often framed in overtly nationalistic terms. At the turn of the twentieth century, fermentation scientists believed that by better controlling the life of microbes' "struggle for existence" within the *moto* (mold-yeast starter), sake brewers could "improve" indigenous industries in the competitive struggle against foreign goods.[3] Shortly after the Japanese invasion of Manchuria in 1931, one of the discoverers of vitamins, Suzuki Umetarō (1864–1943), declared that there would come a time when every kind of food and drink could be made with only coal, earth, air, and water—when "borrowing" the lives of farm animals or microbes would no longer be necessary to address Japan's "resources problem" of nutritional shortage, which had been framed in terms of a national ecological crisis in the industrializing era after World War I.[4]

Technology folded into the myth of history and civilization and became part of national and imperial ideology. One award-winning scientist who was a specialist in Chinese molds wrote in 1945 that the Japanese race's skill and excellent products were proof of their mutational superiority, and that while Korean brewing merely copied Chinese mold preparation techniques, Japanese brewing had received no major influence from China historically.[5] The promise of fermentation—with its focus on local commercial products, and with the idea of microbes as a nutritional and industrial resource—carried over easily into postwar concerns for economic development. *Kōji* has been classified today as an organism unique to Japanese breweries, and in 1957 a biochemist described it as Japan's "national enzyme," an object that could create new industries.[6] A common saying among Japanese microbiologists goes: "Ask microbes, and they will never betray you." And at the Manshuin temple in Kyoto, there is a "microbe mound" (*kinzuka*) to commemorate the lives of the microbes that have died for the advancement of human science and technology.

This book explores why a vision of life as fermentation expanded beyond small-scale traditional manufactures to take special prominence in food, resources, and medicine. It does so by addressing the pivotal role of scientists and technicians in defining the material texture of everyday life as an aspect of political economy through applied science.[7] The scope of the fermentation community itself—which spanned government ministries, the military, industry, and universities—shows the inseparability of science from technology, business, and government in Japan's modern development. The twentieth century was a period of shifting policy paradigms: from industrialization at the turn of the century, to combating resource scarcity and developing the empire after World War I, to autarky and fascism in the 1930s and 40s, and to high-speed economic growth in the post–World War II period. Yet knowledge of microbes—from the earlier concern for improving traditional craft industries at the beginning of the century to the anxiety about resources in the interwar and wartime period, and persisting strongly even in the post-1945 era of big corporations and mass production—lay at the heart of some of Japan's most prominent technological breakthroughs in the global economy. By highlighting the unexpected continuity of a vision of life as fermentation in Japan, I trace the interconnections of the modern state with science, technology, and capital to situate the place of indigenous industry in a modern world.

Here was a modern society that made an alternative technical choice, in which microbes became understood as living workers instead of just deadly germs. A differing set of scientific questions, values, and contributions to biological knowledge resulted. As I will argue, Japanese research in fermentation concentrated on metabolism over genetics; emphasized function over identity; and concerned itself more with what living things can do, or can be asked to do, than with what they are. Moreover, fermentation scientists in Japan came to see biological activity at the level of microorganisms as key mediations in ecologies that bound together far broader processes in society. Over the last century, this fermentation-based vision of life stood in contrast to the dominant trends in the global history of biology and biomedicine.

Microbes other than germs were relatively unseen amid the leading developments in biology and biomedicine in the twentieth century. The rise of molecular biology, which focused on the cell's genetic material and its role in reproduction, eclipsed work on metabolism and physiology in the rest of the cell in capturing scholarly and popular attention.[8] Since the 1970s, microbes' productive capacity has been extensively harnessed in recombinant DNA biotechnology.[9] In the last two decades, however, a new awareness of interactions between realms formerly thought to be separate—reproduction and metabolism, genome and environment, people and other organisms—means that we have come again to pay attention to the role of microbes in nutrition, health, and the environment. These include developments in the fields of epigenetics, the microbiome, horizontal gene transfer, and green chemistry.[10]

These biological debates affect our everyday lives. Environmental scholars calling for a closer engagement with biology have increasingly focused on the importance of microbes, as in discussions of our war on microbes using industrial chemicals, which has changed ecologies that feed back to alter our bodies.[11] Contestations over whether humans should recognize microbial allies alongside enemies have begun to alter the value and culture of labor in the world of craft food.[12] The sea change in biology's view of microbes resonates more broadly with growing ambivalence toward biomedicine's militaristic approach toward microbes, as well as with a widespread rise in the popularity of fermented and probiotic cuisine, and a heightening awareness of microbes' historically positive contributions to human health and the environment.[13] It is a moment when twenty-first-century developments suggest that the traditional eradication-based approach to the microbial world is unsustainable.

Japanese microbiology offers an in-depth look at an alternative world in which microbes were more than just germs, and were central in twentieth-century life sciences. In this book I thus take up an invitation from theoretical work on pluralism in science to conduct "complementary science."[14] The program attempts to recover knowledge that was lost when a technical choice was made in the past, and to thus enrich present-day science by expanding the range of potential approaches for consideration. Rather than pursuing a counterfactual inquiry, however, I follow scientific developments in a lesser-known context that *actually happened*, in order to return to our awareness a sense of the importance of parallel traditions of scientific research. In this case, it was an approach in which microbes became understood and used as much as living workers as pathogens to be fought. My exploration presents a comparable perspective from history's reservoir of possibility. Its aim is to help us reflect on recent dramatic shifts in biology that add to scientists' role as "microbe hunters," trained principally to eradicate disease, the additional role of "microbe smiths" who seek to manage microbial interactions with society.

CRAFT AND SCIENCE

This study's questions go beyond prior investigations of the rise of a military-industrial complex in imperial Japan, by exploring connections between scientists' modern conceptions of microbial biosynthesis and the knowledge within indigenous industries that existed in Japan before the Meiji period.[15] Masayuki Tanimoto has argued for the significance of "indigenous development" in Japan's modern economy.[16] While few will dispute that Western models fundamentally shaped scientific institutions in modern Japan, I follow the questions raised by acknowledging the importance of small-scale traditional industries in the Japanese economy and exploring their implications for science.[17] Recent work has emphasized how the Meiji state built on the initiatives of local industry and government, using a network of middle-level knowledge institutions such as experiment stations and technical schools to improve indigenous industries.[18] Accordingly, my narrative includes fermentation research not just in the science faculties of universities but also in agriculture and engineering, and in national, local, and colonial government research institutes, technical colleges, and small- and medium-scale brewing enterprises, as well as in large companies and the research laboratories within them.[19]

The improvement of indigenous industry in the Meiji period was significant for science far beyond that period. Knowledge of microbes lay at the heart of Japanese scientific and technological contributions to the postwar global economy in the modern pharmaceutical, food, and biotechnological industries (what we would call the recent applied sciences, or the science-based industries), as well as to the life sciences. Rather than following the traditional approach of concentrating on in-depth investigation of a single scientific or industrial field, this book aims at teasing out vital connections across fields. In Japanese fermentation we do not see the persistence of artisanal forms of production free of scientific knowledge and new technologies; instead, we see the knowledge and skills that were linked to the vision of life as fermentation, in scientific and industrial forms that we consider as firmly "modern" and as having global reach. In addition, Japan's modern narrative has key regional implications that remain understudied. This is partly because Japanese scientific work left a legacy in post-1945 Korea and Taiwan as well as in China.[20] But it is also because similarities between indigenous industries in the region suggest the potential for fruitful comparison.

In focusing on the significance of premodern continuities, this study addresses the question of what it means to understand "modern" science in a non-Western context. To reexamine scientific modernity is especially critical in a non-Western setting where our assumptions hide rather than clarify many fundamental changes. "Westernization," and even "domestication" or "vernacularization" in Japanese science, was not merely a question of time, as the "diffusionist" view of the progressive infiltration of modern science from the Western metropole in the non-Western periphery through colonialism and nationalism would imply.[21] When we look beneath the Western-style names of Japan's modern scientific institutions, and look through the eyes and hands of the scientists and technicians themselves, we find that categories of investigation distinctive to that society, which owed their existence at least partly to premodern practices of fermentation, were significant.[22] For example, divisions into conventional scientific or industrial fields preclude investigation of microbial biosynthesis in Japan, despite the ubiquity of fermentation (*hakkō*) in Japanese science and technology.[23]

Fermentation processes in living cells were both biochemical pathways that technical practitioners sought to understand, and manufacturing methods that they tried to design and manage. In centering the narrative on these practical activities of knowledge-making, and tracing how they became so prominent in modern scientific and industrial life, this book attempts to sidestep binaries of theory and practice, as well as those of science and technology and of technology and art, which scholars of the modern American and European context have critiqued.[24] We should not dismiss the practical aspects of knowledge-making in the non-Western context as being "merely" practical, although they are understudied in East Asian history. As Dagmar Schäfer has argued for late imperial China by building on work by Pamela Smith, hybrid figures and "artisanal epistemologies" played a significant role in changes in cultures of knowledge. Even when artisans themselves have left little trace in written sources, understanding the knowledge implied in the objects they have made requires contextualization within how society and various individuals thought about production and labor.[25] Thus, an examination of the role of technology and crafts can potentially uncover an epistemological, social, and cultural cosmos.

This book spans the period from the turn of the twentieth century, when Japan was an industrializing country on the periphery of the world economy, to 1980, when it had emerged as a global technological and economic power. By focusing on a fermentation-based vision of life that took shape in Japan, I examine the role of cultural and technical continuities with the premodern period in sustaining Japan's technological breakthroughs in the global economy. It is not that those continuities made Japan uniquely suited to modernity, but that the premodern context continues to matter when we discuss the modernity of any non-Western country, even in the supposedly culturally neutral and universalistic areas of science and technology. I intend to use fermentation not to essentialize Japanese society or developments in its science, which underwent fundamental changes in each period in my chosen time frame, but rather to point to the importance of the starting point for change that shaped developments while not determining them.

SCIENTISTS AND THE POLITICAL
ECONOMY OF EVERYDAY LIFE

In order to trace Japanese experts' attempts to manage microbial ecologies as well as their social implications, I elucidate scientists' role in shaping the material texture of everyday life as a dimension of political economy.[26] Since scientists were middle managers of production—who neither directed national policy nor were passive agents of it—their visions were important in shaping the specifics of material outcomes, and their approaches revealed persistent patterns across the twentieth century. The time frame of the fermentation story, from 1900 to 1980, is especially telling because it shows how, for these producers, the earlier mindsets from the period of concern for improving traditional craft industries at the turn of the century, and from the period of anxiety about resource scarcity in the interwar and wartime era, absolutely persisted even in the post-1945 high-growth era of big corporations and mass production. The unexpected continuity is what I highlight in this book, and the explanation for it is in the interconnections between state control, business interests, technological innovation, and scientific research. These apparently distinct spheres cannot be separated in the history of the microbe as productive life, nor in modern Japanese development as a whole.

In other words, by focusing on one kind of technology—productive microbes—we see continuities in the approaches of those in the scientific community who were involved directly in production, and in how this "middle level" of society understood its relationships with the state and with consumers and the environment. When considering Japanese political economy, historians have tended to take one of two approaches. The first is top-down, focusing on a developmental state whose policies facilitated growth. Another is bottom-up, adopting the viewpoint of consumers who produced a mass market, or of pollution victims campaigning for redress.[27] Centering those practitioners who were connected to problems of production adds nuance to narratives polarized between the developmental state and social response, and emphasizes that the perspectives of middle-level technical experts are central to understanding the interplay between industrialization and environment in the modern period. How did fermentation scientists conceptualize and navigate ecologies in their material designs that determined what people could consume? Against the perception of Japanese culture as being uniquely harmonious with nature, and in the spirit of previous scholarship that has underlined the multifaceted ecological visions in Japanese society, recent historical work has highlighted the significance of understandings of ecologies in Japanese political life, economic organization, and moral consciousness.[28] Scientists' direct link to industry allows us a look at how modern forms of production were created hand in hand with visions of broader environmental management.

INSTITUTIONAL BACKGROUND

At this point, it will be helpful to describe briefly the relationships between Japanese industry, academia, and government agencies as background to my narrative.[29] In the 1850s, the United States and other Western powers coerced Japan by military force into accepting unequal treaties. Following the overthrow of the Tokugawa (1603–1868) shogunate, the new Meiji (1868–1912) government implemented a series of policies ("Rich country, strong army") to promote national military modernization and industrialization, in order to avoid colonization by Western powers. The Meiji state imported advanced Western technology, temporarily hired foreign experts as teachers and government consultants, sent Japanese students abroad, and established educational institutions to teach Western-style disciplines, including imperial universities to train a Japanese elite for government service. The early Meiji government privileged capital-intensive imported technologies and set up a series of state-owned enterprises which failed commercially and, from the 1880s, were sold off to what became the large industrial combines known as the *zaibatsu*. From then, the state worked to help coordinate already existent initiatives of prefectural governments and local industrialists in promoting traditional industries through a network of middle-level institutions, such as technical colleges to train technicians for industry, and prefectural and national laboratories to conduct research on behalf of small- and medium-scale businesses.

World War I was a massive stimulus to the Japanese heavy and chemical industries, due to the disruption of imports from Europe to East and Southeast Asia. A new awareness of the importance of physical sciences research to the growth of domestic industry and to warfare resulted in the creation of the Institute of Physical and Chemical Research (Riken), which was jointly funded by government and corporate donors, and the institute also came to commercialize its inventions to sustain itself financially.[30] The imperial universities and technical colleges had close and mostly informal links with industry, as certain prominent faculty would train large numbers of students who entered industry upon graduation, while the personal networks they formed might persist long after graduation.[31] In addition, faculty might serve as technical consultants for companies.[32] The Japanese colonial governments in Taiwan and Korea, as well as the South Manchuria Railway Company, supported central research laboratories to promote agricultural and industrial development in the colonies, where various industries also came to compete with those on the home islands.[33] There were other Japanese state institutes for technical exchange with the continent, such as the Shanghai-based Tōa Dōbun Shoin (East Asia Common Culture Academy) and Shanhai Shizen Kagaku Kenkyūjo (Natural Science Research Institute), which are understudied.

In the wake of the Great Depression the Japanese government began to encourage industrial rationalization by fostering cartels, and it became increasingly concerned about self-sufficiency in raw materials and industrial production. Another expansion in the heavy and chemical industries followed the acquisition of Manchuria in 1931, owing to military spending. The outbreak of the Asia-Pacific War in 1937 accelerated trends already in place since the early 1930s. The National Mobilization Law in 1938 gave the state vastly increased powers over the economy, at least in principle, and the state directed research in science and technology through new agencies to coordinate research across government ministries, academic laboratories, and industries. In particular, the new Technology Office (Gijutsuin) oversaw all wartime technology policy.[34]

Allied bombing destroyed much infrastructure and many factories, and the greatly decreased food production was insufficient to feed the population. The economic crisis continued after the surrender in 1945. The American occupation authorities (known as SCAP, for Supreme Commander for the Allied Powers, or GHQ, for General Headquarters) set about demilitarizing Japanese society. They implemented policies to split up the *zaibatsu*, though these eventually regrouped more loosely, and they abolished the Technology Office and the wartime control associations. SCAP substantially expanded the higher education system beyond the former imperial universities, but the latter remained dominant in scientific research. The Japan Science Council was established to promote science and advise the government, with the council being elected by the scientific community. From 1949, the Communist victory in China and the outbreak of the Korean War (1950–53) led to a "reverse course" in occupation policy, after which SCAP strongly supported Japanese industries and made American technology and know-how readily available to Japanese firms.[35]

After the occupation ended in 1952, the Japanese government continued to promote industrial recovery and technology transfer, especially through the Ministry of International Trade and Industry (MITI), which had the power to provide foreign exchange to firms for transactions, though the pharmaceutical industry was under not MITI but the Ministry of Health.[36] Changes in communications technology significantly increased international exchange in Japanese science. In the 1950s and 1960s, a number of large corporations built central research laboratories which were better equipped and funded than most university laboratories. At the same time, informal links remained between academia and industry, and national and prefectural government research institutes continued to perform research. Corporate laboratories supported pure as well as applied science. Movement was relatively fluid, and there were many scientists who took up professorships in universities upon leaving the laboratories. In addition, interdisciplinary research associations bound industrial and academic researchers into a research community.[37]

After the oil shocks of the 1970s, new government policies promoted a broader transition of the industrial structure, from an emphasis on "energy-intensive" heavy industries to "knowledge-intensive" industries. The latter most notably included information technology, but also biotechnology. However, the general changes in economic conditions eventually resulted in an overall drop in research funding, most visibly in private firms.[38]

OVERVIEW

Historians of biotechnology have suggested that fermentation science serves as the main background to understanding developments in Japanese biotechnology as they unfolded after the advent of recombinant DNA technology in the late 1970s and early 1980s. They point to fermentation science — or to one of its most prominent institutions, the discipline of agricultural chemistry (*nōgei kagaku*) — as the backstory for Japanese contributions to areas ranging from drug development to the Human Genome Project to nanobiotechnology.[39]

I build on these insights, and draw on historical studies of individual industries such as sake, soy sauce, alcohols, and pharmaceuticals; scientists' memoirs and biographies; institutional histories (*shashi*) compiled by universities, research associations, and corporations; and accounts of individual scientific fields, as well as technical publications and interviews with microbiologists in both academia and industry. These elements enable an original perspective on twentieth-century Japanese fermentation science. Divisions into conventional academic or industrial fields, with their origins in industrializing Europe and America, have so far obscured the continuity of fermentation science despite its strong disciplinary institutionalization in Japan. I do not intend to give a comprehensive account; instead, I highlight some of the most pivotal changes in research trends across a range of relevant industries.[40] Each chapter in this book focuses on a different class of material goods — the scientists and technicians who designed them, and their effect on the lives of people in Japan — and addresses both contextual developments in Japanese society and comparisons with other national settings.

Chapter 1 looks at how traditional practices of brewing shaped conceptions in microbiological research in Japan from the turn of the twentieth century. It explores a national movement to improve the sake and soy sauce industries as part of a broader effort to upgrade traditional industries in order to modernize the Japanese economy. I trace the adoption of pure culture methods to highlight the role of smaller players and skilled workers in industry against the background of state-sponsored scientific research, especially the *moyashi* makers who specialized in selling dried mold spore preparations to seed the making of *kōji*. The skills and traditions embodied in local industry helped to create a relatively autonomous and lasting scientific practice in Japan of seeing microbes as living workers as much as pathogens.

The next three chapters deal with hybrid sectors between "modern" and "traditional" industries in the interwar and wartime periods, where manufacturing processes relied on a combination of methods from indigenous and imported technological systems. Fermentation work in nutrition and alcohols attempted to address problems expressed in terms of self-sufficiency and the scarcity of agricultural resources. Chapter 2 explores how fermentation scientists intervened beyond the traditional brewing industries in the food and drug industries, bringing their distinctive conception of life to nutrition research. Amid worries about a Malthusian crisis in industrializing Japan, they developed an increasingly ecological vision of nutrition and gave special attention to the role of microbes in mediating the relationships between agricultural and industrial production and everyday consumption. Their material interventions became mass-produced and consumed after the Asia-Pacific War, including synthetic sake, chemical soy sauce, and yeast and vitamin preparations.

Chapter 3 examines the construction of a national "Japanese" fermentation tradition by analyzing an anthropological text authored by an agricultural chemist in 1945. Yamazaki Momoji's *Tōa hakkō kagaku ronkō* (A Study of East Asian Fermentation Chemistry) displays national and nationalistic strategies indebted to Asia, contradicting the notion that Japanese practitioners perceived science to be entirely Western — as implied in Japan's imperialist rhetoric of Western-style scientific modernization — and suggesting instead that modern Japanese scientific identity was defined primarily against Asia rather than against the West. Chapter 4 addresses the practical significance of Asian scientific knowledge in Japan beyond the discursive level. Regional mold technologies were used in the attempt to achieve self-sufficiency in industrial alcohol, especially at the frontiers of empire, and were an important part of Japanese colonial development, particularly in semitropical Taiwan. The all-encompassing wartime drive to develop biofuel as a substitute under the Allied gasoline embargo led to significant expansion of the institutions of fermentation science.

The final two chapters follow post-1945 reconfigurations in fermentation science up to about 1980. Chapter 5 explores the domestication of penicillin production immediately after the war, which was a priority for the Allied occupation government. Penicillin heavily shaped the context of industrial recovery for the pharmaceutical and fine chemical sectors and the resumption of scientific contact between Japan and the United States. It involved the new, imported submerged-culture technology, but intellectual interactions with agricultural science's long-standing approach toward microbes as complex living technologies deeply shaped Japan's medical and antibiotic research. The interdisciplinary restructuring of knowledge, which had wartime origins, played a major role in defining the frameworks of biological research between academia and industry in postwar Japan.

Chapter 6 looks at how the global breakthrough of MSG fermentation from 1956—which made MSG production vastly cheaper, and which was the first technology to be exported out of Japan, in this case to the United States, after 1945—opened up possibilities for microbial biosynthesis far beyond antibiotics. Although laboratories in private corporations led the research, the academy served as an information medium resulting in highly similar lines of research and product development between commercially competing firms, as well as contributions to the international community in biochemistry and molecular biology. Even in the large-scale, science-based food and pharmaceutical industries, the fermentation community continued to rely on local traditions of handling and thinking about microbes.

1

Sake and Shōyu

REMAKING MOLD CULTURES

Whether Eastern or Western, whether old or new—where human culture
flourishes, one finds the technological operations and skills of fermentation.
Yet only in Asia, molds are used to turn starch into sugar before using yeast
to ferment, and these molds create the special properties of the wine of each
brewing region.
—Saitō Kendō, "Higashi Ajia no yūyō hakkōkin" (Useful Fermentation Microbes
of East Asia), 507

Fermentation phenomena, both as life processes and as technologies, hold special significance in Japanese scientific culture. They take pride of place as an area of expertise where the country leads in contemporary biotechnology, are prominent in daily life in producing commonplace foods such as miso (fermented soybean paste) and *nattō* (fermented soybeans) in people's homes, and have been a field of industrial specialization since medieval times in sake and other brewing houses. This chapter illuminates an early period in the creation of this scientific culture by looking at how Japanese scientists in universities, technical colleges, and government research institutes, as well as expert workers in the brewing industry, studied microbes at the turn of the twentieth century in the context of widespread and state-supported campaigns to modernize the indigenous brewing industries.

Japanese workers did not have the concept of "microbes" before the late nineteenth century. However, they had ways of understanding and handling microorganisms—visible en masse as mold formations—as part of essential steps in brewing not only sake and other liquors but also soy sauce and miso. The making of the rice mold *kōji* was the first step in brewing these products, and *kōji* making had been a lucrative monopoly industry since the thirteenth century. By the end of the nineteenth century, the preparation of dried mold spores to seed rice for making *kōji* had become a sector distinct from *kōji* making. These preparations were known as *moyashi* or *tanekōji*, and would be sold either to *kōji* makers or directly to those sake or soy sauce houses that made their own *kōji*. Around the same time in Europe, while bacteriologists developed new techniques for isolating and culturing microorganisms and rarefying their products to make vaccines, experts in brewing developed similar methods that would allow brewers to increase their control.[1] The novel techniques of pure culture were equally important to scientists, allowing them to preserve, collect, and classify individual microbial strains. This chapter focuses on the introduction of the technique of pure culture to the Japanese brewing industries at the turn of the century, and explores its implications for how both brewers—particularly *tanekōji* makers—and scientists worked with microbes.

How experts implemented pure culture in the Japanese fermentation industries opens a window onto the relationship between the modernization of traditional industry and the institutionalization of Western science in the Meiji period. The emergence of a new set of institutional structures for scientific research was driven by the combined efforts of local industrial leaders, prefectural government officials, and the Meiji state to improve Japanese industries, in order to increase their competitiveness both domestically and for the purpose of export.[2] Those industries that had existed in Japan before the Meiji period and continued to exist since—such as textiles, dyes, and pottery, to name a few—are some of the most important and most overlooked areas in modern Japanese science.[3] Among them, brewing contributed by far the highest values of production among the entire manufacturing sector at the end of the nineteenth century, as well as providing the government with its largest source of tax revenue and consuming a sixth of the rice harvested annually.[4] Through pure culture, this chapter traces one aspect of how in the late Meiji period, Japanese experts imported and adapted Western science to take a more systematic approach to processes of manufacturing. As the raison d'être that gave science its rationale in Meiji Japan, this practical application of foreign ideas to Japanese industries underlay the expansion and dynamism of institutions of agricultural and engineering science in the country.[5]

The new categories of "microbe" and "scientist" both relied on a notion of "nature" that had no Japanese-language equivalent in the Tokugawa period—a word that referred to the whole of material reality as something universal, as well as something distinct from "society."[6] As an object of knowledge, the microbe reflected a novel ontological division between cellular life and the environment, which the conception of nature allowed. As an institution of authority, the scientist was a new kind of expert who specialized in nature. Yet both categories were not only linked to European categories by a self-consciously Western refashioning of Japan's political economy in the Meiji period; they were also shaped significantly by conceptions that had emerged amid the vibrant protocapitalism of the Tokugawa era. The procedures and assumptions of commercial brewers that had been developing over several centuries and the pure-culture techniques of academic microbe scientists in turn-of-the-century Japan display a striking, suggestive convergence.[7] Likewise, the role of the microbe scientist within the modern Japanese state was not like that of the early modern intellectual, but was built instead on that of the technical specialist within the brewery: a manager of material production for capital accumulation, now on a national level.

The underexplored narrative of the significance of traditional industry exposes a different side to the formation of modern Japanese science than that seen in the dominant historiographical approach, which portrays the institutionalization of science in Japan primarily as a story of rapid transfer from the West, under the policies of a strong state and constituting an abrupt break from the past.[8] In this chapter, I suggest that local industry helped to shape a relatively autonomous tradition of seeing microbes as living workers as much as pathogens in Japan: a view that, through large and lasting institutions, remained powerful far into the twentieth century.

LANDSCAPES OF EXPERTISE

Brewers in Tokugawa Japan had handled *kōji* molds with specialist skill, understanding them to be essential raw materials in the brewing process, like water or rice. In the first decade of the twentieth century, brewers faced a new landscape of microbe species that scientists had divided into "useful ones" (*yūeki naru mono*) and "harmful ones" (*yūgai naru mono*) (figs. 1.1 and 1.2).[9] Within a species, microbe varieties were characterized by their physiological as well as morphological differences, which corresponded to their role in different industries: the fungi used in the sake industry formed more sugars, whereas the fungi used in the soy sauce and tamari industries formed more amino acids, chemicals associated with protein breakdown and flavor.[10]

By the close of the Tokugawa period, soy sauce, miso, and especially sake brewing accounted for the highest values of production by far in the entire nonagricultural manufacturing sector in Japan, easily surpassing weaving and raw silk. At the beginning of the twentieth century, among those wealthiest people who came by their riches through industrial manufacturing, there were more brewers than any other occupation, and their numbers rivaled those in rising modern industries such as cotton spin-

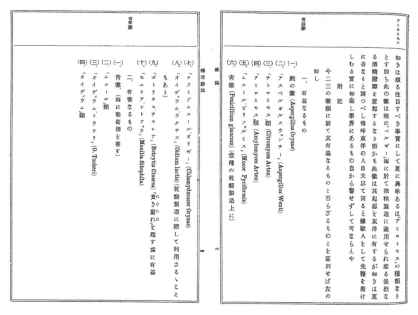

FIG. 1.1. The brewing industry and mold species, in Takahashi Teizō, *Jōzō bairon* (1903), 3–4. Right to left: the category "useful ones" lists ten species beginning with *Aspergillus oryzae*, followed by "harmful ones," which includes four species.

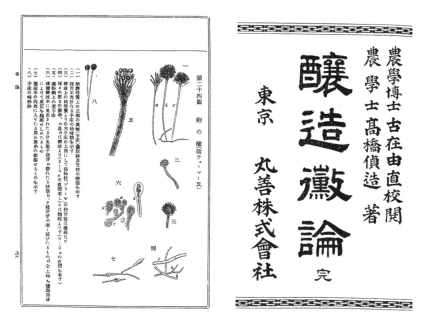

FIG. 1.2. Title page (right), and illustration of the *kōji* mold used in sake and soy sauce brewing (left), in Takahashi Teizō, *Jōzō bairon* (1903), front matter, 249.

ning.[11] The largest brewers had emerged in the second half of the Tokugawa period in Nada (near Kobe, west of Kyoto and Osaka) in the Kansai region of western Japan for sake, and in Noda and Chōshi (east of Tokyo) in eastern Japan for soy sauce. Those breweries, which were located in rural areas, employed dozens of workers from the surrounding vicinities and had become increasingly mechanized.

In the late Tokugawa period, Japan was one of the most highly urbanized societies in the world, with numerous large cities and castle towns that formed part of a national network of consumption and distribution.[12] The leading breweries competed with each other on scale as well as quality to ship their goods to major urban markets, particularly the biggest cities of Edo (renamed Tokyo after the Meiji Restoration), Osaka, and Kyoto. National markets grew first for sake, while the use of commercial rather than homemade soy sauce was initially more common in eastern than western Japan, and miso production was dominated by the home kitchen into the twentieth century.[13] However, brewing was a multilayered industry: outside urban areas, village residents bought mainly from small- or medium-scale local producers. In rural areas there was also widespread home brewing of unrefined sake (*nigorizake* or *doburoku*), drunk early to fuel a day's heavy labor on the farm.[14]

The brewing of rice into liquor, involving the *kōji* mold that grew on the rice, had been known in Japan for perhaps two millennia. The specific origins of *kōji* brewing were not known, though there were clear connections with mold brewing of grains on the Asian continent. In the medieval period, specialist rice wine brewers emerged in Kyoto to supply the aristocracy for ceremonial or medicinal purposes; and by the fourteenth century, commercial sake was also produced in the countryside for public drinking on market days and special occasions.[15] Sake breweries developed especially in urban areas with access to rice, such as port cities in the Kansai region that saw large commercial rice transactions, or nearby temple towns that could sell their products in Kyoto. In the seventeenth century, the establishment of the Tokugawa shogunate based in Edo brought the growth of cities across the country, as samurai were required to live in castle towns to serve their domain lord. Though officially political authorities encouraged commoners merely to farm in order to produce taxes for their lords, over time the domains came to be chronically dependent on prominent merchants for loans and contributions, in return offering privileges such as the recognition of trade monopolies for certain commodities within regional markets. Thus the commerce that flourished in the Tokugawa period came to be dominated by merchants with ties to domain officials, initially wholesalers based in the cities.

By the mid-eighteenth century, the economy experienced a host of changes that historians have documented as a distinctively rural-centered, protoindustrial, protocapitalist transformation.[16] Around the country, there was a dramatic growth in the number of rural households that produced goods for sale in distant markets, typically while engaging in side industries alongside agriculture. Through such small-scale manufacturing activities, as well as moneylending and experimentation with farming techniques, a number of rural elites began to amass substantial wealth. Major urban centers declined or stagnated as "country places" rose in their outskirts, challenged the hold of city merchants, and became centers of vibrant consumption as well as production. Many domain authorities by this time had come to tolerate and even promote commercial growth, since they were increasingly dependent on commoner elites for funds and trading services. As interregional competition for national markets intensified, domain authorities could turn a blind eye when rural elites usurped urban monopolies, or could encourage market-oriented production in the local region by aiding the introduction of new technologies, as well as inviting skilled experts from—or sending observers on trips to—more advanced regions.[17]

In this period, Nada sake brewers, who usually began as largeholders and rural merchants, competed with nearby urban establishments in Osaka, Nishinomiya, Ikeda, and Itami. Nada villages had the geographical advantages of having access to water power that could drive rice-polishing machines, being able to take advantage of a winter wage-labor force when repeated restrictions by political authorities forced the concentration of sake brewing into the agricultural off-season (which incidentally produced a better-quality sake, despite increasing the fermentation time), and being located quite close to city markets as well as growing rural markets.[18] Across the country, refined sake was no longer a luxury reserved for the aristocratic and samurai classes, and it came to accompany meals in restaurants and teahouses as well as rural households when visitors were entertained. Izakaya bars flourished in Edo and along major roads, while many villages hosted small breweries or at least a drinking establishment or two.[19]

Morohaku sake, in which white rice instead of unrefined rice was used not only in the fermentation mash (where white rice had long been used) but in *kōji* making as well, became widespread in the seventeenth century. *Morohaku* brewing encouraged broader changes in labor and technology within the sake industry.[20] The sharp increase in the amount of labor needed for rice polishing meant that rice polishers began to work separately from the brewing workers. Breweries began to rely on external technical specialists who were skilled in the brewing process, eventually leading to the rise of experts called *tōji*. Whereas medieval *kōji* makers had been a separate specialist industry, in the Tokugawa period the *kōji* making process was integrated into the sake brewery. However, *moyashi* making— the making of dried *tane* (seeds) to be used as starters for *kōji* making— gradually emerged as an independent industry.

Brewing differed from other traditional side industries, such as textiles, in that it required relatively high levels of capital and labor. Sake and soy sauce brewery owners needed a certain level of capital because they needed large tracts of land, buildings, and equipment such as tanks for the fermentation process, and because the brew took many months to mature. As for labor, the labor force was not family-based, as it was in textiles. A contract-based, seasonal labor force came to the brewing house to brew in the winter months. The labor force was all male. Until the fifteenth century, sake brewing had been a female-gendered profession, and women had led a number of merchant guilds. By the middle of the Tokugawa period, the rise of religious ideas of pollution linked to women meant that women had been banned from entering the brewery.[21] The brewery owner controlled the production process but did not oversee it himself. In the case of both sake and soy sauce brewing, he left the management of production to the *tōji* (head brewer) and the *kashira* (deputy) whom he hired, and then he interfered little. The *tōji* and *kashira* recruited the workers and oversaw the day-to-day running of the brewing house. It was the *tōji*, rather than the brewery owner, who was master of the knowledge of the fermentation process.[22]

Reflecting the understanding of stages of material change in the sake brewing process, under the *tōji* and *kashira* the employees were divided into *kōji* specialists, *moto* specialists, rice steaming specialists, cooks for all, and day laborers. The space of the brewing house mirrored these divisions, with its different rooms. The process began with the selection of the key raw materials, water and rice. Next was *kōji* making, which was done by adding purchased *moyashi* to steamed and dampened rice, letting it sit in the warm, humid *kōji* room to let it begin to grow, and then putting it into wooden boxes that would be lined up on shelves in the room. It would be left there for a few days, resulting in a fine-smelling green-yellow mold. The *moto* as a material was difficult to achieve well, and it guided the fermentation process. It was made by mixing *kōji* with steamed rice and water in several stages, and required regular stirring with an oar over many days. Then there was *moromi* making, in which *kōji* and *moto* were mixed with steamed water and rice. The ways in which this mash could be prepared were innumerable and characterized the style of the brewery. Finally, the resulting liquid would be squeezed out through a fine press into a large cask and, having been clarified, would be placed into storage. With the first press of the season, the brewers would hang a large green ball of fresh cedar leaves outside the brewing house, which would gradually turn brown to signify the sake's maturation.

To discourage spoilage, *hiire* ("putting in fire," a heating process) would be performed at the time of storage.[23] Soy sauce making too was similar to sake brewing, in that it was based on *kōji* making, followed by applying *kōji* to steamed soybeans, wheat, and water in a variety of styles. Brewery workers believed that gods were responsible for the changes in the materials that made sake, and they aimed to preserve the cleanliness and sanctity of the brewing space. Inside the brewery there would be a shrine devoted to a sake god. The clapping of hands in worship in front of the shrine, the exclusion of women from the space, and the changing into and out of indoor sandals when entering or leaving the brewery were all everyday precautions to ensure a smooth brew amid numerous possible contingencies.[24]

The Meiji period was a difficult time for many brewers, as the new government adapted European and American institutional models for industrial capitalism and military modernization, under the slogan of "Rich country, strong army" (*Fukoku kyōhei*). Having abolished the domains in favor of a centralized system of prefectures, and the status distinctions between samurai and commoners as well as their accompanying occupational privileges and constraints, the government dismantled the early modern guild structure through which the domains and shogunate had controlled production and commerce, lifting all restrictions on who brewed and what amount they brewed. New sake brewers proliferated, especially landlords employing tenant labor, while brewers as a whole experienced unstable fortunes. The central government taxed the alcohol industry heavily, frequently raising taxes in order to meet the demands of military preparation and infrastructure building. Cheap imported alcohol flooded the markets under the unequal treaties, and put further pressure on brewers. The alcohol industry became the Meiji government's greatest source of tax revenue, exceeding the land tax by the end of the nineteenth century.[25]

Private entrepreneurs, including rural elites, and the Meiji state invested in new capital-intensive projects during the state's push for what it saw as the late takeoff of the economy: modern banks, industries, railroads, and other infrastructure. Moreover, at the local level, rural elites were nationalistic and enthusiastic as they spearheaded development efforts in the traditional sector, even though government policies favored export products over sake or soy sauce.[26] In the celebrated brewing regions of Kansai, wealthy brewing improvers published manuals and trade magazines to disseminate scientific principles and new European methods—such as the use of thermometers and salicylic acid—for scaling up production and making goods competitive for export. In other parts of the country, small and medium-sized brewers similarly formed discussion societies, trade associations, and producers' cooperatives. They hoped to survive the competition in local or regional markets during the volatile economic conditions of the period.[27]

In the early Meiji decades, the hopes of rural elites dovetailed with government initiatives to import Western-style knowledge disciplines and factory models, which included establishing educational institutions and large-scale manufacturing enterprises. Rubbing shoulders with both Tokugawa political authorities and now Meiji officials, rural elites—who had sometimes been granted samurai privileges in the past—had high aspirations for the greater role they might play in the new Meiji regime. By the late 1880s, however, not only had many of the state-owned enterprises been sold to the private sector (what would become the industrial monopolies known as the *zaibatsu*) due to commercial failure, and prefectural agricultural schools, training centers, and experiment stations shut down; it was also clear that a goal of state initiatives was to provide a source of employment for former samurai. A majority of the students at government schools or prefectural stations were samurai, not farmers, while high-ranking prefectural officials were generally samurai from other parts of the country.[28]

Led by the rural elite, the local trade associations that had supported improvement-of-industry movements soon turned critical of the government. In the later Meiji decades, the state reversed its policies and, rather than solely prioritizing modern transplanted industries, they began to encourage small- and medium-scale traditional industries in order to make them more competitive, both domestically and for export mainly to Asia. The government consolidated a network of middle-level institutions for disseminating information, training technicians for industry, and conducting experimental research on behalf of small and medium-sized businesses.[29] As historians have more recently argued, rural-based traditional industry continued to play a significant role in the Japanese economy well into the twentieth century alongside the modern factory system.[30] In this way, state institutions built on the existent structures and energy of local rural movements—even where they became overseen by bureaucrats in the Ministry of Agriculture and Commerce and dominated by the scientifically trained, who were predominantly though not exclusively samurai—and consequently "made redundant many of the rural elites' traditional roles," as the historian Edward Pratt puts it, by the first decade of the twentieth century.[31]

It was in this context of industrialization and the improvement of traditional industry that science institutions took shape. As James Bartholomew notes, early on Japan's first imperial university, Tokyo University, was "institutionally innovative" in incorporating not only a powerful faculty of medicine but also strong colleges of engineering (1886) and agriculture (1890) from existing institutions. In government ministries, agricultural research was especially well supported.[32] At the turn of the century, the government worked with local elites to encourage the establishment of higher technical schools as well as agricultural and industrial experiment stations. The national Brewing Experiment Station was founded in Tokyo in 1904 under the Ministry of Finance, as part of a nationwide network of regional brewing experiment stations. Among other things, it surveyed breweries, ran training courses for *tōji*, undertook studies of raw materials, and promoted the use of pure-cultured yeasts in sake breweries in place of *moto* making, centering on yeasts distributed by the related trade association known as the Brewing Society.

In manuals, trade magazines, scientific books, and scholarly journals, microbes (*kin*) began to appear alongside their common names and species names, with sketches of their appearance under a microscope. They became the living forces of the brewing process—the *kōji* mold that turned starch into sugar, or the wild *kōbo* (yeasts) cultivated in the *moto* that transformed sugar into alcohol. In vernacular scientific reports, the microbe that was variously called *baishu* ("mold species," 1878), *shinkinshu* ("mushroom microbe species," 1881), *tōkakin* or *kōjibaikin* ("change-into-sugar microbe" and "*kōji* mold microbe," respectively, 1894) eventually became *kōjikin* ("*kōji* microbe") around 1895.[33] Scientists argued that the quality of the sake was connected to the purity of the *kōji* and to brewers' ability to keep the *kōji* free from contamination.[34] Some argued that the yeasts in the *moto* created specific tastes of sake, but brewers themselves did not accept this simplistic picture; and to scientists it quickly became clear that there were also symbiotic lactic acid bacteria involved in the ecology of the *moto*.[35]

A few of the scientists who named the earliest brewing microbes included the German botanist Hermann Ahlburg (a foreign consultant teaching at the Tokyo Medical School, the predecessor of Tokyo University's Faculty of Medicine) and the German agricultural chemist Oskar Kellner (a foreign consultant teaching at the Komaba Agricultural School, the predecessor of Tokyo University's Faculty of Agriculture), who named *Aspergillus oryzae* as the *kōji* mold of sake in 1876 and 1895.[36] Saitō Kendō, a doctor in botany from the Faculty of Science of Tokyo Imperial University, named *Saccharomyces soja* as a soy sauce yeast in 1905.[37] Yabe Kikuji—official appraiser at the Ministry of Finance, and a doctor in agricultural chemistry from the Faculty of Agriculture of Tokyo Imperial University—named *Saccharomyces sake* as a sake yeast in 1897.[38]

All these scholars (*gakusha*) with doctoral degrees (*hakushi*) in specialist disciplines were considered to be scientists who worked in government ministries, experiment stations, technical schools, or universities, or often in a number of these during their lifetimes.[39] A microbe scientist could also have been trained in a university faculty of engineering as a doctor in applied chemistry (Tsuboi Sentarō, discussed below), or as a doctor in pharmacy (Shimoyama Jun'ichirō, cited above), as well as a doctor from a university faculty of medicine. These trends were a result of the reorganization of investigation in Japan that came hand in hand with the introduction of the microbial entity in modern disciplinary science.

BREWING SCIENCE IN THE IMPROVEMENT OF TRADITIONAL INDUSTRY

In January 1901—roughly two years after returning from a spell abroad at the Versuchs- und Lehranstalt für Brauerei in Berlin (Research and Teaching Institute for Brewing in Berlin)—Kozai Yoshinao (1864–1934), professor and chair of agricultural products in the department of agricultural chemistry of the College of Agriculture at Tokyo Imperial University, delivered a lecture titled "On the Improvement of Sake Brewing" at the Tokyo Chemical Society (Tōkyō kagakkai), the prime meeting place for chemical industrialists and scientists to exchange information.[40] Kozai began by describing to the audience the basic nature of brewing processes (fig. 1.3). What was first needed in the making of sake was *kōji*, the most essential material in the brewing of sake. When the *kōji* was ready, one made *moto*, the material needed to ferment sake. A process of great time and labor that took about eighteen days, making the *moto* ferment through various kinds of manual art was the most difficult part of the *tōji*'s job. Kozai went on to explain these two materials' functions in chemical terms: *kōji* changed starch into sugar by the power of mold, whereas *moto* changed sugar into alcohol and therefore contained abundant amounts of yeast (here called *iisuto*, signifying a foreign term).[41]

FIG. 1.3. *Kōji* making (left) and *moto* making (right). Owned by the Hakushika Memorial Museum of Sake.

Since the various mold and bacterial germs mixed in with the liquid would continue to provoke constant transformation even after refining, the period of storage was when there was particular danger of spoilage, so one needed to be careful and from time to time perform *hiire*. However, in the first place the *tanekōji* used to seed the *kōji* was already not pure. Then during *kōji* making, various bacteria and other kinds of mold from the air would multiply, so the resulting *kōji* would also not be pure. The kind of germs differed with different *kōji*, but they were organisms that were harmful or ineffectual for brewing. Then one would use this impure *kōji* to make *moto*, when more organisms would enter through the air and water and breed. The yeast would gradually win the struggle for existence, but this was not to say that the other plentifully mixed-in microbes would necessarily die out. Some would die when the alcohol was produced, but some would simply be latent, and when their enemy yeast rested in work, which was when the yeast completed the task of fermenting, they would strengthen and cause all kinds of damage. In other words, not only might the sake fall ill from infectious diseases when brewing, but the buds of disease were already in the raw materials. In such cases there was not simply a danger of spoiling; it was *natural* for the sake to spoil. To improve sake brewing, it was necessary to remove every kind of harmful microbe. Kozai encouraged brewers to stop making the complicated *moto* entirely, and instead to use pure cultures of well-chosen yeast. He would have encountered the pure-cultured yeast method at close quarters during his time in Berlin.[42]

By the time Kozai gave this lecture, there had been two decades of movements in Japan to improve the sake industry by the application of scientific principles (*gakuri ōyō*) since the 1880s.[43] The earliest movements had emerged in the traditional brewing centers of the Kansai region with individual wealthy brewing improvers, who published manuals and magazines in the spirit of spreading enlightenment and educating other large brewers like themselves about science. In these early manuals, which disseminated practical methods for achieving high-quality, standardized, mass-produced goods similar to those of their own famed brewing districts, they took knowledge that had previously been secret and experiential, and for the first time publicized it in the language of chemistry. The brewing improvers were well read in the chemical and bacteriological research on brewing that had been published in the 1870s by *oyatoi gaikokujin* (European and American "hired foreigners" whom the Meiji government had brought to Japan as consultants).[44] The movements for enlightenment and education were wealthy brewers' defensive responses on behalf of regional interests against the policies of the new state, as they faced the combined pressures of the domestic alcohol tax and competition with Western imports. They were also the reaction of these regional brewers against the state-led industrial campaigns focusing solely on heavy, chemical, and military-related industries, as well as on certain traditional products such as raw silk, tea, vegetable oils, and pottery that had quickly become important exports shaped to foreign tastes.[45]

By the 1890s, government officials in Tokyo too saw the economic imperative to nurture the larger traditional industries across the country to ease balance of payments, chiefly through export to other Asian countries.[46] Following the reports of Maeda Masana, an official at the Ministry of Agriculture and Commerce, the state shifted its focus from transplanted to traditional industries and began a concerted effort to build on the activities of local voluntary movements. In 1884 the central government issued a set of standards on the formation of trade associations across the nation, at the same time as it left local industrialists and prefectural authorities to oversee and implement the new regulations. Thereafter, the number of trade associations multiplied. Brewing improvers worked together with prominent Japanese scientists at the Imperial College of Engineering and technical advisers in the Ministry of Agriculture and Commerce, as well as local brewing notables, to set up industrial associations along with trade magazines and experiment stations.

While large, established brewers in Kansai sought improvement with a view toward scale-up and export, smaller sake brewers around the country, such as in the Tōhoku region of eastern Japan, subsequently picked up the movement for a different reason. They relied on new forms of technical communication, including brewing manuals, to help them imitate the techniques of the Nada districts and standardize quality. Their concern was to survive the competition against the expansion of larger brewers into local markets, and to push the quality of their product above home-brewed sake. Success was uneven: trade magazines oriented toward small-scale brewers complained about the "stubborn" *tōji*.[47] Results were mixed among large brewers as well. In Imazu in Nada, brewers hired a technical adviser from the Ministry of Agriculture and Commerce to run a series of trials, but the sake spoiled, causing enormous loss. Many Nada brewers became suspicious of the value of Western science for Japanese brewing, keeping "scholars' methods" at arm's length.[48]

Such disappointments left a continuing legacy. Kozai himself was aware of these problems when he gave another lecture in April 1901 before a nationwide association of brewers and brewing experts, upon receiving the Medal of Honor at the Tenth Sake Tasting Meet held in Saitama, outside Tokyo.[49] If the industry had been unable, artificially, to prevent damage by measuring natural influences on the brewing process, it was because knowledge was insufficient and the industry was immature, he said. But now Kozai spoke of the brewers who were paying large sums of money to transport water from the wells of Nishinomiya, one of the renowned brewing districts that was especially famed for its water, and he wondered if there were not sources of suitable water available in a more convenient location. Rather than investing in water or rice, Kozai pointed to microbial materials as the most important for brewing good sake (*junshu*). His pure yeast method was lifted directly from the model of beer and Emil Hansen's extensive research on yeast varieties.[50] He explained that different types of yeast (*kōbo*) had different physiologies, and only by selecting a superior kind of yeast could one guarantee a good brew.

Scientists saw themselves as playing a key role in raising production levels by encouraging a high level of knowledge across the entire country. Since the end of the Sino-Japanese War in 1895, when the Japanese government raised the alcohol tax to fund further military preparation, peripheral eastern brewers in Kantō and Tōhoku had become the most vocal in calling for the government to establish a national Brewing Experiment Station in Tokyo, hoping that the research and training it could provide to brewers would alleviate the challenges they faced vis-à-vis the large breweries in Kansai.[51] Small business owners in Tōhoku were hiring highly skilled *tōji* from Tanba in Kansai, for example, to help themselves survive.[52] Kozai argued that science-based technical training was superior to relying on the uncommunicated power of skilled craftsmanship that was held by the *tōji*. He, along with other sake improvement advocates, saw a number of changes as part of the tide of modernity. He extolled the advantages of labor- and cost-saving technologies and the ability to produce a uniform sake from a uniform yeast. In addition, he thought that professionalism in the industry should rely on academic knowledge as well as practical experience, as was the case in the professions of medicine and law. Science advocates hoped to improve the sake industry by attempting to "first improve [*kairyō*] the heads of *tōji*," as one opinion in a trade magazine put it.[53] Brewery owners were often preoccupied with account books and abacuses, entrusting technical matters entirely to *tōji*, so what was the point of lecturing to the owners?

Brewing was similar to other "traditional" rural industries in that most producers were too small-scale to support scientific experimentation, so that experts believed government institutions to be necessary to shoulder the development of better materials, techniques, and tools.[54] Around this time, while Tokyo Imperial University's agriculture and engineering colleges prepared students for work in government ministries, new technical schools trained many technicians for such sectors in industry as pottery or textiles. The largest of these were the higher technical schools, for which the government created a set of national standards in 1903 in order to encourage local elites in major manufacturing centers to establish such schools, though some cities had done so already.[55] The government issued a similar ordinance defining national standards for industrial experiment stations, which conducted surveys of local industries and hosted training courses. In the trade journal *Jōkai* in 1902, Takayama Jintarō, director of the Tokyo Industrial Experiment Station, opined in an article titled "The Necessity of a Brewing Experiment Station" that it was clear Japan needed a government-run institution for brewing improvement, as it should not be left entirely to the tradesmen to take up the initiative for an industry so important to the national income.[56]

From the perspective of chemical industry, Takayama explained, there were two different kinds of industries. On the one hand, there were industries that had their origins in the distant past and were unique. On the other, there were "imitation" industries taken from the West after the Meiji Restoration. The former included porcelain, lacquerware, sake, and soy sauce; the latter included matches, beer, cement, glass, soap, Western-style paper, and alkali products. The vitality had been tamed in the former, and naturally these industries tended to adhere to old customs and hesitated to seek improvement. But among them, Takayama reminded his brewer readers, there were industries such as porcelain and lacquerware that had attained worldwide renown for their manufacturing methods, and the shape and de-sign of their products. It was well known that these products had won great acclaim at the international exhibitions of Paris in 1867 and Vienna in 1873.

Takayama did not remind brewers how badly sake—a product possess-ing "not a single merit"—had been received at the same exhibitions. In the 1870s, the majority of Japanese writers had thought sake to be unhealthy, in contrast to the benefits of wine and beer. The imported liquors that "civi-lized" people drank flooded into the cities of Yokohama and Kobe under the unequal treaties, and could supposedly make farmers and soldiers work more efficiently.[57] Takayama reflected now, three decades later, that while the Meiji state campaigns under the slogan of "Develop industry, promote enterprise" (*Shokusan kōgyō*) had successfully transplanted Western indus-tries—not only stemming imports but even exporting products overseas—Japan should not lose unique products like sake. Takayama urged the ap-plication of modern scientific principles to the sake industry, even though sake brewing was mostly a household-scale enterprise.

Microbes made frequent appearances in trade magazines in the early 1900s. In the Q&A columns of *Jōkai*—a Tokyo-based periodical that aimed to disseminate enlightenment to medium and small-scale breweries—brewers asked questions such as, What kind of soy sauce fermentation mi-crobes existed? Moreover, what were the great enemies of sake brewing, "*bakuteriya*"? A technical expert replied that a multitude of different types of bacteria existed in the raw materials, and not all of them were bad or strong, nor was it easy to distinguish between them either by outward ap-pearance or under a microscope. He described and illustrated the appear-ance under a microscope of common bacteria that caused the main types of spoilage: incomplete fermentation (too sweet), nonvolatile acid fermen-tation (lactic acid fermentation), and volatile acid fermentation (butyric acid fermentation).[58]

One lecture published in the magazine explained the importance of hy-giene in brewing: some still believed that sake was the work of gods, the speaker began, for science had not completely opened the country. But disinfection by new chemical means was essential, and those hands who worked in the brewing houses must also keep clean clothes and clean bodies, as if they were in a sacred place; it was no coincidence that good sake was produced in places that adhered to such customs.[59] Another lec-ture explained methods for inspecting the number of bacteria in apparatus, or in materials such as *kōji* or *moto*, without a microscope, for microscopes were extremely expensive items, though they were certainly handy if one used them all of the time. Bacteria were invisible to the naked eye, but could be seen if they were propagated as colonies. The speaker explained the sampling and culturing methods, which required other specialized equipment such as flasks, filter paper, test tubes, petri dishes, and pipettes.[60]

The Ministry of Agriculture and Commerce and the Ministry of Finance finalized the decision to establish a national Brewing Experiment Station in Tokyo in July 1902, and soon it was under sole jurisdiction of the latter, sake brewing being the largest source of tax revenue in Japan. Among those in the initial investigation committee were Tejima Seiichi (principal of the Tokyo Higher Technical School), Takayama Jintarō (director of the Industrial Experiment Station), Kozai Yoshinao (professor at the College of Agriculture of Tokyo Imperial University), and Yabe Kikuji (official appraiser at the Ministry of Finance, who had isolated the first sake yeasts in 1893).[61] Medical bacteriology made the most glamorous and highest-visibility achievements in Japanese science in the early 1900s, and pathogenic thinking also dominated the earliest microbial ideas in brewing science. At the same time, the large-scale efforts of government officials, scientists, and industrialists to improve traditional industry—with their origins in movements from below at the initiative of local governments and rural industries—constituted one of the most dynamic, powerful, and wide-reaching scientific trends of the period.

PURE CULTURE PRODUCTION IN
THE *TANEKŌJI* INDUSTRY

By the turn of the twentieth century, to improve the brewing process using the study of microbes meant applying the techniques of pure culture. Pure culture methods first made their way into the brewing world not through the yeasts of *moto*, despite the pure-cultured yeast campaigns of academic and government scientists, but rather through *kōji*, via private-sector *tanekōji* suppliers (fig. 1.4). It is likely that *tanekōji* companies that cultured *kōji* microbes pure were in the minority in the early twentieth century. For example, Kōjiya Sanzaemon in Kyoto, which sold *moyashi* under the label Biokku and claimed lineage from a *kōjiza* ("*kōji* group") licensed by the Ashikaga shogunate (1338–1573), adopted pure culture technology much later, in 1951.[62] In the post–World War II period, however, the number of *moyashi* companies shrank. Compared with the hundred or so that existed at the beginning of the twentieth century, by the last decades of the twentieth century there were fifteen *tanekōji* companies, of which six distributed

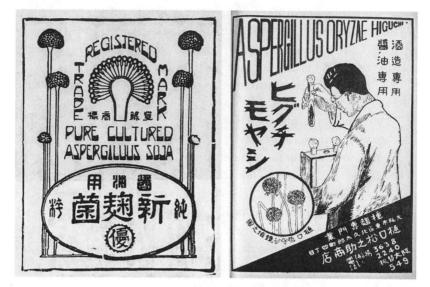

FIG. 1.4. Advertisements of *tanekōji* makers. Left: Konno Shōten, in *Jōzōkai* 15 (autumn 1924), 50. Courtesy of Akita Konno Shōten. Right: Higuchi Moyashi, in Shudō Nagatoshi, *Jōzō daijiten* (1931). Courtesy of the Society for Biotechnology, Japan.

nationally. Though the companies were all small-scale, with fewer than fifty employees each, the concentration of the industry implies the level of technology needed to stay competitive.[63]

Konno Seiji, the original founder of today's *moyashi* companies Akita Konno Shōten and Kobe-based Konno Shōten in the Meiji period, was a pivotal figure in bringing pure culture of *moyashi* into the brewing industry (fig. 1.5).[64] Born in 1882, Seiji was one of the sons of a brewing family in Kariwano in Akita Prefecture, the oldest after his elder brother died when Seiji was five. In the snowy town on the Sea of Japan side of Honshū, Seiji's father was the kimono-clad *tōji* of the family's soy sauce factory. However, a fire entirely destroyed the factory when Konno Seiji was young.[65] After Seiji completed his studies at Akita Middle School, he left the cold northern prefecture to study the scientific principles of brewing. At the time, the only college in Japan that had a brewing department was the Osaka Higher Technical School, located in the heart of the metropolitan merchant capital.[66]

Surrounded by the traditional brewing districts of Kansai, the Osaka Higher Technical School's Brewing Department trained technicians from breweries all over the country. The Brewing Department had been established in 1897 in response to calls from brewers to create an independent

FIG. 1.5. Konno Seiji. Courtesy of Akita Konno Shōten.

subject for brewing, unlike at the Tokyo Higher Technical School where training in the use of microscopes, for example, was under the Applied Chemistry Department. It was rumored that the manager of Osaka Beer (the predecessor of Asahi Beer) prodded the prefectural government's decision by buttonholing a high-ranking official after a nationwide meeting of the Association of Sake Brewing.[67] At the school, Konno Seiji studied under Brewing Department head Tsuboi Sentarō (1861–1921). Tsuboi, a chemist and microbiologist who had graduated from the Imperial College of Engineering (later the Faculty of Engineering at Tokyo Imperial University), saw the department's research as bringing scientific ideals and actual practice (*jicchi*) close together.[68] At the time, Tsuboi's laboratory was working on the pure culture of *tanekōji* as well as yeast for application in industry. Tsuboi's advertisement in the back pages of *Jōkai* in 1902 asking brewers to buy pure-cultured *moyashi* made by his college laboratory joined those of established commercial *moyashi* makers licensed by the Ministry of Agriculture and Commerce, who variously claimed that their particular pure-cultured *tanekōji*, the fruit of laborious research efforts and enthusiastically tested technology, drew high praise in the "twentieth-century brewing world."[69] In the picture that both Tsuboi and commercial *moyashi* makers painted in their advertisements, the application of science placed their product at the cutting edge of the industry.

Konno Seiji graduated in the spring of 1905 and entered Kawamata Shōyu, one of the largest soy sauce companies in western Japan, whose factory was part of the chimneyed cityscape of Sakai, south of Osaka. As chief technician of Kawamata, Konno was busily occupied with the mechanization of the factory.[70] He was a man so obsessed with the precision of watches that he would make charts of how late each one ran to record its reliability, checking its performance in horizontal and vertical directions.[71] Apprentices remember that he kept the factory very clean.[72] While directing the newly opened Kawamata Shōyu Brewing Experiment Station, that autumn Konno Seiji isolated an excellent *kōji* microbe, "*Kawamata kin*," which the company began to use for its soy sauce. In 1909 Seiji isolated a microbe suitable for sake; and the following year, while keeping his position at Kawamata, he and two brothers, Shigezō and Kenkichi, opened a shop called Konno Shōten in Kyoto, and began selling "*sake moyashi Konno kin*" and other microbes as pure-cultured *tanekōji* to *kōji* makers and to sake, miso, and soy sauce companies. The shop soon moved back to Osaka and opened another department for selling tools and machinery, many of which Seiji played a leading role in developing and patenting at Kawamata. Konno Shōten also published the trade journal *Jōzōkai* (Brewing World), and later opened a further soy sauce *moyashi* branch in Sakai and a sake *moyashi* branch in Nada.[73]

These developments were underway well before the national Brewing Experiment Station in Tokyo developed a method for the pure culture of *kōji* microbes, on which the related Brewing Society published its first report in 1911.[74] By then, other companies were already rapidly adopting the use of pure culture. The largest soy sauce companies in Kantō, such as Noda Shōyu (later Kikkōman), Yamasa Shōyu, and Higeta Shōyu, also began to make *tanekōji* in-house by the 1910s.[75] In fact, the head technician at Higeta Shōyu had interned under Konno Seiji at Kawamata before he first began isolating and pure-culturing *kōji* microbes for soy sauce *tanekōji* at Higeta in 1912.[76] Subsequently, new specialist *tanekōji* companies appeared in the late 1910s and 1920s to supply smaller soy sauce makers.[77]

Unlike the yeasts that the Brewing Society worked to maintain, the distribution of *kōji* microbes was already under the private monopoly of *tanekōji* companies, which specialized in preparing what were dried microbial spores that would seed *kōji* making elsewhere. The *tanekōji* sector had distant roots in the medieval *kōjiza* that held lucrative monopolies over *kōji* making and thereby controlled the source of the entire medieval brewing economy. In the medieval and early modern period, the shogunate and domains usually banned *kōji* making by unlicensed houses, partly to regulate tax collection but also to minimize brewing activity in order to suppress wastage of valuable rice. Where the monopoly system weakened, as had happened in the past, specialist *kōji* makers continued to supply brewing houses that preferred not to make *kōji* in-house.[78]

In the Meiji period as the sake improvement movement grew, more and more brewing companies requested *tanekōji* from specialist makers rather than making the starter in-house, and by the end of the period almost no sake brewers in Nada were making *tanekōji* themselves, though in-house manufacture was still prevalent in regions on the periphery of the sake economy.[79] Now it was common practice to shake off the spores of the *kōji* from a good brew, dry them, and use them as seeds in the next round of *kōji* making; the spores in this case were called *tomokōji*. However, the original spore starter—at this point sold by *tanekōji* houses distinct from specialist *kōji* makers—was tricky to generate.[80] In theory, if a sake or soy sauce company built a *kōji* room, it was possible to make original starter spontaneously by putting steamed soybean and ground wheat (to grow suitable microbes for soy sauce) or steamed rice (for sake) in an open *kōji* box, and then leaving it on the shelf of the *kōji* room to wait for mold to enter from the air. In a long-standing *kōji* room, plenty of good *kōji* microbes should have settled there and be floating in the air. But how could one get to that point, and where did *kōji* microbes come from? Moreover, how could one maintain good *kōji* after finding it? With successive culturing, any good *kōji* would become old, produce fewer spores, and become contaminated by other molds like *kekabi* or *kumonosukabi*. The color of the spores would darken and turn black, the mold would have a strange smell, and the taste of the sake or soy sauce would worsen.[81] Secrecy helped to preserve the *tanekōji* sector.

Konno Seiji's nephew Konno Kenji became head of Akita Konno Shōten much later, by which time the main branch of the company had moved back to Kariwano due to rice shortages in Kansai during World War II. Kenji has a childhood memory of watching an apprentice of Seiji, Ueno Shiejirō, "making *genkin*," or generating the original starter microbes (fig. 1.6).[82] Ueno had not attended a technical school, and had learned these methods under Konno Seiji during the war. Starting with a mass of spores floating in water, Ueno used a syringe to deposit a droplet of the spore mixture into a container of pure water, repeating until he had a very dilute mixture. Then he drew a mark on the cover glass and looked at the ster-

FIG. 1.6. Konno Shōten at the time of the company's founding in 1910. In the front row, fourth from right, is Ueno Shiejirō. Courtesy of Akita Konno Shōten.

ile colored liquid through the microscope, to see whether there was only a single spore on the dish. If there was, he sterilized a piece of filter paper by splashing alcohol on it and used it to suck up the spore. He expanded the single spore into a colony by culturing it on rice grains, in other words making *kōji* within a flask, using wide-bottomed flasks that Konno Seiji had specially designed to increase the area for culturing. Then he subjected the pure colony to testing, investigating properties such as the formation of proteases, amylases, acid-resistant amylases, and so on. Repeating the procedure with hundreds of single spores taken from the same original sample, Ueno selected the strongest resulting colony; and then, taking the spores from that colony, he repeated the entire process. In this manner, by thoroughly investigating weak and strong microbes using the single-spore method, only the microbes with the very best qualities would be propagated and made into *genkin*. Droplet by droplet, taking spores wrapped in single droplets, one could cultivate them and bring up their descendants.

The single-spore method was also crucial to preserving the selected microbe. Otherwise, when successively cultured, the strain would quickly degrade.[83] What protected the products of *tanekōji* makers who employed pure culture methods was partly their reputation, but also the fact that other makers did not have the technology to maintain the strains even if they physically possessed them (fig. 1.7). The expense of maintaining high-

FIG. 1.7. *Kōji* molds being cultured in wide-bottomed flasks on a lab bench at Akita Konno Shōten, February 2012. Photo by author.

quality strains also meant that breweries increasingly preferred to purchase *tanekōji* from specialist makers.

The reason for the swift adoption of pure-culture technologies in the *tanekōji* industry was that they were upgraded versions of technologies that *tanekōji* makers had already been using. Since the products they sold were dried microbial spores, they had long-held practices for identifying and isolating "good" cultures, mainly relying on sensory means. By inspecting the color of the *tane*, one could tell what kind of mold the *kōji* consisted of, as well as how old the *kōji* was, as the resulting yellow or yellow-green *tomo-kōji* tended to darken to brown with successive rounds of *kōji* making and with time. From the smell, one could tell how dry the *moyashi* was and the method of production.[84] This varied widely between makers—for example, in the geographical source of the ash used, the way they stacked the *kōji* boxes during *kōji* making, and the way and degree to which they dried the *tane*.[85] If the *moyashi* maker put the *tane* in his mouth, he could make similar distinctions through their taste and hardness. He could also check them for irregularities, now understood as growths of bacterial colonies. Finally, the maker could actually make *kōji* and see how smoothly it went, and then ask breweries to try out the *moyashi* and see how the sake, soy sauce, or miso tasted. By these means, the *moyashi* maker could select the best spores.

Since the late thirteenth century, makers had also attempted to store, maintain, and propagate the mold cultures as purely as possible by adding special ash.[86] In a 1903 report Tsuboi Sentarō noted that if one went to the places where *kōji* was made, sometimes the workers would first sprinkle camellia ash onto the rice; and then, after they mixed in the *kōji*, on which were stuck all kinds of microbes and bacteria, and brought the whole thing into the *kōji* room, somehow only the mold microbes would multiply.[87] New scientific methods allowed *tanekōji* makers to fulfill the same aims with a much higher degree of control.

Most importantly, *tanekōji* makers and academic scientists did not only share common tools and techniques; they also shared similar intellectual concerns. University laboratories, government-run experimentation stations, and the thousands of brewing houses that specialized in *tanekōji*, *kōji*, sake, or soy sauce shared concerns for isolating, identifying, and preserving individual microbial strains, and investigating their properties. Academic scientists depended upon *tanekōji* makers and other brewers to provide them with their objects of study, the microbes, which they then studied and preserved in a similar fashion. It was the intellectual concerns shared between academia and industry that helped to drive the adoption of new technologies in a dynamic private sector, a sector which in turn shaped the way academic researchers thought about problems and the research objects they used.

NOTIONS OF *KIN*

As academic experts adapted the techniques of Western microbiology to the microorganisms they found in Japan, processes of indigenous brewing shaped the concepts at the foundations of their science. For example, one, *kōji*, two, *moto*, three, *moromi* (brewing mash): these were the steps by which sake brewers organized their work to effect transformation in the raw materials. Scientists organized their categories of microbial inquiry along the same understandings of labor and chemical change.[88] Saitō Kendō (1878–1960) and Takahashi Teizō (1875–1952) later became key figures in building microbial strain collections in Osaka and Tokyo, and in establishing a tradition of microbial studies. From the material context that local industry provided for their investigations, they came to attach meaning to microbes as useful organisms as well as living beings with complex physiologies. Their approach to microbes had an important influence on fermentation science in the interwar period.

Much of the botanist Saitō Kendō's work was pioneering because "at the time, the kinds of wild fungi produced in Japan were completely unknown."[89] Back in 1902, the Investigation Committee for the Establishment of a Brewing Experiment Station asked Saitō to investigate the microbes around a possible site for a modern brewing laboratory in Takinogawa Village in Tokyo.[90] Saitō was a fresh graduate from the Faculty of Science at Tokyo Imperial University, and a mold specialist. Since at the time nobody in Japan was an expert on "fermentation microbes" (*hakkōkin*), he copied methods from German, Danish, and Japanese books on brewing science. The Tokyo Tax Office and Inspectorate published a collection of Saitō's scientific reports in 1905. His research was primarily taxonomic in aim, with the goal of elucidating and classifying new microorganisms.

Saitō specifically looked for organisms with properties that mapped onto operational stages in the brewing process.[91] Scientists knew that during soy sauce brewing, there must be microbes with starch and protein-decomposing ability that caused dramatic changes in the raw materials during *kōji* making and as the *moromi* matured; microbes that produced the acids, particularly lactic acid, present in high amounts in a mature *moromi* and in commercial soy sauce; and microbes that produced the small amounts of alcohol in the mature *moromi* that were likely to be responsible for soy sauce's distinctive aroma. Even if he did not find microbes with these three properties, Saitō reasoned, microbes that could survive in *moromi* with such a high concentration of table salt must have interestingly complex functions.

By seeking microbes as functional steps within the soy sauce brewing process, Saitō's study yielded one new species of yeast and two new species of lactic acid bacteria. Rather than transferring droplets of *moromi* directly onto the colloid-based culture medium for investigation, Saitō created an intermediate culturing step with a medium that more closely mimicked the conditions of the soy sauce itself—*kōji* water with 17 percent table salt (the same concentration as commercial Yamasa soy sauce from the Chōshi region), in which he first let a drop of *moromi* blossom for a few days. He argued that other microbes played symbiotic roles with the main *kōji* microbe, *Aspergillus oryzae*, and the soy sauce yeast, *Saccharomyces soja*; and that if brewers mismanaged this symbiosis, the brewing would proceed sluggishly.[92] On the other hand, he also found a host of new species of molds and bacteria that were "useless" in soy sauce brewing.

Saitō found more new species as he sampled microbes in the air and water in a sake brewery in Kumagaya in neighboring Saitama Prefecture in December, the month when sake brewing typically began.[93] Organisms and spores in the air settled onto petri dishes that Saitō placed around the rooms twice a day: the hot, humid *kōji* room that incubated *kōji* mold, where *kōji* microbes were plentiful, and the cool fermentation room where *moto*, rice, and water were left to transform into *moromi*, and in which various molds were present. But he discovered that some of the yeasts in the air were not good yeasts, because when they were cultured in *kōji* water, they easily suppressed sake yeast (*Saccharomyces sake*). At the same time, he found species of lactic acid bacteria that would multiply—seemingly symbiotically—when cultured with sake yeast in *kōji* water. Afterward he spent several years extending his studies on "microbes floating in air," knowing that their number and variety changed vastly with the seasons as well as day by day with the weather, and that different crowds and spaces drew different microbes.[94]

Over time, Saitō came to view microbes as sensitive and localized organisms that were broader than simply pathogens (*byōgenkin*), germs (*baikin*), or bacteria (*saikin*). Likewise, the Japanese term for microbe, *kin*, came to refer as much to fungi (*kinrui*, or more specifically for molds, *shijōkin*). For Saitō, Japanese brewing microbes were a part of the diversity of "useful fermentation microbes" (*yūyō hakkōkin*; figs. 1.8 and 1.9) produced in Asia, which Europeans had first introduced to the world through their expeditions in the 1880s.[95] In a 1906 review, Saitō arranged his descriptive survey by product, because a particular kind of microbe was responsible for each product. Beginning with *Aspergillus*, he described *A. oryzae* for Japanese sake, *A. wentii* for the soy sauce of Java, *A. luchuensis* for *awamori* of the Ryūkyū islands, and *A. batatae* for the sweet potato wine of Hachijō island. He moved on to *Monascus purpureus* for Taiwanese "red *kōji* wine," followed by the "Chinese yeast" that consisted of various *Mucor* and *Rhizopus* fungi (commonly called *kekabi* or "hairy mold" and *kumonosukabi* or "spider web mold" respectively, as they were often found in spoiled vegetables) used in China, Cochinchina, India, Java, and Taiwan, which operated as mold starters like Japanese *tanekōji*. He had studied some of these himself, using samples sent to Tokyo from the Chinese quarter in Kobe.[96]

Saitō came to see microbial collections as being like gardens, and along with the gathering and classification of tropical plants, he thought that there should be botanical gardens that corresponded to the microbial world, which would gather specimens from the tropics to the north and south poles and allow research on their theory and applications. He regretted that

第六圖　麴　菌　屬

1.　梗子單條ナルモノ.

2.　梗子分歧セルモノ.

3.　頂嚢ノ細長瓶子形ナルモノ.

4.　多數ノ被子器ヲ生セルモノ及ビ子嚢ト嚢胞子ノ縱

　　横兩面ヲ示ス.

FIG. 1.8. *Aspergillus*, in Saitō Kendō, *Hakkō kinrui kensa benran* (1929), fig. 6.

only Dutch and Japanese researchers were interested in useful rather than pathogenic tropical bacteria.[97] In 1911, Saitō Kendō left the Brewing Experiment Station to take a position at the South Manchuria Railway Central Laboratory. In 1922 he became director of the entire laboratory, where he gathered microbial strains used in the local fermentation industries for the laboratory's collection, until he moved back to the home islands in 1927 as lecturer in the Brewing Department at the Osaka Higher Technical School (the school later became the Faculty of Engineering, Osaka Imperial University).[98] He sent many new specimens to the Centraalbureau voor Schimmelcultures (CBS) type culture collection in the Netherlands.[99] In the twentieth century, Saitō Kendō was one of the most important contributors to Japan's microbial culture collections.[100]

Another of the earliest large-scale microbial culture collections was at the Brewing Experiment Station, which opened in the Takinogawa district of Tokyo in 1904 (fig. 1.10). The collections held strains contributed by Saitō Kendō, Nakazawa Ryōji (discussed in chapter 3), and the agricultural chemist Takahashi Teizo.[101] Takahashi Teizō was a nonregular staff member at the Brewing Experiment Station alongside his position as assistant professor in the Department of Agricultural Chemistry in the College

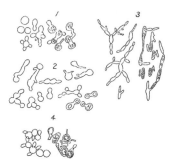

第 二 圖 けかび屬

1. Mucor Mucedo. (Monomucor)
   イ 胞子囊柄　ロ 胞子囊　ハ 接合胞子
2. M. racemosus. (Racemomucor)
   イ 胞子囊柄　ロ 胞子囊　ハ 芽子
   ニ 酵母狀發芽　ホ 接合胞子
3. M. javanicus. (Cymomucor)
   イ 胞子囊柄　ロ 胞子　ハ ニ 芽子
   ホ 酵母狀發芽　ヘ 接合胞子

第十三圖: Zygosaccharomyces.

1. Z. Soya.
2. Z. japonicus.
3. Z. farinosus.
4. Z. Pastori.

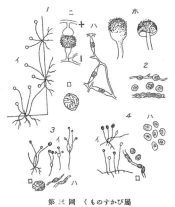

第 三 圖 くものすかび屬

1. Rhizopus nigricans
   イ 菌叢一部，胞子囊柄，假根，匍匐枝ヲ示ス.
   ロ 胞子.　ハ 接合胞子形成.　ニ 接合胞子.
   ホ 中軸.
2. Chlamydomucor Oryzae.
3. Rh. tonkinensis.　イ 胞子囊柄　ロ 胞子　ハ 芽子
4. Rh. chinensis.　イ 胞子囊柄　ロ 芽子　ハ 胞子

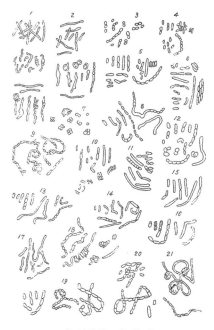

第二十四圖　乳 酸 菌

1. B. Delbrücki.　2. Bᵃ. lactis acidi.　3. Bᵐ. lactis acidi.　4. B. Haydncki.
5. B. Wortmanni.　6. B. brassicae fermentati.　7. P. Liudneri.
8. B. Liudneri.　9. B. Leichmanni I.　10. B. panis fermentati.
11. B. Buchneri.　12. B. cucumeris fermentati.　13. B. Wehmeri.
14. Sb. Pastorianus.　15. B. Aderholdi.　16. B. Maerkeri.
17. B. bulgaricus.　18. B. Leichmanni II.　19. B. Beijerincki.
20. B. Listeri.　21. B. Leichmanni III.

FIG. 1.9. Top to bottom, left to right: *Mucor, Rhizopus, Zygosaccharomyces* yeasts, and lactic acid bacteria, in Saitō Kendō, *Hakkō kinrui kensa benran* (1929), figs. 2, 3, 13, 24.

FIG. 1.10. The National Research Institute of Brewing, formerly the Brewing Experiment Station, Tokyo, December 2011. A bust of Yabe Kikuji, who named the first sake yeasts, is in the foreground at the left. Photo by author.

of Agriculture at Tokyo Imperial University (Tōdai). A student of Kozai Yoshinao, at Tōdai he took over Kozai's seminar in agricultural products, until his new seminar in brewing science and microbial physiology (later renamed fermentation science) split from it in 1924.[102] Takahashi copied strains from the Brewing Experiment Station to develop a smaller, parallel collection in Tōdai's Department of Agricultural Chemistry for his research and teaching.[103] Because of institutional developments that had separated the Department of Agricultural Chemistry from the Department of Agricultural Science, agricultural chemistry (*nōgei kagaku*, "chemistry for the agricultural arts") had come to bear the role of transmitting basic chemistry to Japanese agriculture, and had a much more basic orientation than one might expect.[104] Moreover, the department had begun to include substantial microbiological teaching focused on brewing microbes.[105]

Takahashi's research questions on yeast were simultaneously problems of practical relevance to the brewing industry and ways to distinguish between different varieties for classification, as he continued Kozai Yoshinao's project to encourage the use of pure-cultured yeast in sake brewing. He studied methods of yeast preservation developed in Europe, and confirmed which methods worked best for sake yeast. In extensive studies, he and other members of the Brewing Experiment Station isolated sixty-two strains from *moto* (now technically called *shubo*, a term probably derived from the neologism for yeast, *kōbo*) collected from various brewing districts, and characterized each strain's morphology, whether or not it made spores or a film, what kind of colony it formed, and—more relevantly to brewing—its ability to ferment sugar, what kinds of acids it produced, how much it assimilated amino acids, and its ability to liquefy gelatin.[106] Takahashi also conducted investigations of sake disease-causing *hiochi* microbes (*hiochikin*), the name originating from the term for spoilage when sake "dropped" the fire put in during *hiire*.[107]

Kōji microbes only slowly came to take more prominence in Takahashi's collections, for studies on them were greatly outnumbered by yeast studies for which the investigative techniques of isolating, culturing, testing, and preserving had already been developed in Europe. Takahashi isolated strains of *A. oryzae* from different *tanekōji* samples for sake, and from *kōji* for soy sauce and tamari (a type of rich soy sauce).[108] He found that a variety of strains existed even within one kind of sake, but the physiological differences were most striking between industries: the fungi used in the sake industry formed more sugars, and the fungi used in the soy sauce and tamari industries formed more amino acids, chemicals associated with flavor. It was clear to him that the precise configurations of amylolytic (turning starch into sugar) and proteolytic (breaking down protein) enzymes made by these different *kōji* microbes were related to how brewers had selected and propagated particular kinds of microbes for creating specific good products in the breweries.

In fact, unlike *tanekōji*, pure-cultured yeasts were not very successful in the brewing industry in the early twentieth century, though Takahashi visited numerous factories to oversee attempts to use them.[109] In 1906, a new sake trade association, the Brewing Society (Jōzo kyōkai), opened on a site adjacent to the Brewing Experiment Station and began to distribute pure-cultured yeast strains to sake factories.[110] These "Kyōkai yeasts" were neatly organized as numbered strains for the nation's brewers to select and order, and they had been isolated and chosen from samples of *shubo* or *moromi* sent from breweries or local tax inspection offices across the country.[111] In 1909, scientists at the Brewing Experiment Station developed the "fast *moto*" method whereby pure-cultured yeast would simply be added to *moto* to speed up the process. However, the *moto* foamed too quickly and produced sake that tasted poor—a fact that brewing scientists would later explain by the complex and at the time hazily understood role of lactic acid bacteria in the ecology of the *moto*.[112] Japanese sake brewing traditionally had not preserved yeast-containing samples from the last brew to seed the next, as German beer brewers had or as they themselves did for *kōji*; instead, wild yeast entered spontaneously and multiplied during the long *moto*-making process. Industrial surveys show that until the late 1960s, the majority of brewers in Nada were using the natural *moto* method.[113]

In 1925, Takahashi Teizō's research laboratory at Tōdai took a sharp shift from practical studies of brewing to fundamental studies of organic acid fermentations, especially of molds. There was material continuity with earlier studies, as he concentrated on molds, such as *Rhizopus*, that were already employed industrially in Japan or other Asian countries and were present in his culture collections. There was also intellectual continuity in the vision of microbes as complex objects for the study of life and as effectors of chemical change. Mostly, Takahashi's investigations were biochemical studies to determine what acids were synthesized from or converted into other acids by microbes.[114] His students also worked on problems related to the fermentation of organic acids, which were relatively unusual topics worldwide at the time.[115] Their work established a pattern of fermentation research at Tōdai that lasted into the post–World War II period.[116]

Agricultural chemistry as a discipline encompassed a variety of fields, with an especially strong focus on microbiology and natural product chemistry, especially of microbial metabolites. When Sakaguchi Kin'ichirō (discussed in chapters 3 and 6) entered the Tōdai department in 1919, for example, Kozai Yoshinao held the chair of agricultural products, Suzuki Umetarō the chair of biochemistry, and Asō Keijirō the chair of soils and fertilizers.[117] Departments of agricultural chemistry also existed in the other imperial universities, with differing emphases; the department at Hokkaidō Imperial University, for instance, was especially famed for its microbiological studies of dairy and marine products and *nattō*, while the department at Kyoto Imperial University took a biochemical focus. By the post–World War II period, agricultural chemistry with its integration of microbiological and chemical studies had become a mammoth discipline, an umbrella for work not only on brewing but on pharmaceuticals, food, and fine chemicals, as well as for basic work in microbial physiology and genetics.[118] The discipline provided an institutional frame in which fermentation would continue to shape the development of both pure and applied research.

THE BOUNDARIES OF THE CELL

Tokugawa-era Japanese had lacked the language to convey directly the European notion of bodily "nature" as autonomous mechanical necessity.[119] In the Meiji period, Japanese microbial scientists adopted European conceptions of cellular identity, which relied on a new notion of nature and enabled scientists' role to serve state intervention in the lives of people and industries. The work of the naturalist and folklorist Minakata Kumagusu (1867–1941) during the same period throws the conceptual foundations of Japanese academic microbiology into sharp relief, by way of contrast. Minakata held no degrees and worked entirely outside the academy. Famed internationally for his studies of slime molds, he sought to use these microbes' variety and life cycles to interrogate contemporary understandings of the identity of living organisms.[120] Slime molds were cryptogams (spore-reproducing plants) that held an ambiguous status between the animal and plant kingdoms. In reaction to academic mold studies, Minakata's studies staked out a position that directly opposed the emerging purposes and methods of microbe scientists: standing against the taxonomic fixation on species, against the assumption that microbes should be understood as bounded entities, and against the notion that scholars should be functionary experts instead of tackling more wide-ranging philosophical questions.

As numerous letters to Japanese acquaintances testify, behind Minakata's fascination with slime molds lay his interest in epistemological and metaphysical questions about life. Is there such a thing as a species? What is the nature of life and death? Minakata came to believe that there were no true species, and that new species and new varieties of slime molds were nothing more than states of variation.[121] Even more radically, for Minakata, the slime mold as protoplasm—that is, as a substance that was formlessly alive, as opposed to a discrete, bounded entity such as a cell—challenged the independent identity of the organism's life.[122] The life cycles of slime molds included a plant-like reproductive phase in which they grew spore structures that were visually appealing and morphologically well defined, and an animal-like protoplasmic phase in which they moved in an amorphous form and ate rotten wood and dead leaves. In Europe, biologists had come to characterize life as the autonomous interior maintenance of an organized individual, taking animals rather than plants as their primary objects.[123] The notion of life as a self-directing ensemble of functions to resist death went beyond medical or animal studies, since cell theory drew on the same ideas.[124] As Christina Matta has shown, as early as the 1860s these ideas had laid the philosophical foundations for a taxonomic system of bacteriology that divided bacteria into distinct, stable species. Such taxonomic systems were based on physiological identifiers where morphological properties were not sufficient, and had emerged from a tradition of cryptogamic botany that sought to apply experimental methods from the physical sciences instead of relying on morphology alone.[125]

Minakata, however, drew on Buddhist ideas of life and death as well as the naturalist tradition of morphology to critique the very concept of the organism's autonomy, centering on the protoplasm. A person observing the slime mold under the microscope, he described, could see that the life of the slime mold's reproductive form, which resembled the shapely plants that his fellow naturalists appreciated, was different from the life of its protoplasmic form, which resembled the metabolic animals that physiologists vivisected. Yet the reproductive form's apparent birth, and its budding and growth—so celebrated by naturalist observers—relied on the protoplasmic form ceasing to eat or move, and effectively dying. The two lives shared an interconnection that could not be seen from the vantage point of either kind of scientific observer, illustrating that individual life itself was a superficial phenomenon.[126] In contrast, for scientists who imported the foreign concept of the microbe and adapted it for the purposes of rapid national industrialization, the existence of species was not a question in itself. Similarly, as Olga Amsterdamska has argued, for medical bacteriologists the stability of species was a "grounded assumption" that allowed efficient research strategies in laboratory investigations on disease, even when scientists knew there were reasons to question the assumption.[127] In Japan the same was true for studies on industrial processes. The new conception of "nature" enabled the cellular division between the independent living force and the mastered environment, and made scientists managers of production. Though the microbe came with a set of philosophical assumptions, microbe scientists separated the material from epistemological and metaphysical problems.[128] While microbes became living workers, the modern scientist became a technical expert for the state.

CONCLUSION

The late Meiji-period adaptation of Western science to the indigenous brewing industries left an impact on microbiological research far beyond the Meiji period. Technologies to generate and propagate *genkin*, the original starter microbes with which *kōji* mold and the brewing of all sake, soy sauce, or miso began, were refined long before microbiology emerged in the late nineteenth century. For *tanekōji* makers, their ability to control the forms of life that constituted their product was critical to their reputation and survival. The standardization of new, precise procedures of pure culture for the microbes of the Japanese brewing industries relied on tinkering by both Western-trained Japanese scientists in technical colleges and universities and expert workers in the brewing industry, and on exchange between them. National government-supported institutions placed an emphasis on "Western" techniques even in the traditional brewing industry, while local industrialists and the technical colleges that served them focused on upgrading the practices they knew best. Because of this, the *tanekōji* sector provided a way of modernizing the brewing industry in the area of microbiology, and the small scale of these companies belied the level of specialization and technology they possessed. The dynamism of smaller players in industry has been neglected historiographically in favor of efforts initiated by government officials and university-trained scientists, such as yeast pure culture. In this respect, the pure culture of *kōji* was one of the most important and pervasive technological changes in the brewing industry, and it echoes the continuity between tradition and modernity encapsulated in the common motto of the *tanekōji* industry, *Onko chishin* ("Find new wisdom through cherishing the old").

State policies to promote a capitalist economy co-opted the local initiatives of long-standing experts in industry to improve manufacturing. Newer scientific experts had an even clearer role to play in the political economy of the Japanese state. The government took an interventionist approach to what it saw as Japan's capital-poor, late-developer economy. Modern science was institutionalized at the same time as a network of imperial universities, technical colleges, national and prefectural experiment stations and research programs, scholarly and trade journals, and industrial associations took shape around government policies. This resulted in a social constellation of scientific expertise in Japan that did not make a strong distinction in definition or hierarchy between science and medicine, engineering, or agriculture. As highlighted by the contrast between scientific questions and the goals of enquiry of the nonacademic naturalist Minakata Kumagusu, the new conception of "nature" enabled the cellular division between the independent living force and the mastered environment, and made scientists managers of production. Thus, the scientist did not have special authority to speak on metaphysical and epistemological issues, unlike the early modern intellectual. The modern scientist became a technical expert for the state.

Techniques of handling and studying microorganisms were standardized in the exchange between industrial experts and scientists. However, looking beyond standard techniques such as pure culture, local material traditions deeply shaped the practice of microbiological research in Japan.

Scientists saw their categories of investigation through the organization of processes by which brewers in local industry operated. Indigenous industry helped to create a relatively autonomous microbiological tradition in Japan that treated microbes as useful resources to be manipulated in spite of their variation and sensitivity as living objects, and which saw microbes as effectors of chemical change and complex objects for the study of life. The significance of these processes in other countries in Asia and in the Japanese empire doubled their meaning to Japanese researchers, especially as they were working in part to develop traditional products for export to Asia. What was left after this era of the scientific improvement of traditional industry was a particular way of seeing microbes not only as pathogens but as living workers. This was distinctive and powerful, because of the way scientists came to know and use microbes.

2

Nutrition

NO LONGER A LAND OF PLENTY

The mash barrels of that era were open at the top, and so the strange,
unpleasant odor from the amino acids crept along the floor of the brewing
house and leaked into the outside street, but nobody complained because
of the war.
—Kawamata kabushiki gaisha, *Murasaki: Sakai no shōyuya Kawamata, Daishō
200 nen no ayumi* (Murasaki: Two-Hundred-Year History of Kawamata-Daishō,
Soy Sauce Brewer of Sakai), 143

Between 1918, when rising rice prices triggered massive riots across Japan,
and the end of World War II, Japanese fermentation scientists came to see
microbes as a nutritional resource. The interwar period saw the invention
of new food technologies such as synthetic sake, chemical soy sauce, yeast
preparations, and vitamin synthesis that scientists believed would help to
distribute agricultural reserves more efficiently. This view of efficiency was
not simply economic: it encompassed growing concern for the health of
the citizenry. Nutrition science emerged as a new field of expertise at the
same time as the Japanese state came to value the population more and
more as a source of military manpower and industrial labor. Scientists focused their research on the specific dietary and resource problems that the
Japanese population faced, assuming that they could not rely on European
ideas because Japanese bodies and dietary trends differed from those in
Europe. Unlike the European diet, the Japanese diet consisted mostly of
grains, especially rice, with a small amount of meat. Microbial activity featured prominently in their thinking since a number of fermented goods
such as sake, soy sauce, and miso were dietary staples across the country.
This chapter looks at the products designed by fermentation scientists to
redistribute agricultural and industrial outputs of nutritional and economic
value.

Malthusian visions of resource shortage cast a prominent shadow over
the interwar period, and a notion of Japan as a "resource-poor" nation arose
from the 1910s onward.[1] Japan's industrial economy took off in the final two

decades of the nineteenth century. From 1880 to 1900, the population rose from roughly 35 million to 45 million, and millions left agricultural work in rural areas to migrate to towns and cities. An inflationary boom-and-bust cycle during and after World War I, stimulated by manufacturing investment to fill the vacuum left by the halt in European exports, anticipated recurrent economic crises throughout the 1920s. Technologically driven increases in agricultural production stalled, coinciding with an era of social instability and political violence.[2] As the country emerged from the revoking of its semicolonial status in the form of the unequal treaties, from successful light—and the early phases of heavy—industrialization, and from the acquisition of a new empire following victory in two wars, interwar Japan nonetheless, in the words of the historian Mark Metzler, "saw no mass-market consumer revolution and no roaring twenties."[3] In an era of political liberalism, one common hope among Japan's leaders and populace was that the country's socioeconomic "deadlock" (ikizumari) might be broken by consolidating its empire, and that Japan could thus take up a place among the first-class nations of the world.[4]

In reality, such ambitions were unsustainable. As Jordan Sand has argued, for example, women's magazines and media-sponsored home exhibitions called for national middle-class ideals that included Western furniture, the reform of household living and entertaining habits along modern lines, and scientific cooking practices; yet the budgets of most white-collar wives, even, could barely accommodate the cultural standards required to maintain their families' fragile social position.[5] As both of the two major political parties attempted to stabilize the economy, swinging between policies of fiscal expansionism and austerity, their aspirations to maintain Japan's imperial and industrial expansion manifested in what were sometimes naive campaigns for the reorganization of everyday life. For instance, in the wake of severe retrenchment policies implemented by the ruling Minseitō party following the worldwide depression of 1929, which hit rural areas especially hard, the message to ordinary urban women from the Ministry of Finance was this: "To rationalize Japan's macroeconomy, the 'rationalization of the kitchen' was necessary"—for apparently it was the morally decadent "bloating of a kitchen economy near destruction" that had been to blame for the country's inability to gain a strong foothold at the upper levels of the international order.[6]

Fermentation scientists played an "in-between" role that mediated between the state's visions for political economy and the material construction of everyday life. In other words, as these early Japanese nutrition scientists created goods to intervene in the food that people bought and ate, their concerns centered on the desires of the state more than those of the individual consumer. There was no middle-class, market-driven "vitamania" akin to that in the United States in the same period.[7] While fermentation scientists concentrated on designing goods for consumption mainly in the urban market, the problems of economy and self-sufficiency that they dealt with were widespread. During the 1930s depression, the monthly magazine *Ie no hikari* (Light of the Home) encouraged its mostly rural readership to make do with what they had, asking them to make soy sauce in the household rather than buy it.[8] In such a context of industrialization, the language of food chemistry was comparable to the moralizing and rationalizing legacy of economic chemistry in late eighteenth-century France. Emma Spary argues that French chemists, in their artificialization of nature, had aimed to fashion an ideal consumer whose behavior might align with the goals of the state. Similarly, early twentieth-century Japanese fermentation scientists sought to define the distinctions between luxury and necessity, or between unacceptable and acceptable processing, and engaged with flavor as a part of their interventions.[9] The main difference was that their statist efforts were expressed in the recent language of Taylorism, eugenics, and the latest biochemistry, which was shared by contemporary industrialized nations.

The prominence of fermentation science within Japan's nutrition research, however, was distinctive. Studies of nutritional metabolism and fermentation were linked, for scientists understood fermentation to have meaning beyond the microbial decomposition of chemicals, and to refer more broadly to processes of chemical change in living organisms.[10] All of the experts considered in this chapter had a background in fermentation science (table 2.1). Their microbial thinking was both national, as a consequence of their attention to local material problems, and nationalistic, emergent with their self-construction of identity. While British experts investigated the nutritional and economic efficiency of margarine, Japanese did so for sake and soy sauce.[11] The constellation of statist concerns in nutrition science that bound together Malthusian anxieties with national strength and war was not unique to Japan, though Japanese scientists played a leading and underappreciated role in this early period of the field's development.[12] But the close connection between fermentation and nutrition science in Japan led to a particular focus on the role of microbes in mediating the distribution of vitamins, amino acids, and other nutritional constituents. As a result, a strongly ecological conception that the chemical

TABLE 2.1. Fermentation scientists and their inventions examined in this chapter

Scientist	Training in fermentation science, with doctoral degree	Career trajectory in research institutions	Food technologies examined in this chapter
Tsuboi Sentarō (1861–1921)	Graduated in applied chemistry from Imperial College of Engineering (predecessor of Faculty of Engineering, Tokyo Imperial University).	Head technician at chemical company Nippon Seimi Seizō. Taught in Brewing Department of Osaka Higher Technical School (predecessor of Brewing Science Department, Faculty of Engineering, Osaka Imperial University). See also chapter 1.	*Daiyō seishu* (synthetic sake); Katsuryokuso ("activation element") yeast preparations
Suzuki Umetarō (1864–1943)	Graduated in agricultural chemistry from Faculty of Agriculture, Tokyo Imperial University. Studied abroad with Emil Fischer at University of Berlin.	Discovered vitamins (as "oryzanin") in 1910. Chaired biochemistry seminar in Department of Agricultural Chemistry, Faculty of Agriculture, Tokyo Imperial University. Directed laboratory at Riken.	Rikenshu (synthetic sake); synthetic vitamins
Hashitani Yoshitaka (1888–1975)	Graduated from Tōhoku Imperial College of Agriculture (predecessor of Faculty of Agriculture, Hokkaidō Imperial University). Trained in agricultural chemistry.	Worked in industry, including beer company Dainippon Beer (predecessor of Asahi Beer and Sapporo Beer).	Ebiosu yeast preparations
Takada Ryōhei (1898–1978)	Graduated in industrial chemistry from Faculty of Engineering, Kyoto Imperial University.	Worked at Institute for Nutrition Research. Taught in Brewing Science Department, Faculty of Engineering, Osaka Imperial University. Also taught in Industrial Chemistry Department, Faculty of Engineering, Kyoto Imperial University. Founded Vitamin Society of Japan.	Microbial and fish technologies, including mold vitamin B_2 production by *Eremothecium ashbyii*, and Korean *mentai* (Alaska pollock) "eyeball boom" vitamin B_1 preparations

makeup of everyday diets and foods was a part of the national agricultural and industrial economy as a whole emerged in Japan: a conception that was uncommon among other societal contexts in that era.

The new "economizing," "rationalizing" food technologies that scientists had designed for managing the nutritional economy eventually became widespread during the Asia-Pacific War. If there was any public distaste for chemical adulteration or a fear of the degradation of product quality in Japan, as there had been in the early nineteenth-century British brewing industry, the fact was that the state under total war backed the austerity technologies that fermentation scientists had created.[13] Worsening conditions of material deprivation in the 1940s, especially, left little room for the complaints of ordinary consumers. These scarcity technologies became even more widespread in the 1950s, filling crucial gaps in the food and drink industries until the postwar economy began to recover.[14] In the process, they permanently transformed the industries. By exploring the design of new products both synthetic and fermented, this chapter traces some of the major attempts of fermentation scientists to intervene in political economy through material culture. In nutrition science, fermentation scientists went beyond traditional brewing to transform key approaches at the core of Japan's food and drug industries, creating a knowledge of microbes as methods of national resource management.

CIVILIZATION AND FERMENTATION

Like chemists in nineteenth-century Germany and Britain, chemists in Japan at the turn of the twentieth century first made their way into the fashioning of mass dietary consumption by associating closely with producers and creating a shared scientific culture, via the brewing industry.[15] Tsuboi Sentarō, head of the Brewing Department at the Osaka Higher Technical School, was one such chemist, who held opinions on a range of scientific subjects that would become vigorously debated after his time.[16] His Meiji-era product innovations anticipated post-World War I nutritional technologies of synthetic sake and yeast tablets, but contrasted with them in meaning. Tsuboi's creations reflected an early moment of transition in the years immediately before the emergence of nutrition science as a scientific discipline, when the conceptual connection between individual health and national-level resources, which in turn implicated industrial production, did not yet exist.

When Tsuboi contributed to a book titled *Lectures by Twenty-One Experts* published in 1908, the title of his chapter was "Theory That Meat-Eating Will Ruin the Country." He stated that he had many reasons to choose an outlandish subject that diverged from his area of specialization in brewing.[17] With the wind of Western civilization blowing across Japan in all its material and spiritual aspects, many people were jumping hastily to the conclusion that whatever Westerners did was the most excellent, and that it would be most pleasing to bring a superficial Western appearance to their actions, their home, and their food. In the midst of the Western "infection," the most detestable thing was meat-eating. Tsuboi argued passionately that meat-eating was the wrong way for the nation, both from the principles of animals, plants, and nature, and from the economic perspective.

Tsuboi stated that plant protein and animal protein were clearly distinguishable from each other in analytical studies, and animal protein was more concentrated. But on the basis of the authority of present-day science, plant protein was not inferior to animal protein. He referred to an analogy between the human body and a boiler. Adding a small quantity of British coal to a boiler would not generate steam power. In order to get steam power, one needed to add the amount of coal that could be burned, even if it were Japanese coal or Fushun (Chinese) coal (both varieties known for their impurity compared to British coal).[18] Continuing the analogy, Tsuboi argued that if the stomach were a boiler, the intestines would be the chimney, and neglecting to clean the chimney would cause fire. What he meant was that eating animal matter dirtied the chimney and caused abnormal defecation patterns. Some Western medical experts believed that beriberi and rheumatism came from eating rotten fish and meat, but nobody claimed that one could fall ill by eating plants.

Tsuboi was facing fashionable views that Japanese had weaker physiques than Europeans due to a diet that lacked meat, and because the protein in animal matter was qualitatively superior to that in vegetable matter. Such ideas focused on Japanese national deficiency, with the implicit reasoning that Japanese needed to compensate for it in order to contest Western empires in Asia. These views persisted into the early twentieth century, even though the chemist Justus von Liebig's 1840s theory that meat was the sole fuel for muscles had long been overturned. Physiologists in the 1880s had shown energy equivalence between different foods (in terms of the "calories") in bodily metabolism, based on applying the principle of conservation of energy to living organisms.[19] In addition, Japanese research in the 1880s had upended previous protein standards set by German research, and argued that achieving health did not require duplicating a European diet—a notion further supported by the Japanese victory over Russia in the war of 1904–5.[20]

More than being scientific arguments, however, Tsuboi's views on metabolism as an exchange of material between living things were moral arguments.[21] To him, protein intake was an ability that was natural to plants, but animals did not have such an ability: they needed to choose between taking proteins from other animals and taking them from what had accumulated in plants. Which of the two was most appropriate for human beings according to natural principles? One could immediately see that humans—as the lords of all things—must be blessed with the most loving heart. When the Creator made humans, He surely did not say that they should eat other animals in order to live. Buddhism, too, forbade one to kill life and stopped one from eating meat. Tsuboi's appeal to religious ideas as well as to worries about meat poisoning by intestinal autointoxication echoed arguments among American vegetarian physicians of the period.[22] Like other Japanese critiques of meat-eating at the time, they were, above all, cultural critiques that aimed to reject the decadent aspects of Western civilization.[23]

That Tsuboi Sentarō, a brewing expert, could even write about meat-eating was thanks to the fact that nutrition science did not yet exist in Japan in the first decade of the twentieth century. This is not to say there was no significant scientific work on nutrition. The physician Takagi Kanehiro had implemented dietary reforms in the Japanese navy from 1884, following experimentation with rations to solve the problem of beriberi (*kakke*), which afflicted more than one third of sailors. His modified diet, which replaced rations of white rice with a mixture of barley and rice, lowered beriberi incidence to 13 percent.[24] Takagi's treatment was effective and based on modern epidemiological reasoning and experimentation, and yet it was dismissed by his critics in medicine, all of whom had been trained in bacteriology. Army physicians did not modify white rice rations until after the Russo-Japanese War, and both the Sino- and Russo-Japanese Wars saw heavier losses of soldiers from beriberi than from combat. As Christian Oberländer argues, physicians rejected Takagi's nutritional work because it lacked laboratory methods.[25] Bacteriology as a laboratory science had raised the standards of causal explanation in medicine, and they would not be matched in the field of nutrition until the work of Suzuki Umetarō after the Meiji period.

The "substitute sake" (*daiyō seishu*) that Tsuboi Sentarō created, using alternative raw materials to rice, would look similar to Suzuki Umetarō's later invention, but with a different meaning. Like Suzuki, Tsuboi claimed that it was part of the progress of human intelligence to alter natural products and make them into artificial (*jinkō*) products. But Tsuboi's purpose was to contract the long years and months of nature's power into short hours and days. His aim was to improve mass production efficiency in traditional industry to make sake competitive against imported alcohol, and to combat the mid-Meiji-period trend in which brewers made great volumes of "mixed sake" (*konseishu*) by adding imported alcohol to sake.[26] Tsuboi's method was to use a cheaper raw material than rice, such as starch or a

FIG. 2.1. Advertisement for Katsuryokuso in the *Asahi shimbun* (October 5, 1910), morning edition, 6.

starchy crop. It would be broken down first by *kōji* and acid-producing microbes, and then by acid hydrolysis, into a predetermined constitution of essences and alcohol—thereby avoiding the process of protein decomposition during maturing that caused rotting or a "stale liquor" smell—and then adding ethanol. The "sake" would be placed into a cask for several weeks to add a pleasant wood aroma. Another method of his for "improved sake" (*kairyō seishu*) in 1918 involved a synthetic ethanol made from carbide. The procedure was to hydrolyze refined rice with dilute sulfuric acid, and then heat and liquefy it into dextrin, neutralize the mixture with calcium carbonate, and add sugars, ethanol, and tartaric acid for taste.[27]

Yeasts are commonly thought to have emerged as a drug in Japan from the 1930s onward, but Tsuboi Sentarō's yeast preparations were available as early as the 1910s in the form of powdered and bottled tonic, and later as pills (fig. 2.1).[28] He gave them the name Katsuryokuso ("activation element"). Tsuboi's nephew ran the business to manufacture and sell the tonic, pitching it as a patent drug for a bourgeois market, and it thrived through the 1910s and 1920s.[29] In wartime Germany, yeast became an ersatz food in the 1910s, in contrast to in Japan, where the marketing materials for Katsuryokuso used no appeal to vitamins nor national deficiency.[30] According to 1920s advertisements in the magazine *Fujin gahou* (Women's Graphic), the drug could stop the depletion of intracellular plasma, strengthen the ability of white blood cells, purify the blood, and be active against various ailments including chronic intestinal diseases, tuberculosis, heart disease, kidney disease, beriberi, constipation, irregular menstruation, rheumatism, hysteria, asthma, and lack of sex drive, as well as being able to serve as an antifungal for soy sauce. Such claims were an extension of Meiji-era "vegetarian nationalism," in straightforward opposition to the meat-centered diets associated with Western civilization.[31]

SYNTHESIS

By the 1910s, the stakes had changed. Agricultural chemists at the Institute of Physical and Chemical Research (Riken) began to think seriously about the "food problem" (*shokuryō mondai*), which they also called the "population problem" (*jinkō mondai*). Scientists framed the problem as a universal one along the lines of Malthusian logic, in which unlimited population growth would meet saturated food production due to limited cultivation area; but in Japan itself many people felt the problem acutely. Serious disturbances known as "rice riots" spread across the country in response to rising rice prices in 1918, which were triggered by the economic recession and restricted rice imports from Southeast Asia amid World War I. Because of the scale of the problem, the scientists envisioned synthetic food to be the ultimate goal of chemistry. Two elements came to the forefront of their attention: the vitamins that people were unable to synthesize and needed from their environment for growth and balance, and the amino acids that built protein in the body, and which also contributed to the flavor of foods. The agricultural chemist Suzuki Umetarō hoped to free humankind from dependence on agricultural production by making amino acids chemically from mineral substances; and he created synthetic sake and synthetic vitamins.

Suzuki had discovered vitamins in 1910 with his isolation of a component ("oryzanin," or vitamin B_1) in rice bran that could cure beriberi. Though the priority is arguable, whether or not Suzuki truly discovered vitamins is less pertinent than the fact that his work, more than anyone else's, channeled the confluence of multiple forces in Japan, and founded nutrition as a scientific discipline there.[32] When nutrition did become a science in Japan, it came with laboratory-based animal experimentation on micronutrients such as vitamins. Above all, the field's emergence was inseparable from the rise of statist considerations about national resources, which implicated industry. It was statist concerns about resource scarcity that ultimately motivated support for laboratory research on nutrition, in *agricultural* science, outside the discipline of medicine. It led to another path of microbiology in Japan in the agricultural study of fermentation, beyond medical bacteriology.

Riken was very different from the network of higher technical schools and experiment stations that focused on aiding medium and small-scale manufacturers in traditional industry. Founded in 1917 as a private institution with funds mainly from business, and with some support from government, Riken aimed to promote advanced fundamental research in the physical sciences with the aim of achieving national self-sufficiency.[33] However, agricultural concerns continued to play an important role at Riken. The physical chemist Ikeda Kikunae (1864–1936)—best known for isolating the umami component monosodium glutamate (MSG) from kelp in 1908—advised one of the members of his Riken laboratory that the problem humankind now faced was the population or food problem.[34] For Japan, food synthesis (*gōsei*) was one solution, but it was still far from achievable, and it was necessary rather to increase food production. Ikeda thought that refrigeration technology would be the problem of the future. Civilization, he thought, had begun in the warm regions of Egypt, India, and Mesopotamia and then moved north thanks to the development of heating devices; but in the future humankind would once again return to the south and the tropics where plant growth was ten times faster than in the cooler regions, and because of this, refrigeration technology would be essential.[35] He thus presented the problem to be simultaneously a universal limit to empire and an obstacle to Japan's destiny as a first-class nation.[36]

Suzuki Umetarō had another answer to the food problem, which was synthetic sake. Suzuki's work would tie the food problem directly to the local conditions of Japan's agricultural economy and diet, and especially to the national requirements of war. Along with his large laboratory at Riken, he ran a laboratory at Tokyo Imperial University's Faculty of Agriculture in Komaba, where he chaired the biochemistry seminar in the Department of Agricultural Chemistry. His engagement in the apparently unestablished, unsystematic fields of nutrition and agricultural science made him an oddball among the physical scientists and engineers at Riken. Nonetheless, Riken had the commercialization of his laboratory's invention, vitamin A preparations (under the name "Riken Bitamin"), to thank for saving its struggling finances and keeping the institute afloat from the early 1920s.[37] Since beriberi had become an especially important problem in the Japanese military, Suzuki thought of nutritional resources as a matter of war.[38] To Suzuki, the Russo-Japanese war had been a "vitamin war" that would have turned out differently had the Russians known about the vitamin C in the soybean that was plentiful in northeast China.[39]

In Suzuki Umetarō's mind, it was not in the spirit of a peasant that he had sought to master agricultural science.[40] (He was probably gesturing at tensions with medical experts, as he was a professor in the agricultural rather than the medical faculty, and was famously from the commoner and not the samurai class.) Rather, the questions he was interested in were: When fertilizer was applied to crops, what chemical changes did it go through as the ammonia became proteins? What intermediate substances were formed? These were the chemical pathways that he had mastered agricultural science in order to study. It was linked to a fancy he had nurtured since his student days to see whether food could be made artificially by the hand of a chemist, and not from livestock or grains.[41] Suzuki had done work on proteins under Emil Fischer in Germany. As in the chemical changes that occurred when fertilizer was taken up by crops, when people ate meat, the proteins in the food underwent change.[42] The complex molecules of protein were gradually decomposed, and would be absorbed as only some twenty types of amino acids. It was a faraway dream for a chemist to synthesize meat in the event of a protein shortage as the world population grew. But chemists instead could synthesize such simple things as amino acids, as was happening in his lab at Riken.

Synthetic sake design reflected key features of the Taylorite ideological movement, which rose in Japan as early as the 1920s. As William Tsutsui explains, Taylorite thinking would later become widespread in the Depression-era "industrial rationalization movement" in the 1930s, and especially in the mobilization campaigns of World War II.[43] In the concern with biochemical processes of metabolism, in this case the breakdown of proteins and amino acids, and in locating them in a constellation of problems of national self-sufficiency and socioeconomic inequality, Suzuki's work was similar to that of nutrition scientists working in Britain in the same period.[44] In Japan, however, nutrition scientists themselves played a leading role in designing new products to intervene in consumer trends, unlike in Britain where it was the food industry, such as margarine manufacturers, that took the initiative in developing new products as a defensive reaction to findings in nutrition science.[45] The stronger technocratic position of scientific experts in Japan was reflected in the fact that the leading national scientific institution, Riken, manufactured and sold commercial products including vitamins, unlike in Britain and Germany where pharmaceutical companies produced vitamin preparations.[46] This technocratic trend resulted in discernibly Taylorite features in the design of everyday products.

One such feature was scientism, or a belief in progress premised on science, which was usually expressed in terms of quantitative estimations of efficiency.[47] The 1918 rice riots spurred Suzuki to begin his research on synthetic sake (*gōseishu*) in 1919 in order to economize on rice consumption in brewing. The key to sake as a unique product was in the amino acids, he argued. The nature of rice protein, as opposed to any other kind of protein, was determined by its amino acid makeup. In brewing, *kōji* and yeast microbes gradually broke rice protein down into amino acids, which in turn underwent secondary changes; the resulting complex series of intermediates created the unique flavor of sake.[48] But among the amino acids, only two or three types were strongly linked to the creation of the flavor. Other amino acids did not contribute to flavor, or were even related to bad aromas in the sake. The most important flavor component was alanine, in addition to leucine, tyrosine, valine, and phenylalanine. Suzuki stressed in his talk at the Food Meeting in the House of Peers in 1922 that the amino acids that produced the flavor of sake were no more than 4 to 5 percent of the rice protein, which in turn was 7 percent of the weight of the rice. To brew sake was effectively to consume the entirety of the rice in order to obtain amino acids that made up only three to four thousandths of the rice. This was not economical.

At Riken, he said, researchers were attempting to industrialize synthetic processes for making the amino acids that were relevant to sake flavor, one by one. If these amino acids were decomposed by chemical methods, mixed together with acid-saccharified (broken down into sugar) starch, fermented with yeast, and then added to alcohol at the end, one could obtain a product extremely similar to sake without consuming any rice.[49] In a review of the state of fermentation science in 1926, Suzuki wrote that flavor—the biochemical problem of understanding the intermediates in the decomposition of proteins, such as when amino acids changed into sugars and fats—was the most important problem in fermentation science.[50] At the time, analytical methods were not accurate enough to measure chemical equivalence with fermented sake, and so it was done by the conventional brewer's test: tasting the synthetic sake.[51] The question of flavor was thus driven less by advances in chemical instrumentation than by Taylorite concerns of agricultural resource efficiency. Suzuki was loath to use agricultural produce at all in making the synthetic sake components, and he quantified the cost in terms of land acreage. For making the amino acids, he suggested acetylene. For making the alcohol that would be added at the end following the fermentation, he advocated researching carbide, for then the only acreage needed for sake production would be for the starch component: it would mean using fifty thousand instead of two hundred thousand *chō* of paddies.[52]

A second feature of synthetic sake typical of Taylorism as a whole was a belief in apolitical expertism. The benefits of applying scientific objectivity to increase efficiency were meant to accrue to all, whether urban or rural, working-class or middle-class, tenant or landlord. The role of the expert, then, was statist—to work to help the state in its management of efficient resource consumption—but apolitical. The ultimate aim of technocratic design was not only cost-cutting; it was idealistic. On the shop floor it was intended to improve labor/management relations.[53] For food, similarly, designs to make more available more cheaply were supposed to ease social conflict in an era of mass protest and violence, by helping (and disciplining) both capitalists and consumers. Riken might have been founded on *zaibatsu* capital, accrued through the World War I boom and bust that had devastated so many on the lower rungs of the socioeconomic ladder, but its scientific experts were the university elite who saw themselves as working on problems of the greatest national importance. Amid the perception of deadlock in the 1920s and 1930s, the idea was that the application of science would provide the necessary civilizational breakthrough to accompany Japan's first industrial revolution.[54] The real breakthrough that Suzuki Umetarō aimed for—like many other Japanese experts in the period—was empire.

That was why, to Suzuki, agricultural shortage was a long-term problem that would not go away even after the era of the rice riots. In 1927, he lamented Japan's annually rising reliance on rice imports from Korea and Taiwan, as well as the country's continuing increase in population.[55] (By 1925, Japan's population was about 60 million.)[56] As population increased, sake brewing would increase; as the level of culture rose, people's lifestyle would improve; and the shift from a wheat-based to a rice-based diet would continue, he reasoned. Food was an international commodity, and so it seemed that if a nation only had the money it could import food from other countries. This was fine in times of peace, but not in times of war. He reminded readers that Germany had lost in the Great War because of food shortages.[57] He remarked that sake was a luxury item, and it had a very particular taste that people were fussy about. Because of this, synthetic sake would not be embraced unconditionally by the public. However, people's tastes could gradually change: consider beer, which people at first had thought

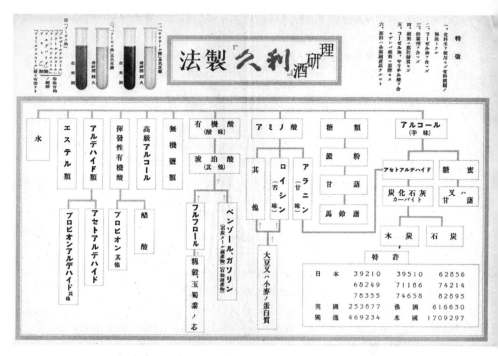

FIG. 2.2. "Method of manufacturing Riken sake 'Riku.'" Promotional chart showing the different constituents—alcohol, sugars, amino acids, organic acids, minerals, and others—that go into making synthetic sake. The notes and illustrations along the top highlight the absence of fusel oil and salicylic acid. Courtesy of the National Museum of Nature and Science.

was too bitter, but had now come to be widely loved, and which if it lacked the bitter taste was not thought to be real beer. He recommended paying some attention to marketing, and with time the food problem would compel the Japanese people to change their tastes.

The results of Suzuki's "alanine fermentation" method for synthetic sake did not go down well, according to the tasting session involving large sake and drug companies that had contributed funds to tests at the factory (at a new company called Yamato Jōzō, to which Riken had sold the patent).[58] Rather than fermenting amino acids, Suzuki's laboratory decided to try instead to simply assemble all of the relevant chemicals in the right proportions (fig. 2.2), which included organic acids (such as succinic acid, fumaric acid, and lactic acid), amino acids, glucose and other sugars, and aromatic compounds such as esters. In effect, he was using all of the chemicals in minute amounts as flavorings, and Suzuki was optimistic that traditional brewers too could employ them to adjust the taste of a poor product.[59] The trickiest part was how to manufacture them cheaply and in bulk. Suzuki mobilized a number of his peers at Riken and Tokyo Imperial University to help him in researching methods for the different chemicals. Especially, Yabuta Teijirō managed to obtain succinic acid—which had been very expensive to derive from amber—from furfural decomposition.[60] Other researchers worked on methods for making amino acids, while lactic acid supply was entrusted to the pharmaceutical company Sankyō.

In an early attempt at a marketing strategy, Suzuki boasted that synthetic sake was healthier than ordinary sake due to the lack of harmful chemicals in synthetic sake, since the combinatorial approach meant that they could simply leave harmful chemicals out. Synthetic sake had no preservatives such as salicylic acid, and best of all, one did not get hung over because the manufacturers would not add fusel oil. This was not unlike the approach of British food companies who claimed that their vitamin-fortified foods were healthier than unprocessed foods.[61] The idea that a combinatorial approach to synthetic food could make the food better—that is, the idea of analytically resolving food into its elements, and making an artificial version using only the "good" elements—probably came from the way experiments were conducted by nutrition scientists, including the many experts who worked at Suzuki's Riken lab. As Suzuki explained, nutrition scientists answered questions such as: What food makes us live? What food makes us die?[62] To do systematic research on food as a nutrition scientist, one needed to purify each constituent, combine them in appropriate amounts, and feed the mixture to laboratory mice to see what would happen.[63]

Aspiring to empire in Asia meant competing with the West. What Taylorite scientism in Japan articulated above all were comparisons with Western industrialized countries, and a constant worry about Japanese national inferiority and lack of efficiency, expressed in numbers. Just as household reformers made quantitative estimations of the time wasted per year, per capita, by the populace sitting down on and getting up off tatami mats instead of chairs, Riken experiments, too, bore similarities to nutrition research in other national contexts, but with a focus on cultural practices specific to Japan.[64] We find research on such typical Western objects as butter and sugar, but also concerns about a nationally characteristic diet based on vegetable and cereal proteins and marine products. As Suzuki described, from these experiments it was clear that plant proteins were inferior to animal proteins for growth, and that rice was inferior to fish or meat.[65] Why were Japanese people's physiques small? Researchers at Riken had discovered a crucial amino acid, "oxyaminobutyric acid," that was absent in the proteins of rice and wheat. When it was added to the rice fed to mice, the mice would show a huge improvement in growth. Butter was the best type of fat, rich in vitamins and hormones (though Japan had a shortage of butter), whereas fish oil was not only useless but had strong toxins.

Mixing feed out of 65 percent starch, 20 percent protein, 10 percent butter, and 5 percent minerals, and adding vitamins in appropriate amounts, the mice would grow perfectly. But if one replaced some of the starch with 5 percent of refined sugar, the mice would not grow; they would wither, sugar would appear in their urine, and later their urine would be bloody. However, if one added 4 to 5 percent of yeast or yeast extract, such damage from overconsumption of sugar would not occur. Suzuki recommended supplying vitamin B supplements (such as the yeast preparations available on the market) to children, as they were prone to eating too much sugar. From similar experiments, researchers also knew that the vitamin B complex could undo some of the harm from overconsumption of alcohol, though that overconsumption was nowhere near as harmful as overconsumption of starch or sugar.[66]

Lecturing to a lay audience at the mothers' meeting hosted by the Ministry of Education in 1933, Suzuki argued that it was simply not enough to eat what one could afford and survive, as people had done throughout history.[67] He used the story of a Japanese horse in the early Meiji period being displayed at an exhibition in Europe, to the amazement of the crowd. They wondered at the rare sight of a horse with a physique so inferior that it showed a complete lack of artifice in terms of modern breeding and feed improvement, in contrast to the artificially excellent European horses they were used to. It was the same with humans: one must not think that present Japanese bodies were fixed and could not grow larger. He thought eugenics was too idealistic, but that in food alone there was room for improvement and for Japanese people to fulfill their genetic potential through "rationalization." One problem, for example, was the bias toward plants in the diet, and another was vitamin B deficiency.[68] Speaking at the Manchuria Medical University in 1937, Suzuki related how, based on a 1925 nutritional survey, the government had a thirty-year plan to promote the livestock and marine industries in order to increase the proportion of meat to 25 percent of total protein in the Japanese diet by 1957. But in Europe, 50 percent of protein in the diet came from meat, and Japan's thirty-year goal did not even match the 30 percent of Italy at that time. He noted that a vitamin shortage at least would not be a problem in Japan in the future, since basically all vitamins and even some hormones had been crystallized, structurally elucidated, and chemically synthesized for use as drugs or supplements; the vitamins would be available at a cheaper rate in one or two years.[69]

The promise of new access to fossil fuel resources in Manchuria, following Japan's invasion in 1931, buoyed Suzuki's dream of chemical food. His was a kind of stage-theory thinking shaped around resources, born in the notion that agricultural resources were limited and mineral resources were unlimited. In his keynote lecture at the first meeting of the Manchuria branch of the Industrial Chemistry Society in 1932, Suzuki opened with the multitude of possibilities that came from a side product of coke manufacturing, benzol.[70] When organic acids and amino acids were made from benzol, a part of the coal would become food. From benzol one could derive fumaric acid. From fumaric acid one could derive succinic acid. (Researchers had been making these acids at his Riken lab for synthetic sake.) From fumaric acid, one could also derive malic acid, which was a better flavoring for soft drinks than tartaric acid or citric acid; Japan could stem imports of tartaric acid by making malic acid instead. Suzuki's laboratory at Komaba had found that the sodium salt of succinic acid, made from benzol or gasoline, was chemically identical to essence of clam and gave the same unique flavor. Neutralizing malic acid gave a substance that could be used as an alternative to table salt, and as a flavoring for ill people with low salt tolerance. Aspartic acid derived from fumaric acid had few uses as yet, but could be used to make amino acids. In the future, food would be entirely chemically manufactured. Suzuki had no doubt that eventually there would come an era when every kind of food and drink could be made with only "coal, earth, air, and water."

In synthetic food, Suzuki envisioned a fundamental energy transformation of the economy, with agriculture to tide society over only temporarily. He saw agriculture as a dead end that could not achieve further intensification of production.[71] The question of resources for production, framed in terms of national self-sufficiency, directed the heart of his lab work—unlike the profit-driven taste research that created Tang, the 1950s American "space food," and in contrast to the medical concerns for an individual's maintenance of their own biological body that led to Yakult, the Japanese commercial invention of the 1930s. As Suzuki explained, chemistry was not quite there yet, and so there was no choice but to continue research on agricultural production. Opining on the best uses of soybean in Manchuria, he thought it would be better to use it to nurture a livestock industry than to eat it. Chemistry was not yet good enough to compete with the abilities of chickens and cows, and chickens were still better than chemists at making eggs. If soybean lees were fed to animals, the animals would forge it into better protein (milk protein, instead of soybean protein); for now, it was better to borrow the power of animals than to try chemists' poor tricks.[72] But he left the ecological implications of agricultural resource research to others, and for his own part focused on developing a futuristic chemistry.

THE ECONOMY OF YEASTS

Vitamin supplies, even alongside the vitamins sold by Riken, were mostly yeast preparations in the prewar period. The success of yeasts in the pharmaceutical market relied on the new notion of vitamins, as well as international research on yeast metabolism. It marked the rising status of agricultural chemistry in the medically dominated area of pharmaceutical development, by way of nutrition science. Hashitani Yoshitaka (1888–1975), a pioneering entrepreneur in the yeast business, maintained that his company was the first to sell yeasts in Japan around 1929 to 1930 (with no mention of Katsuryokuso) as digestive and medical supplements.[73] The yeasts came out of the beer industry. When Hashitani entered Dainippon Beer (the predecessor of Asahi Beer and Sapporo Beer) in 1915, the company was seeking a way to get rid of excess yeast instead of flushing it down the river and polluting the surrounding paddies. He decided to develop yeasts as a drug after reading in the British journal *Institute of Brewing* about the extremely high amounts of vitamin B purported to be in yeast. In Europe and the United States, researchers were working on problems of yeast metabolism, especially the effect of "bios"—which was eventually shown to be part of the vitamin B complex—on yeast growth. In 1930 Hashitani started selling bottles of yeast preparations, under the brand name Ebiosu, as powder and as tablets (fig. 2.3).

The drug Ebiosu's advancement into the pharmaceutical market was gradual, and in 1932 it was finally listed in the *Nihon yakkyokuhō* (Japan Pharmacopoeia), the official catalogue of drugs. According to Hashitani, things changed in the 1930s with the explosive progress of vitamin science: the isolation of vitamin B_2, the structure determination of both B_1 and B_2, and the synthesis of B_2, along with work on the physiological mechanisms of vitamin action in the body. Commercial offerings of yeast on the market multiplied to dozens of brands, and Ebiosu was joined by Wakamoto, Entsaima, and others. Japan's swimmers took yeast drugs at the 1932 Olympics in Los Angeles and performed well, and subsequently yeast drugs became widespread in the Japanese sporting world. They were a feature of the age in the 1930s. Japan differed from other national contexts not in the prominence of yeast drugs and vitamins, but in that these products were developed by agricultural chemists who came from a background in fermenta-

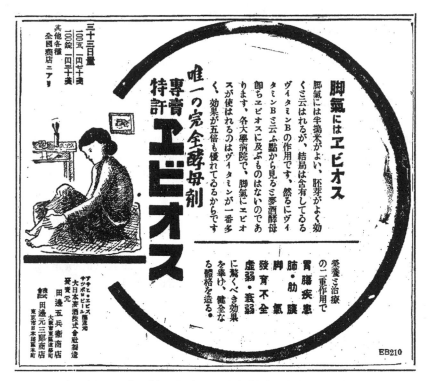

FIG. 2.3. Advertisement for Ebiosu in the *Yomiuri shimbun* (July 10, 1932), morning edition, 1. Reproduced with permission of the *Yomiuri shimbun*.

tion and brewing science, not medicine.[74] Hashitani, who had trained in agricultural chemistry, explained the slow reception of yeast drugs in the pharmaceutical field by observing that until after World War II, agricultural chemistry had a low status in the medical world, and medical journals generally did not cite agricultural chemical research.

The premise of Hashitani's design was an economy of yeasts: an excess supply from the beer industry could be revalued and sold in the pharmaceutical industry. But by centering the yeast microbe itself, his design also implicated ecological thinking, or a perspective on resources that took into account the interrelationships between industries, specifically as mediated by yeast metabolism. By contrast, the Japanese physician Shirota Minoru's design of the probiotic lactic acid drink Yakult in the 1930s showed no sense of the macrolevel resource redistribution that concerned scientists working in agricultural chemistry. Inspired by Russian theories of immunology, Shirota's ecology related to the competition between good and bad bacteria in the human body, not to a national resource economy.[75] While Hashitani's was essentially an economic view, the later thinking of other scientists, especially Takada Ryōhei, was an actualization of the ecological line of reasoning already implied in Hashitani's yeast design.

The work of chemical engineer Takada Ryōhei (1898–1978), a leading figure in the fields of vitamin science and seasoning (*chōmiryō*) science in mid-century Japan, gives us a closer look at nutrition research as it developed across the transition into the Asia-Pacific War. Takada Ryōhei—whose main body of work came slightly later than Suzuki Umetarō's—analyzed agricultural and industrial production as parts of a single national landscape of nutrient distribution. Unlike in interwar Germany, where ecological thinking on food was associated with vitalist currents and lay challenges to expert authority, in Japan ecological thinking resulted from sheer scientism in managing the nutritional political economy.[76] It is evident in the arguments over the limits to industrialization. In the thick of wartime, Takada criticized Suzuki's proud chemical products, remarking that vitamin synthesis to compensate for wastage of natural resources was not conquest of nature, but "treason against nature."[77] Distinctively, Takada's conceptions brought the role of microbial activity to center stage in mediating the nutritional ecology.

A less well-known figure than Suzuki, Takada had a career that reflected the choices available to a younger transwar generation of fermentation scientists. His trajectory spanned prominent government institutions and university departments in the fields of food and chemical research, and he founded the Vitamin Society of Japan. It was a period when broader technocratic ideologies of building a fascist, self-sufficient bloc in East Asia upon total, effective, Japanese-led resource management were at their height.[78] The government had been unable to contain the various sources of violent energy that were present across the Japanese empire and which had supported radical change since the Meiji Restoration. This culminated in a series of high-profile assassinations of politicians and capitalists in the early 1930s, the invasion of Chinese territory in Manchuria by the Kwantung Army (Japan's imperialist arm in northeast China) in 1931, and the end of parliamentary rule in 1932, along with increasing militarization of the cabinet. Deficit spending then aided the country's relatively rapid recovery by financing the heavy industrialization of Manchuria and Korea, effectively keeping Japan's economy afloat by military spending in the 1930s, up to the outbreak of full-scale war with China in 1937.[79] Though Takada's specific product innovations ultimately did not become important commercially, the approaches in his work would become widespread after the war.

When Takada Ryōhei graduated from the Faculty of Engineering of Kyoto Imperial University in 1922 with a specialization in industrial chemistry, he first entered the Institute for Nutrition Research (Eiyō kenkyūjo) as an assistant engineer. Synthetic sake was not the only scientific consequence of the 1918 rice riots; the riots also left a major scientific legacy in the form of the Institute for Nutrition Research, established under the Home Ministry in 1920 to coordinate long-term responses to problems of diet. On its founding it was one of the only large-scale institutions in the world dedicated to human nutrition, and it aimed at both fundamental and practical research.[80] Along with the results in vitamin science spearheaded by Suzuki Umetarō at Riken in the 1930s, work at the Institute for Nutrition Research would play a central role in informing strategies for efficient resource management as they related to the critical area of food during the Second Sino-Japanese War. Takada worked at the Institute for Nutrition Research until 1931, when he became faculty in the Brewing Science Department at Osaka Imperial University's Faculty of Engineering. In 1943 the chemist Kita Gen'itsu invited Takada to move to the Faculty of Engineering of Kyoto Imperial University as a professor of industrial chemistry, and he remained there for the rest of his career.[81]

While Takada's youthful views on the "food problem" were similar to those of other nutrition scientists when he began working at the Institute for Nutrition Research in the 1920s, he brought his own microbiological expertise to bear on the issue. He wrote in 1929 that along the lines of Malthusian thought, unlimited population growth would meet saturated food production because there was limited land available for cultivation.[82] Like the dominant chemists of the period, he believed that the most "rational" (gōriteki) method of food production was synthesis from nonagricultural matter because it required no land area, and that humankind was making gradual scientific progress to that end: sugars and carbohydrates had been synthesized in Europe, as were even polypeptides, which were a step toward protein synthesis. Vitamins would be synthesized in the future once their chemical nature was elucidated. However, he thought, the ideal of chemical synthesis was still far away, and at this intermediate stage, the two most promising possibilities were the saccharification of industrial waste products and the use of microbes as food. The former would supply calories, the latter, proteins and vitamins. After a year of reading the contemporary literature, at the end of 1923 Takada began research on turning yeast into food, a subject receiving much attention in the international scientific journals of the time. Takada recounted that his research focused on experiments on *Saccharomyces sake*. However, sake yeast later become well known for its low vitamin value in contrast to beer yeast.

Various technical difficulties, as well as his frustration that fresh yeast was not in any way tasty—meaning that he was increasingly reluctant to attempt to replace fish or meat with yeast—led Takada to seek other types of microbes that might be more suitable. In 1925 he abandoned yeast research and would not revisit it until ten years later. In the meantime he began investigating *kōji* microbes as a possible food. *Kōji* microbes drew his attention because they were easy to culture and process, but also partly for cultural reasons: since *Aspergillus oryzae* was Japan's unique product, he reasoned that inventions and food deriving from it would become a source of pride for the people. Moreover, very little research existed on the chemical makeup and nutritional value of molds, and nobody had yet suggested using molds for food.

In the course of the late 1920s and through the early 1930s, the focus of Takada's research expanded to the nutritional content of traditional brewed seasonings more generally, especially that of miso and soy sauce.[83] This research concentrated on vitamin content in the product as the measure of nutritional value. Because microbes were known to consume vitamins as they propagated as well as synthesize them, the changes in levels of vitamin B or B_1 in the fermentation fluid during the brewing process were complex and fed back to affect the growth of the microbes. Takada's research on measuring the changing vitamin levels in the fermentation fluid led him to link brewing processes with processes inside the cell: how vitamin B_1 disappeared from the fluid to accumulate inside the microbes, how microbes biochemically synthesized vitamin B_1, and how these processes affected microbial growth and the nutritional value of the fermented product.

With the increasing anxiety about resources that followed the 1927 Shōwa financial depression and the 1931 invasion of Manchuria, Takada responded in the mid-1930s by directing his attention to amino acid and protein production from industrial waste, in the context of the new amino acid seasoning that was becoming available in the market of the brewing industry. In the late 1930s his attention turned entirely to problems of nutrient supply during the restrictions of war. He was especially devoted to the more rational processing of fish catch as a solution. In the course of his career, he never touched chemical synthesis of food from mineral matter, and overall he focused on the full use of very constricted resources—both agricultural and marine products, as well as industrial by-products—from the perspective of a nutrition scientist. To Takada, this meant viewing the industrial and agricultural landscape of Japan in terms of relationships mediating nutrient distribution, primarily as an ecology of vitamins.

From this nutritional perspective, Takada later recalled, "the Greater East Asia War was a vitamin famine," when food imports stopped and the demand for vitamins in the military grew.[84] The food shortages in the years after the beginning of the Second Sino-Japanese War in 1937 apparently came as a surprise. In January 1939, Suzuki Umetarō wrote in the *Yomiuri Shimbun* that Japan had nothing to worry about in terms of food, however long the war continued.[85] Writing in an academic journal in 1941, Takada Ryōhei said that three years earlier, nobody could have predicted the country's unprecedented and comprehensive food crisis. The time when they believed that their country was blessed with an abundance of food compared to that of the Western civilized countries was firmly in the past.[86] Suzuki was more upbeat about the state of nutrition science in 1941, stating that thanks to dramatic progress in the previous two decades, they had reached the point where they could now specify standards for the rational distribution of food.[87] Takada was less optimistic, writing later in 1944 that the body needed not only to meet a minimum amount, but to be saturated with each kind of vitamin in order to be truly healthy, and that, though researchers had established the amounts for daily vitamin requirements in the United States, there was little research on the appropriate amounts for Japanese people.[88]

The "rationalization" (*gōrika*) of the resource economy first emerged as a goal of nutrition and food science with the resource anxieties of the 1930s. After the war began, however, rationalization of resources became the overwhelming focus. In a 1941 article titled "The Food Problem from the Industrial Perspective," Takada Ryōhei criticized root assumptions about the entire Japanese diet.[89] Even in the prewar period, Takada said, the 2 percent of food that was imported had been due not to a shortage of food production but to the poor use of domestically produced resources.[90] He began with staples. Presenting tables of the amounts of nutrients in various starchy crops, Takada argued that rice was high in both calories and proteins, even though it did not surpass sweet potato for the former and soybean for the latter, and that in that sense it was not erroneous for rice to be the staple in Japan in peacetime. In times of shortage, however, it was dangerous to rely only on rice. He complained that bread and udon were being called "substitute food" (*daiyōshoku*) for rice; bread and udon were splendid foods in their own right. "Economic rice" (*setsumai*), which referred to mixing rice with daikon or wheat, was almost as bad: the method entirely disregarded the special qualities of these foods, since wheat could be milled into powder, and daikon was best eaten raw. Mixing them with foreign rice was even worse; foreign rice truly was substitute food.

Moving on to seasonings, what seriously bothered Takada was the practice of soy sauce brewing, which in contrast to sake brewing had not drawn any attention for being a waste of food. He had for years investigated soy sauce in terms of the rate of usage of the nutritional constituents in the raw material, soybean—and the rates were low. The protein lost in soy sauce brewing each year was enough to feed tens of millions of people in that year; the calories lost could feed a million. And then, a million tons of *katsuo* (skipjack tuna) were caught each year, but most of this excellent protein source went into fertilizer or animal feed. This was "irrational" (*fugōri*), especially when one considered that a shortage of animal protein characterized the condition of the nation. In the northern part of Korea, the fish oil industry dumped huge amounts of waste water into the sea, which perhaps carried off 70 percent of the fish matter. If they could take this water and condense it into half its volume, it would result in a seasoning with the same nitrogen level as soy sauce (chemists took nitrogen level to be the quantitative measure of flavor concentration in soy sauce). He stressed that there was enough waste water produced in the fish oil industry each year for this hypothetical seasoning to replace half of the entire soy sauce market.

Powerfully, Takada argued in a 1943 article that Japan was a "vitamin resource kingdom."[91] He drew detailed tables of the domestic production of each kind of vitamin, which he calculated by adding together the vitamins in every kind of agricultural produce (fig. 2.4), and compared them with the estimated amount of the vitamin required daily for the health of every person in Japan. He showed numerically that there was enough of every kind of vitamin produced each year for Japan to be far more than self-sufficient in vitamins, with the exception of vitamin B_2. Just as in the microworld of his laboratory he saw vitamins moving out of the culture fluid and accumulating inside the yeast microbes, on the macroscale of national industries he saw the vitamins in each crop being harvested with expense and labor, and then being destroyed by being boiled and steamed in industrial processes, streamed into livestock feed or fertilizer, or simply discarded into rivers and seas along with the "inedible" part of the crop, or in refinement or distillation processes.

This, Takada explained, put Japan's vitamin situation in a critical condition. He suggested that microbes such as yeast be cultured on some of

我が日本は世界に誇るビタミン資源王國である。

日本全土1箇年に生産する食糧資源中に保有するビタミン量を算出して見る。

食糧名	年産額 (1000t)	A	B_1	B_2	C	D
米	13,520	9.5	48.7	40.5	—	—
麥類	4,832	3.4	19.6	10 8	—	—
雑穀	1,365	1.3	8.2	1.1	16	—
大豆	816	9.0	4.1	2.0	82	—
其他豆類	426	3.2	1.8	1.2	42	—
甘藷	5,214	36.5	5.7	5.0	573	—
馬鈴薯	2,600	0.5	2.3	0.5	468	—
其他芋類	701	1 1	0.8	0.4	72	—
大根,蕪	2,728	—	2.3	0.6	962	—
其他根菜類	940	21.0	0.8	0.4	106	—
葉菜類	1,818	96.6	1.4	1 4	1130	—
茶葉	194	34.0	0.3	0.3	281	—
果菜類	2,135	12.0	1.1	1.0	416	—
果實類	1,790	1.8	1.8	1.0	716	—
藻類	622	18 6	3.2	2.0	746	—
鳥獸肉	298	2.0	1.5	3.0	24	—
乳及卵	462	11.4	1.2	0.8	5	—
魚貝類	6,228	182.7	7.0	5.1	333	14.0
水産動物	783	7.8	0.5	1.6	47	1.2
總計	—	452.7	107.4	78.6	6019	15.2

FIG. 2.4. From Takada Ryōhei, "Bitamin shigen yori mita Nihon" (1943), 289. The table shows that the total amount of each vitamin produced annually in Japan is a surplus relative to what is needed to keep every member of the population healthy, with the exception of vitamin B_2. Reproduced with permission of the Chemical Society of Japan.

the waste products as a way to "recover" or recycle (*kaishū*) the vitamins that would otherwise be discarded, and he also mentioned his work on producing B_2 using the mold *Eremothecium ashbyii*. He praised the stunning results of recent work in vitamin synthesis as the conquest of nature at the hand of Japan's chemists. But he worried that the brilliance of synthetic processes blinded people to their flawed logic: to use tens of different types of industrial chemicals and large amounts of raw material to synthesize one kind of vitamin did not constitute "turning nothing into something." While recycling discarded vitamins could not be compared to the scientific achievements of synthesis, surely one should first use the natural resources available, and then rely on synthesis as a supplement where nature fell short. He worried that people's blindness led to a tendency to take the abundant vitamins in Japan's natural environment and turn them into nothing.

In 1944, Takada Ryōhei published a short book summarizing his thoughts on rationalization, titled *Wartime Food Measures*. He wrote that the drought in Korea, Japan's "granary," in the summer of 1939 — which was followed by the first rice rations — was over in a year, but the food problem had steadily gotten worse. Who would have imagined that people would be queuing to buy daikon, and that they would need ration tickets to buy sardines? In 1939 the government had established two principles for food policy in response to the drought: to increase food production, and to economize further on food consumption. That year there was still leeway: people were debating whether hotcakes should be called snacks or substitute food, and hair permanents were the issue of the day.[92] Now it was different. People were desperate simply to ease the painful feeling of an empty stomach, which they usually accomplished with water. According to the latest newspapers, it would be possible to make bread out of straw, and then factory workers and soldiers could eat their fill on it. But Takada argued that food problems could not be treated in this way: one must think from the nutritional perspective of proteins and vitamins.[93]

Where Suzuki Umetarō had imagined a step change enabled by fossil fuels, to which empire unlocked access, Takada saw a closed system of agricultural resources to be the only pragmatic way of envisioning the country's immediate problems. There was no more room for economizing on food, he argued; even in peacetime, people had not eaten in excess of the daily essential amount, and now they were at the limit of economization. War power would decrease, and they would fall like Germany in World War I if they went below the minimum.[94] As for increasing food production, the government thought that if they could return production to prewar levels, that would solve the food problem. So would they dig out wild fields, farm barley in strawberry fields, and grow vegetables along the roadside? It was impossible.[95] There was room to use resources more rationally, something research laboratories had worked hard on. In peacetime, only 81 percent of the food resources produced actually went into food, so there was room in the other 19 percent. Takada dismissed imperial visions of food resources expanding into limitlessness, with each additional country to come under the flag of Japan and into the Greater East Asia Co-Prosperity Sphere; those expectations were dreams of wartime, he wrote. And either way, any leading power such as Japan in East Asia could not maintain its dignity if it depended on other countries for its own food.[96]

Highlighting his perspective of the vitamin ecology that lay in the chemical makeup of agricultural products, Takada wrote that it was not just the quantity of food but the quality of food that was important. Vitamins were required to truly replenish a person's health. Here Takada noted that the United States had never been self-sufficient in the fish liver oil that supplied vitamin A and D, and that it had relied on Japanese exports. In this way, in the prewar period, Japan had contributed steadily to strengthening the enemy's bodies; but, he reassured readers, vitamins were not stored for long in the body and so their efficacy should be lost by now. Takada slammed the idea of synthetic sake being more economical than brewed sake. For him, the culprits of resource wastage did not lie in sake (especially now that it consumed less than a fifth of rice supply under the new restrictions), nor in *mochi*, candy, or miso. They lay in rice vinegar, which represented an extreme waste of calories and proteins, and in industrial products such as alcohol and butanol, which used potatoes and destroyed or threw away all of their vitamin C and B.[97] Whether alcoholic drinks were essential to daily life was outside of the scope of his book, but he would proceed on the assumption that they were somewhat essential. Previously, due to rice shortages, sake had been at the center of the food problem debate; but now every kind of food was scarce. The main ingredient of synthetic sake was alcohol, which used sweet potato as the primary raw material. It would be different if it were molasses, which was not easy to convert into food. But the nutrients of sweet potato—proteins, vitamins, and minerals—were simply flushed away with the distillation waste liquid in the manufacturing process. Because of this, synthetic sake should not be called the darling of the age.[98]

Takada directly addressed flavors alongside nutrients as analytical components for rationalization, specifically through the consideration of seasonings. He recognized that East Asian people had loved umami as the "taste of *dashi*" from long ago, and their pioneering seasonings were their pride: rice vinegar, *mirin* (sweet rice wine), miso, soy sauce, kombu, *katsuobushi* (dried skipjack tuna flakes), *niboshi* (dried baby sardines), and so on. To satisfy people's craving for tasty things was one of the three main purposes of food (the other two being to fill an empty stomach and to obtain nutrients). It could not be denied that, even if one should restrain one's greed for tastiness in a situation as strained as that at present, eating tasty foods encouraged the secretion of digestive juices, and the absorption of nutrients. Of course, under the pressures of the war it was most important to consume food for nutrients, but one still wanted to eat tasty foods where possible. Eventually, Takada believed, only salt would remain, but the food situation was not yet quite as desperate as that.

He revisited his idea of using fish such as herring and *katsuo*, which were caught in large quantities in northern Korea and Hokkaidō, and using the waste water from the fish oil industry to make a product so concentrated and flavorful that it could substitute for half the market of the highest-grade soy sauce. He lamented that the fish catch had declined so much in recent years (due to overfishing, as well as the conscription and commandeering of vessels) that he had had to abandon tests on his idea.[99] Moreover, he was fascinated by the fact that the amino acids in *katsuo* created an umami flavor different from those of all other seasonings. Conventional seasoning—whether it was soy sauce, amino acid seasoning, or powdered seasoning—relied on glutamic acid for the umami flavor. But *katsuo* seasoning was based on inosinic acid, which created the aroma of *katsuobushi*.[100]

In the amino acid ecology, too, Takada saw the redistribution of resources including nutrients, minerals, labor, and energy. The costs of chemical work in making amino acid flavoring incensed him in particular, as had vitamin synthesis. It was true that in soy sauce brewing the *moromi* traditionally took a year to mature, which meant that much of the carbohydrate was lost to the production of carbon dioxide gas. But reducing the fermentation time by pressing the soybean before brewing could alleviate this problem. Amino acid flavoring, on the other hand, was not an ideal alternative to soy sauce. The carbohydrates were lost in decomposition by hydrochloric acid, and the process involved using such expensive industrial chemicals as hydrochloric acid and sodium carbonate to produce common table salt. Along with the bad smell of the seasoning, these were the reasons why his own laboratory had abandoned work after years of research on the topic. Likewise, it angered him that crystallized glutamate salt (MSG) as a seasoning would often simply be mixed with soy sauce, wasting all the labor and energy that had gone into its refinement.[101]

As an alternative to agriculture for producing nutrients and flavors, Takada's work highlighted three resources in particular: fish, yeasts, and molds. His new pet project during the wartime period was the development of marine products. In his book he devoted many pages to the "irrationality" of the industry's use of fish matter, and during the war he succeeded in developing a method of using the eyeballs of *mentai* (Alaska pollock, caught and consumed in Korea) as vitamin B_1 preparations. The method was used in the military and incited a temporary "eyeball boom" in Korean restaurants as well.[102] Fish took up no land for cultivation and required only the minimal resources (manpower, fuel, and boats) to obtain, so Takada perceived them to be truly a bounty for Japan in terms of proteins and vitamins, which he argued should be better used.

Throughout his career, Takada repeatedly turned to a different kind of natural resource, microbes, as a strategy for recovering nutrients from agricultural products that were discarded as industrial waste. Like many other nutrition scientists, he was irritated by the state's decision to encourage *genmai* (unrefined rice) in the military as a source of vitamin B_1.[103] He had hoped to produce vitamin B_1 much more efficiently by growing yeast on the husks discarded from rice refinement. He had also experimented with inoculating yeast into the distillation waste water in alcohol and butanol manufacture. In this manner, he reasoned, one could recover not just the vitamin B_1 but also the carbohydrates and nitrogenous matter dissolved in the water, which the yeasts would absorb into their cell constituents as they propagated, and which could be harvested by collecting the yeast. He suggested that the same method be applied to the waste water from pulp making, and he was developing a similar process to recover vitamin B_2 from butanol fermentation. In this way, he was hoping to do what Suzuki Umetarō had suggested earlier, but replacing soybean with waste fermentation fluid, and cows with microbes: he wanted to harness the biochemical pathways of a living organism to transform and concentrate nutrients from a raw material.[104]

Takada's laboratory turned to a mold to plug the gap in vitamin B_2 that represented what he saw to be Japan's only real poverty in vitamin resources. Traditionally, miso was the nation's main supplier of vitamin B_2, since the *kōji* microbes, yeast, and bacteria that contributed to miso fermentation all synthesized vitamin B_2. But it was not quite enough vitamin B_2 to supply the entire country.[105] Because of miso's nutritional qualities, Takada argued that it should be classed as a food, not a seasoning. After many years of miso research, his laboratory had turned to the possibility of employing the peculiar properties of molds to produce sufficient vitamin B_2 to make up for the national shortfall. *Eremothecium ashbyii* came from a cotton plant in Africa. The mold had been named by Alexandre Guilliermond in France, who had sent a sample to Saitō Kendō at Osaka Imperial University, where Takada had picked it up and found that it could make surprisingly large amounts of vitamin B_2 when cultured under the right conditions. In partnership with Wakamoto Seiyaku, Takada's laboratory had the mold mass-manufactured at Wakamoto's factory in Itami, and later in other factories in Japan, Korea, and China. The vitamin B_2 thus produced went entirely to the military. However, the factories stopped making the product after the surrender, when military demand vanished.[106] To Takada, vitamins were not drugs; they were nutrients and were like food. They should not be given only to weak and ill people.[107]

Takada Ryōhei offered a brief explanation of the meaning of much of his life's work in 1956, in a description of the link between fermentation and vitamins that he wrote for a compilation titled *Mold-Utilizing Industries*.[108] His microbe-centered thinking emphasized ecology and symbiosis, and he inverted the traditional hierarchy of the living world as captured in the Great Chain of Being.[109] Vitamins, he said, were essential elements animals needed that could not be made in their bodies. Animals had to obtain them externally, generally from higher plants, which had the ability to synthesize vitamins inside the plant body. Microbes, which were lower plants, had the ability to synthesize vitamins, but with restrictions and conditions that differed for each kind of microbe. The lowest kind of microbes which represented the closest proximity to animal-type conditions were bacteria, which could synthesize a few vitamins but mostly needed vitamins from the outside world. Yeasts were closer to plant conditions than bacteria, and many vitamins were synthesized within their cells, but often growth would be promoted when vitamins were supplied from the external world.

Molds, which were even closer to higher plants, had the functions in their cells to synthesize the majority of vitamins. The fact that each microbe had the ability to synthesize various vitamins was significant for the fermentation industries because, Takada wrote, the activation centers for the various enzymes that participated in fermentation were the vitamins. Mutual supply of vitamins primarily accounted for the symbiosis between microbes seen in fermentation processes. From that perspective, the vitamins that microbes made were the source of the fermentation industries. There existed microbes that made certain vitamins in especially large quantities, and they could be used industrially, but due to the progress of synthetic chemistry, all the vitamins apart from B_{12} had been chemically synthesized, so the microbial production of vitamins had not become established commercially.

AMINO ACIDS AND VITAMINS IN REDISTRIBUTION

Some of fermentation scientists' most lasting nutritional technologies— namely, amino acids in the form of chemical soy sauce and synthetic sake, as well as vitamins—were first mass-manufactured and widely consumed during the Asia-Pacific War. These changes were not market-driven developments; mass production did not thrive until the wartime conditions and economic planning of the late 1930s. At the same time, neither was it the state, any more than consumers, that entirely decided what kind of products should exist. Both groups set constraints that scientists and technicians—those people who were directly involved in production in between—found a way to navigate materially. As a result, scientists incorporated flavor as much as nutritional components into their conceptualization of the national resource landscape and into their pioneering designs for tools of national management. In contrast to the sweeteners, margarine, and vitamins prominent in Europe and the United States, in Japan efforts to develop material interventions concentrated on the massive seasoning and sake markets.[110] Scientists and technicians hoped, as Suzuki Umetarō had put it, that consumers would "change" their taste preferences in a time of great resource pressure.

Within nutritional technology in Japan, flavor research led to a distinctive focus on amino acids, which first entered the Japanese food and drug market as seasoning. In the late 1920s, around the same time as Suzuki Umetarō's synthetic sake laboratory at Riken was acid-hydrolyzing waste silk threads in the attempt to obtain amino acids similar to those in rice protein, research on the acid decomposition of the proteins in soybeans was underway within soy sauce companies and government institutes. The goal was to increase the speed and scale of soy sauce brewing and thereby to raise profit margins in traditional industries, much like the original aim of Tsuboi Sentarō's substitute sake. Since the Meiji period, soy sauce companies had made continuous efforts to mechanize and create new "fast brewing methods" (sokujōhō) for this purpose.[111]

The method of using acid hydrolysis to decompose proteins had emerged out of flavor manufacturing, and had been employed originally to create amino acids in Ikeda Kikunae's monosodium glutamate (MSG) process, patented in 1908. It was only in the mid-1920s that acid hydrolysis began to be applied in soy sauce brewing.[112] In Kansai the Yamashita Chemical Laboratory and the Osaka Municipal Technical Research Institute collaborated to study the acid decomposition of pressed soybean, and their product was commercialized in liquid form as Yamashita Chemical Laboratory's "amino acid seasoning" (aminosan chōmiryō). Compared to brewed soy sauce, which took six months to a year, amino acid seasoning took three to seven days to make.

Following the Shōwa financial crisis of 1927, numerous sites, including the national Brewing Experiment Station, the Tokyo Industrial Experiment Station, and soy sauce companies, began similar work. They were motivated by depressed economic conditions to develop cheaper chemical methods that better conserved agricultural raw material. Amino acid seasoning was also a vendible by-product of MSG manufacturing, and so when the Suzuki Chemical Company's patent on MSG expired in 1929, a number of companies began to produce both MSG and amino acid seasoning.[113] The invasion of Manchuria in 1931 spurred new anxieties that raw materials for soy sauce might not be plentiful for much longer. Soy sauce manufacturers had been relying on Manchurian imports for soybeans, but on US and Canadian imports for wheat. In Manchuria, too, the South Manchuria Railway Central Laboratory produced soy sauce using acid hydrolysis of pressed soybean. Self-sufficiency thus provided a powerful incentive to develop amino acid seasoning.

Agricultural chemistry at Riken (that is, Suzuki Umetarō's lab) had already made flavor a part of its research agenda in the political economy of nutrition. However, whether amino acids could be a *nutritional* substitute for protein, in terms of their effect on human growth, was investigated but secondary to considerations of flavor substitution in the technological development of amino acid manufacturing. In any case, it was clear that amino acids and proteins were not equivalent in nutritional value, any more than the Riken sake lab had found them to be so in taste. In the 1930s Takada Ryōhei, then at the Department of Brewing Science of Osaka Imperial University's Faculty of Engineering, worked in conjunction with the soy sauce company Kawamata Shōyu to research the nutritional qualities of amino acid seasoning. It did not lead to growth in mice if it were used as the only source of protein, but when combined with other sources of protein, it did have a supplementing effect, and studies showed no harm, at least, when it was 10 percent or less of the feed mixture.[114]

Quantity became more important than quality after the beginning of the Second Sino-Japanese War in 1937, when amino acid seasoning became widely used in soy sauce. In 1938 the Japanese government effected the National Mobilization Law. As the United States levied sanctions on Japan, the government attempted to achieve self-sufficiency through rationing, cartelization, and manufacturing control associations, including those related to soy sauce and miso.[115] The supply of raw materials began to dry up. Amino acid fluid had a higher rate of use of the main raw material, soybean, than did brewed soy sauce, and so it was considered a more economical use of the remaining resources. It became common practice to mix amino acid fluid with soy sauce. As the war went on, the proportions in the mixtures changed until they consisted mostly of amino acid seasoning, also called "chemical soy sauce" (*kagaku shōyu*) or "substitute soy sauce" (*daiyō shōyu*). At Kawamata Shōyu, as its institutional history recalls, the acid-decomposed amino acids had a peculiar smell. The proteins for acid hydrolysis came less and less from soybean, shifting to anything that was vegetable matter: pressed vegetable lees from oil factories, the raw lees of fallen petals, cotton fruit lees, or coconut lees from Micronesia.

Amino acid soy sauce flourished best in the 1950s, reaching its highest production levels around the end of that decade. The US occupation authorities encouraged the manufacture of chemical soy sauce as an important solution to the continuing resource shortage. Noda Shōyu (later Kikkōman), which had developed "New-Style Soy Sauce No. 1" in 1944, went on to release "New-Style Soy Sauce No. 2" in 1955.[116] In 1948, researchers at Ajinomoto converted waste water from MSG manufacture into a seasoning product, selling it as Esusan Mieki ("S-Three Flavor Fluid"), or simply Mieki. The mixing and interchangeability between amino acid seasoning and brewed soy sauce remains in manufacturing processes today.

The expansion of synthetic sake production followed trends similar to those in the production of chemical soy sauce. It was Riken that led research, rather than private companies and government experiment stations, due to the relative technical difficulty of the flavor problem in sake. Research was thriving at Suzuki's Riken and Komaba labs by the late 1920s, with many of the researchers at Riken especially coming from a background in protein chemistry.[117] But the drink's production levels only shot up in the late 1930s. Yamato Jōzō began to sell it in the mid-1920s, and the Industrial Experiment Station in Hokkaidō and the Taiwan Government-General Monopoly Bureau began to manufacture it in 1924 and 1925 respectively. Riken set up an internal pilot plant (fig. 2.5), and in 1936 it was able to have the patent rights returned to it from Yamato Jōzō. Synthetic sake manufacturing expanded across Japan after 1937, and by 1943 the drink was made in fifty-two factories by twenty-seven companies, with a producers' association coordinating the distribution of rationed raw materials. Technicians from companies flowed through the Riken laboratory seeking guidance on making synthetic sake in their factories.[118] The product was sent across the Japanese empire to troops, especially in colder regions; and to Suzuki's delight, production amounts soared. He hoped to see them climb to a million *koku* in his lifetime, though when he died in 1943, they had reached over seven hundred thousand *koku*.

After the surrender in 1945, the occupation authorities seized the finances of the producers' association and ordered it to disband. When it regrouped in 1948 as the National Synthetic Sake Association, synthetic sake

FIG. 2.5. Alcohol fermentation tanks at Riken. Courtesy of the National Museum of Nature and Science.

production (and sake production in general) was at an all-time low, and synthetic sake had again come under the suspicion surrounding its quality. In 1949, however, rations on potatoes were annulled, encouraging the alcohol industry to begin to revive, and in 1951 Riken received a license to use rice for synthetic sake. When in 1953 the Tax Department of the Ministry of Finance began offering financial assistance to the drinking alcohol industries, which would continue until 1958, synthetic sake production reached its peak at 790,000 *koku*, surpassing wartime levels. Continued shortages of rice and other raw materials, as well as further research and technological refinement, meant that synthetic production flourished through the 1950s. It was only from the early 1960s that the brewing industry was able to recover and the production of synthetic sake wound down—although, much like the relationship between amino acid seasoning and soy sauce, the interchangeability between synthetic and brewed sake in industrial production continues to the present day (fig. 2.6).

For vitamins, the technological means for truly mass-scale manufacture did not exist until late in World War II. Wartime policies therefore centered instead on providing vitamin-rich foods, the scientific knowledge of which was well established. It was in the army that vitamins first shifted from being solely a drug to being employed in foods. According to memoirs, Japanese

FIG. 2.6. A bottle of Riku synthetic sake recently produced by Asahi Breweries, on display in the office of the Riken archives, June 2011. Photo by author.

nutrition scientists had been preparing for the harsh conditions of fighting in far northern Chinese and Russian lands, and were somewhat surprised that the battles were concentrated in southern tropical areas where fresh food was more plentiful. Nonetheless, they developed many vitamin-rich foods for the battlefield, mixing dried egg yolk with vitamins A and D supplied by Riken (the preparations were made at the Institute for Nutrition Research), adding *kanpō* (Chinese-style) drugs to sake to refresh airplane pilots, or using *mentai* eyeball powder for vitamin B_1 or *mikan* fruit skin for vitamin P. These scientists knew that it was important for the foods to be tasty to maintain the morale of the military men. Vitamin fortification of common foods was a later postwar phenomenon.[119]

This chapter has traced the development of technologies to manage the nutritional resource economy, such as synthetic sake, chemical soy sauce, and yeast preparations from the aftermath of World War I to the beginning of the post–World War II period. In doing so, it has looked at how fermentation science expanded beyond the traditional brewing industries into the broader food and drug industry through engagement with nutrition science. Although the new technologies bore similarities to products appearing in the late Meiji period, they were motivated by entirely new questions that had emerged in the 1910s. In the late Meiji period, the push to improve and modernize traditional industry drove the bulk of innovations in Japanese science, and questions about health and diet were inseparable from the question of Western culture and the novelties it introduced to Japan. By the late Taishō and early Shōwa periods, however, the stakes had changed. Although the Malthusian conception of the "population problem" was not unique to Japan, scientists such as Suzuki Umetarō investigated nutrition as predominantly a local question that tied science to Japan's peculiar local diet and agricultural economy. In particular, it was bound up with the practical, national requirements of war, and Riken scientists came to be less concerned with the advancement of traditional industry than with the improvement of the Japanese race by the science of nutrition. At the same time, even amid Suzuki's most ardent chemical dreams of food synthesis from nonagricultural matter, connections with biological processes marked the entry point of his investigations. His interest in synthesizing amino acids as building blocks for proteins came from observing how proteins and amino acids broke down during fermentation processes, which he believed mapped onto similar pathways of chemical change inside any living organism.

By the 1930s the goal of nutritional work had narrowed to focus on the rationalization of resource use in Japan. In this period, scientists such as Takada Ryōhei continued to prioritize the local dimensions of their work. In his bid to develop microbes and especially molds as a means to provide nutritious food, Takada turned to "unique" Japanese microbes and drew on East Asian culture's propensity to use molds in industry. Like Suzuki, Takada tied the microprocesses in fermentation—in his case, how vitamins were distributed in the microbes and culture media in the course of brewing—to the large-scale distribution of food resources in a national industrial ecology. Unlike Suzuki, however, in the constricted economy of the 1930s he returned to cheaper agricultural rather than nonagricultural substances for raw material, and he emphasized the use of industrial waste in particular. Just as Suzuki suggested "borrowing" the power of livestock to forge better animal protein from inferior vegetable protein, Takada attempted to employ microorganisms' physiologies to select and concentrate, or even synthesize, the desired nutrients in their bodies. The approaches that his work represented became important commercially in the longer term, even though his specific innovations did not.

During this time, various new technologies became widespread. The technique of acid hydrolysis of proteins was a common process that became employed across the brewing industry, for products from synthetic sake to chemical soy sauce. Both technologies—synthetic sake and chemical soy sauce—remained, albeit in a more refined form, in the postwar food and drink industry as a way to economize on resources and to supplement brewing processes by allowing chemical adjustment of flavor. Amino acids and especially vitamins stopped being associated solely with medical substances or luxury drugs, and became a part of ordinary food consumption. These shifts took place because of the heavy strain on resources during the Second Sino-Japanese War, which drove the intensive study of resource use and the reorganization of industry to economize on resources. The local dimensions of these forces contradict the notion that nutrition science in Japan was primarily a story of transformation by Western science and culture. Moreover, because similar conditions continued in the immediate postwar years—the food scarcity, hunger, and shortage of raw materials for industry—postwar innovations such as glutamic acid fermentation bore resemblances to prewar fermentation work in their motivation and design.

Fermentation scientists' nutritional research tied their laboratory understanding of microbial physiologies to their grasp of the relationship between national resources and consumption, specifically in terms of the agricultural components of flavor, nutrition, and industry. In this sense, flavor was seen to be no different from other fundamental abstractions of food constituents like proteins, vitamins, or calories; people in the nation needed to consume them, and would; consumption of all these things needed to be managed efficiently. The practical laboratory-based work of fermentation scientists suggests that scientism in modern Japan lies in large part in these everyday technologies. The notion of progress premised on science is clearly visible if one considers the national agricultural landscape itself as a Taylorite shop floor in the interwar period. Scientists and technicians had an in-between role as managers of production: material developments could not entirely conform to the ideals of political economy set out by state actors at the highest level, but neither were these technologies purely market-driven developments. Everyday technologies were a site where experts could intervene in political economy, not through regulation but through production.

In the alcohol industry similar forces drove innovation, centering on the need for resource self-sufficiency linked to war, and the diffusion of new technologies in that field were also driven by resource concerns during the Second Sino-Japanese War. But unlike in nutrition science, much of the research in alcohol production took place outside the main islands of Japan in the Japanese empire, and the innovations went beyond indigenous methods to combine them with Asian continental traditions. Because of this, despite the local dimensions of nutrition work, the study of alcoholic production methods enhanced a sense of Japanese uniqueness as well as cultural solidarity with Asia to a far greater degree than did early nutrition research. It brought about a broader consciousness of fermentation not merely as a technology, but as part of Japan's self-conception of its history and culture.

3

Nation

ASIA'S MICROBIAL GARDENS AND JAPANESE KNOWLEDGE

Microorganisms are a part of culture, like language and religion. They are a part of history.
—Komagata Kazuo, professor emeritus, Faculty of Agriculture, University of Tokyo, in discussion with the author

Tōa hakkō kagaku ronkō (A Study of East Asian Fermentation Chemistry), published in March 1945, is an anthropological work which presents the history of fermentation technologies as a reflection on the abilities and culture of civilizations.[1] Its author, Yamazaki Momoji (1890–1962), was a Japanese agricultural chemist who had worked in Shanghai from 1914 to 1927 and subsequently was based in Utsunomiya, a prefectural capital in eastern Japan, from 1930. He won the prestigious national Suzuki Prize in Agricultural Sciences, the highest honor of the Agricultural Chemistry Society of Japan, for his pioneering book in 1945. As early as 1906, the botanist Saitō Kendō had called Japanese brewing microbes part of the diversity of "useful fermentation microbes of East Asia," and had thought of collections of such microbes as "gardens" that could allow research on their theory and application, from the tropics to the poles.[2] The work of these prominent Japanese microbial strain collectors—who were at once fermentation scientists and, in the case of Yamazaki, writers of anthropological study— reveals how Asia's microbial gardens looked to Japanese eyes, and how they were used in, and contributed to, Japanese knowledge.

This chapter and the next examine the construction of a national "Japanese" fermentation tradition by considering Yamazaki Momoji's 1945 book against the background of industrial alcohol production in the formal and informal empire. This chapter analyzes the intellectual structure of the understanding of fermentation history, and the next chapter traces the material development of alcohol production technologies. For these scientists, fermentation techniques were traditional technologies, possessing qualities including uniqueness and regionality—which, in turn, were qualities used by other Japanese writers and anthropologists in a multiplicity of ways in their depiction of cultural traditions at the time.[3] The frequent portrayal of Japan as modern *and* Asian at the height of 1940s Japanese pan-Asianism, for example, relied on particular ideological and rhetorical uses of sameness between Japan and Asia. That is, sameness was often constructed around the region's cultural uniqueness vis-à-vis the West, while it was deployed to highlight a *temporal* contrast between an advanced Japan and a more primitive Asia.[4] As the historian Robert Tierney explains, Japanese imperial culture is best understood as a "triangular structure," in which "the West was always the (implicit) third party."[5]

Science and technology have rarely been considered ancient cultural traditions in Japan, whether by intellectuals in imperial Japan or by historians today. Existing scholarship has assumed those things to be Western, universalist, and modern, something of the "now." Thus, a focus on intellectual discourses of science and technology has turned up only the latter assumption. To practitioners in fermentation science, on the other hand, microbial technologies in Japan were seen not as Western transplants but as Asian technologies. The historian Hiromi Mizuno has argued compellingly that the term "science-technology" (*kagaku gijutsu*, a now ubiquitous and seemingly innocuous phrase which can also be read as "scientific technology") was coined by wartime technocrats to mean a self-sufficient system of technological development led by Japanese research and based on raw materials from Asia, in order to support the imperialist vision of a Greater East Asia Co-Prosperity Sphere (*Dai Tōa kyōeiken*).[6] Practitioners in fermentation science, however—who had remarkably little to say on such abstract subjects as definitions of science or technology—would have struggled to reproduce even the technocratic imperial discourse that guided wartime policies in their own field: the rhetoric of Asia as representing merely resources, and Japan as offering science-technology. For Japanese scientists and technicians working in the empire, how could scientific and technical knowledge's traditional nature, as well as its debt to Asia, be reconciled with the imperialist ideology of scientific modernization?

It was a question at the heart of Japanese scientists' deepest ambivalence about national identity. *Kōji* today is the unique "national microbe" (*kokkin*) of Japan that makes the geographically distinctive products of sake, soy sauce, and miso; yet it is known to originate from China.[7] The phrase *Tōa* (East Asia) itself, which Yamazaki used in the title of his work, was so closely associated with the imperialist ambitions of the wartime state that after the country's defeat in 1945 it largely disappeared from common parlance. Yet until 1945 the identity of Japanese fermentation traditions and ultimately their function within modern Japanese political economy could only be defined in relation to *Tōa*—a fact of which Japanese scientists themselves were overwhelmingly aware. Here I read the Japanese accumulation of microbial strains and technologies within time-specific layers of visions of *Tōa*. Maps of *Tōa* were simultaneously mirrors of Japanese identity, but they nonetheless show how Asia's microbial gardens contributed to Japanese knowledge.

The regional reality was starkly clear to fermentation scientists, for several different reasons. First, local technologies—beyond Japanese or European technologies—were needed to develop economic products from the specific raw materials available in the formal and informal colonies. To this end, Japanese scientists working in the empire needed knowledge of local products. Alcohol fermentation research took place primarily in the Japanese empire rather than on the main islands, and it vividly reflected the use of regional knowledge in practice. Taiwan, especially, was important because of its sugar industry, which generated large quantities of molasses as a by-product, which were then used as media to grow microbes. Second, it was debatable and not obvious that Japan had historically leapt ahead of China and Korea in fermentation technology. Therefore, those scientists promoting the Japanese state's goals in the empire needed to draw on regional knowledge to make an explicit case for Japanese superiority. Both regional achievements and the historical debt to Asia needed to be reconciled with empire's ideological justification in terms of a civilizing mission, amid aims for practical economic integration in the empire on the part of statist elites.[8]

The practical problem became more urgent after Japanese forces broke away from the international consensus on empire in China by invading Manchuria in 1931. Consequently, the Japanese state's rhetoric shifted from presenting Japan as a Westernizing influence in Asia to emphasizing Japan's exceptional suitability as a leader in Asia vis-à-vis the Western powers due to Asian regional uniqueness.[9] Fermentation, in particular, was a body of scientific and engineering knowledge that grew to massive proportions during wartime, under the Allied embargo on gasoline exports to Japan. Chemical sites across the ambit of the Japanese empire, with synthetic or fermentation capability, turned to the problem of alcohol production for fuel substitution.[10]

Examining the work of practicing fermentation scientists, then, highlights the debt to Asia owed by Japanese scientific knowledge and, even more significantly, Japanese scientists' clear awareness that Asian contributions were centrally important *even* in science and technology. Yamazaki

Momoji's scientific work on history constructs a particular conception of the Japanese nation that must be understood in terms of comparison—not between Japan and Europe, as conventional accounts of modern science and technology have assumed, but between Japan and other Asian countries including Korea and China, even in the context of modern imperialism. In this chapter as well as the next, I thus respond to Andre Schmid's call (here paraphrased by Taylor Atkins) to "assess the effects of empire building on Japan's successful modernization process, rather than viewing the empire as the *outcome* of that process."[11]

Between 1945 and the years that followed lies a chasm in the way scientists verbalized their memory of Asia's importance, and a discontinuity in scientists' deep awareness of Asia. Immediately after Japan's defeat in the war, government officials, textbooks, and public discourse as a whole enforced this break by practicing what Carol Gluck has described as "a kind of 'transliterated history'": "The words changed, as if they were writing the past in another script. The Great Empire of Japan vanished from utterance, even in negative mention."[12] For the new generation of scientists who came of age during Japan's economic miracle, and amid the silence regarding Asia's significance in Japanese modernity, *kōji* accordingly became Japanese alone, while Japan offered technical assistance to countries in Southeast Asia as a major foreign aid donor with a record of modernization that was perceived to be exceptional.[13]

When recovering the positions inherent in scientists' accounts of Asia during the imperial period, it must be remembered that we expect *no conceptual coherence*. Since other structures of power were at play, ideologies produced by middling statist elites such as scientists and technicians were not required to do all or even most of the work involved in stabilizing colonial institutions. "On the contrary, duality, not to say duplicity, characterizes Japanese writings about the colonized," to put it in the words of the historian Robert Tierney. "Depending upon time and place, Japanese played up affinities between themselves and other peoples of Asia or stressed differences, claimed to represent Western civilization or asserted their absolute uniqueness. Accordingly, the Japanese rhetoric of sameness was not invoked consistently."[14] This is true for writings about both the formal and informal empire. The most striking feature of Yamazaki Momoji's 1945 text is schizophrenia. With the enormous effort displayed in Yamazaki's attempt to survey regional knowledge on fermentation, the very assertion that regionality was not important, which he felt compelled to make repeatedly throughout the work, is a reflection of Asia's monumental importance for Japan in his eyes.

Yamazaki Momoji's *Tōa hakkō kagaku ronkō* (A Study of East Asian Fermentation Chemistry) and the circumstances of its production and publication offer a glimpse into a pre-1945 vision of Asia (*Tōa* in Yamazaki's title) within which Japanese fermentation scientists worked. It was a world that all but vanished from public discourse after Japan's surrender in August 1945. The book was published in March 1945, immediately before the watershed in public memory. In more recent times, the *Study* is cited as a key source in *longue durée* technical works on the history of Asian fermentation, but the book is also remembered by older scientists who lived through the war as not unlike a symbol of the world lost. Whatever these scientists' intentions or hopes were regarding other Asian countries at the time, it is clear that they were keenly aware of Asia's contribution to Japanese modernity.

On a day in 1985, as part of a featured series on Japan's former higher agricultural and forestry schools (after the war, the higher school system was dismantled and many of the specialist schools were incorporated into universities), editors from the *Nippon nōgei kagaku kaishi* (Journal of the Agricultural Chemical Society of Japan) interviewed faculty and faculty emeriti of the Utsunomiya University Faculty of Agriculture, which was until 1944 the Utsunomiya Higher Agricultural and Forestry School (from 1944 to 1949 the Utsunomiya Agricultural and Forestry College).[15] There Yamazaki Momoji had taught since returning from China in 1930 until his retirement in 1953, first as a lecturer in agricultural products and applied microbiology, and then as the first head of the new agricultural chemistry department from 1945, which opened in the chaotic period preceding Japan's surrender; his title was changed to professor in 1950 after the school became a university. The higher school in Utsunomiya had been especially well known for its strength in foreign studies, teaching languages including not only English and German but also Russian, Chinese, and Spanish.[16]

The main gate of the university campus opened onto an old French-style garden, which students curiously called the English garden, leading to the entrance to the Faculty of Agriculture, where the interview was conducted.[17] As the agricultural scientists reminisced about former teachers, Komagata Kazuo, an agricultural chemist serving as interviewer from the journal, raised the topic of Yamazaki Momoji's 1945 *Study*, of which he had come to possess a copy. Even to that day it was the only work that had investigated Asian matters, he observed (here, of course, Komagata used the contemporary term *Ajia* for Asia, rather than *Tōa*). One of the interviewees, an agricultural chemistry professor who had himself graduated from the college in 1948, recalled Yamazaki Momoji cutting a figure as a distinguished scholar even though he was only in his early fifties: dressed in Chinese clothing, having just published the book, having won the Suzuki Prize. To defeated Japanese, the interviewee remembered, the book was a work of epic grandeur in its ambitions to unite Asian and Western culture.[18] Later, the *Study* was introduced by the leading Japanese agricultural chemist Sakaguchi Kin'ichirō in the PR magazine of the publishing house Maruzen, where the work was called "an illusory masterpiece among the war damage." It was illusory because most of the three thousand printed copies had been destroyed in the air raids and not sold in bookstores, making physical copies of the book relatively uncommon.[19]

A highly "individual" personality who was interested in many things, from the experimental brewing of Shaoxing wine to the locating of raw materials in Japan for making tsatsai (a pickled mustard featured in Sichuanese cuisine), Yamazaki was, as one interviewee put it, a "famous local product" (*meibutsu*) himself, giving loudly applauded public lectures in town dressed in Chinese clothing and bellowing at the top of his voice. After the war he styled himself "Dr. Nattō," after the strong-smelling fermented soybeans distinctive to Japanese cuisine, and wrote pamphlets lauding the nutrition value of *nattō* and other protein sources: "Eat *nattō*! My daughter raised on *nattō* is more gigantic than me." He could no longer walk by then, and would send his student (one of the interviewees) to kindergartens to do *nattō* taste research on his behalf.[20] Students and guests at his home were entertained with homemade vodka along with Chinese preserved duck egg and Chinese fermented bean curd.[21] When Utsunomiya University began selling some of their experimental agricultural products to the local residents of the city as an outreach initiative, a popular product was one of Yamazaki's creations: a lactic acid drink called "Milkis," which resembled the commercial drink Calpis.[22] His preserved duck eggs and Shaoxing wine were also said to be sold in the city.[23]

Considering Yamazaki's work in the empire as an agricultural chemist brings us further down the layers of memory into that pre-1945 world in which the modern construction of a national "Japanese" fermentation tradition took place. Yamazaki was a microbial strain collector, and his contributions are materially embedded in the Japanese culture collections. After graduating from the agricultural chemistry department of Tokyo Imperial University, he moved to Shanghai in 1914 and conducted research on Chinese agricultural products while based at the Tōa Dōbun Shoin (East Asia Common Culture Academy). He published a study titled "On Shaoxing Wine" in the *Nihon jōzō kyōkaishi* (Journal of the Brewing Society of Japan) in 1917, and completed his doctoral thesis for the Tokyo Imperial University Faculty of Agriculture in 1925, titled "Research on Chinese-Produced Fermentation Molds and Fermented Goods."[24] In between, Yamazaki returned for a time from the Tōa Dōbun Shoin to work in his mentors' laboratories at the agricultural chemistry department of Tokyo Imperial University. He isolated and investigated mold strains from the Chinese *kōji* materials he had brought with him, and his systematization of strains in the *Rhizopus* genus became the basis of his doctoral thesis.[25]

Like Yamazaki's contributions, the history of Japanese microbial type culture collections as a whole is quietly entangled with the history of Japanese empire, and with it the *material* construction of a national fermentation tradition. Around the same time that the botanist Saitō Kendō was gathering strains at the Central Laboratory of the South Manchuria Railway Company, Nakazawa Ryōji (1878–1974), a graduate in agricultural chemistry of Tokyo Imperial University, built an extensive culture collection in Taiwan. From 1911, Nakazawa worked at the Taiwan Government-General Research Institute, at first as a technician, and then from 1916 as head of the Brewing Science Department, later reorganized as the Fermentation Industry Department, Division of Industry, Taiwan Government-General Central Research Institute.[26] Finally, Nakazawa became head of the entire Division of Industry in 1937. From 1930 he was appointed professor in the Faculty of Science and Agriculture, Taihoku Imperial University. When Saitō Kendō and Nakazawa Ryōji returned to the home islands in 1927 and 1939 respectively, copies of strains from the Taiwanese and Manchurian microbial type culture collections followed them.[27] Hanzawa Jun's (1879–1972) laboratory in the Department of Agricultural Chemistry, Faculty of Agriculture, Hokkaidō Imperial University was another prominent center of strain collection on the frontier.[28]

The imperial heritage of Japan's culture collections was a noted fact. Sakaguchi Kin'ichirō (1897–1994) remembered being a third-year undergraduate in the agricultural chemistry department of Tokyo Imperial University who was charmed by Takahashi Teizō's talk of microbes while dining at his house one evening, and how he thus decided to write his thesis with Takahashi, using as his starting point Felix Ehrlich's study of fumaric acid production by *Rhizopus*. Yamazaki Momoji's isolation of many *Rhizopus* species from China at the time offered Sakaguchi the possibility of using experimental materials different from the strains Ehrlich was using, and of thereby doing original work.[29]

As Komagata Kazuo remarked in the 1985 interview, *Rhizopus* strains that Yamazaki had collected were still preserved and used by other researchers at the Institute of Applied Microbiology at the University of Tokyo.[30] To return to Saitō Kendō's metaphor, if Japan's national microbial culture collections were like gardens, then they were built from other gardens and were consciously celebrated for doing so. They did not encompass only newly discovered strains. They drew on Asia's microbial gardens, in turn built knowledgably by Asia's industries, which had curated strains of useful molds and yeasts for locally distinctive manufacturing.

In Shanghai, Yamazaki Momoji worked during a time of increasingly difficult relations between China and Japan from the late 1910s through the 1920s. He was in Shanghai in the first place due to scientific exchange being a part of cultural diplomacy with China. The Tōa Dōbun Shoin, where he worked as a professor, was a semiofficial institution, since it was sponsored by the Tōa Dōbunkai (East Asia Common Culture Association), which in turn received subsidies and grants from the Japanese Foreign Ministry. Taking its name from the slogan *Dōbun dōshu* (Common culture, common race), the Tōa Dōbun Shoin opened in the early 1900s and served several purposes including intelligence and empirical surveys for the Foreign Ministry, as well as the education of both Chinese and Japanese in the name of idealism and friendship.[31]

The term *dōbun dōshu* had been used by Japanese politicians and journalists since the late nineteenth century to describe the belief that China's relations with Japan should be different from relations with the Western powers because China and Japan supposedly shared a special cultural affinity. After Japan's victory in the Sino-Japanese War of 1894-95, Japan gained privileges equivalent to those of the other treaty powers in China, and around this time there also emerged the idea of the "Ōkuma doctrine," or the notion, as the historian Peter Duus explains, "that the Japanese, in repayment of their cultural debt to China, should take an active role in pulling China up the steep path toward 'civilization'" and against Western encroachment in Asia.[32] Along with the Western powers, however, Japan operated in the uneasy equilibrium of informal empire in China within the structure of the unequal treaty system. Following Japan's acquisition of rights in Manchuria after the Russo-Japanese War of 1904-5, Japanese thrived in this informal system, and on the eve of Japanese forces' invasion of Manchuria in 1931, not only did Japanese residents in China outnumber those of all the other treaty powers put together, but Japan had displaced Britain as the dominant foreign economic power in China.[33]

Tensions heightened with the growth of Chinese nationalism after the 1911 revolution and especially in the 1920s. Chinese movements called for the broader dismantling of the treaty structure. There were four different major anti-Japanese boycotts in China, all of which included Shanghai, between 1919 and 1928.[34] While the British were willing to contemplate a gradual withdrawal from empire in China, Japanese leaders were concerned that the stakes they held in China were particularly wide-ranging and of acute economic and strategic importance, even though in the 1920s they were strongly divided on the question of foreign policy in China. The Japanese Kwantung Army's assassination of the Chinese warlord Zhang Zuolin in 1928, followed by its invasion of Manchuria in 1931, finally pushed domestic political forces toward a "strong" China policy that departed from the international consensus on informal empire in China.[35] In 1933, faced with international condemnation, Japan withdrew from the League of Nations.

When it came to cultural diplomacy, as the historian Sophia Lee puts it, all the treaty powers "followed the same prescription: relief work and medical services for the masses, and, more important, education, especially higher education, for China's future elites."[36] In the decade after Japan's victory in 1895, Chinese students flocked in the thousands to schools in Japan. But these numbers, as well as Chinese student numbers at the Tōa Dōbun Shoin, collapsed after the Japanese government's issuing of the Twenty-One Demands to the Chinese government in 1915, which was a naked display of Japanese leaders' imperialist intentions in China.[37] Moreover, by the mid-1910s Chinese students had increased opportunities to study in modern schools elsewhere, both abroad in Western countries and at home. Japanese educators in China in the 1920s, such as Yamazaki Momoji, felt themselves to be on the defensive regarding the value of education for Chinese at a Japanese school or university, compared to education at a school in the United States or Western Europe, or at a Western (mostly American)-run missionary school in China. As Lee describes: "In 1918, one Diet member declared that the difference between the [Japanese-funded] Dōjinkai hospital in Peking and its neighbor, the [American Rockefeller Foundation-funded] Peking Union Medical College, was akin to the difference between the houses of common Japanese and the mansions of the Mitsui and Mitsubishi dynasties. He fretted about what the Chinese thought of the obvious disparity."[38] More Chinese studied in Japan in the 1920s and 1930s than in any other foreign country, but it was widely perceived that a Japanese education did not confer prestige.[39]

By 1923 the Japanese government, following the other treaty powers, was providing backing for cultural initiatives in China using Boxer Indemnity remissions, creating a China Cultural Affairs Office (Tai-Shi Bunka Jimukyoku) in the Foreign Ministry to oversee programs, followed by a binational advisory committee in Beijing in 1925. (The China Cultural Affairs Office was quickly renamed the Cultural Projects Division, or Bunka Jigyōbu. This was due to Chinese opposition to the unilateral nature of the words *tai-Shi* [toward China], as well as to the use of *Shi* representing the Japanese phonetic character compound *Shina* for China, instead of *Chūgoku*, or the character compound meaning Middle Kingdom, which was used by China itself. However, the term *tai-Shi bunka jigyō* [toward-Shina cultural projects] remained widely used by Japanese until 1945.) The Cultural Projects Division subsidized Chinese students in Japan and Japanese hospitals in China, as well as the founding of two new research institutes. These were the Pekin Jinbun Kagaku Kenkyūjo (Peking Humanities Institute), which opened in 1927 and became "principally a Chinese organization headed by a Japanese sinophile," and the Shanhai Shizen Kagaku Kenkyūjo (Shanghai Natural Sciences Institute), which opened in 1931 and in the end turned out to be "strictly speaking a Japanese organization."[40] After Japanese troops incited conflict in Jinan in 1928, all Chinese members withdrew from the advisory committee and the Nanjing government no longer recognized either institution, though both institutions continued informal engagement with Chinese scholarly and scientific associations.[41] Of course, the Manchurian Incident in 1931 ended all possibility of negotiating a new agreement with the Nanjing government on the matter.

Yamazaki Momoji's activities in Shanghai in the 1920s included advocating for the establishment of the Shanhai Shizen Kagaku Kenkyūjo. His publications from the time demonstrate the ways in which he articulated the significance of scientific (especially agricultural) research for Sino-Japanese relations, and vice versa. Much of the rhetoric produced for a *Japanese* audience, such as an essay on negotiations for the new institute that appeared in the *Dainihon nōkaihō* (Agricultural Society of Japan Report) in 1924, is predictable as well as predictably jarring to post-1945 ears. That is, he invoked the trope of cultural sameness between Japan and China vis-à-vis the West, as well as of temporal difference between Japan and China. The essay was an apologetic attempt to justify Japan's bid for a special leadership role vis-à-vis the West in Asia, in the name of both national and universal welfare.

For example, Yamazaki stated in the essay that he had long been in Shanghai for the sake of developing Asia (*kōa*) and for the sake of humankind, devotedly conducting investigations and research on China's (*Shina*'s) agricultural products, believing in the necessity of opening the way to utilization and welfare.[42] For Japan to explore each of the various locations of Asia that were scientifically unknown, and to achieve widely framed comparative research with them, was a pressing need for the advancement of academic research and industrial development.[43] Yamazaki believed in making the world an ideal state of mutual love between humankind, but also believed that there was a proper order to things; first, that Sino-Japanese cooperation should be strengthened and used as the root axis to establish Asianism (*Ajiashugi*) among Asian people (*Ajiajin*), and that only then should there be movement to a world of mutual love among humankind. Cultural projects, he believed, should be managed along this principle. Thus Yamazaki welcomed the participation of certain Western scholars in the new natural sciences institute, but believed that it should be mainly Sino-Japanese scholars undertaking the research.[44]

Speaking of his own promotion of agricultural science within plans for the new institute, Yamazaki stated that the new institute should determine its priorities for topics with the following three questions in mind: On which topics could Japan offer to the world reference materials from scientific research on China that were qualitatively and quantitatively superior? Which topics were likely to produce the best research results if the research was assumed to depend only on cooperation between Chinese and Japanese people? Which topics were most essential to China then and in the future? In all these areas, agricultural science must obviously come first, Yamazaki argued. He turned again to the special role that he believed Japanese research played in Chinese science, asserting that results obtained in Japan served as important reference materials in the Chinese context, due to China and Japan most resembling each other in their crops, farming tools, climate, and agricultural economy.[45] Here he referred to the Japanese archipelago as extending "from Hokkaidō in the north to Taiwan in the south" (formal colonies such as Taiwan were usually spoken of as being part of "Japan"), implicitly emphasizing its climatic range despite its small size compared with China. Finally, Yamazaki made a moral appeal to agriculture using the rhetoric of Japanese affinity with China in the realm of ethics, drawing on sweeping generalizations that were clichés of the period. The West's way was the quest for hegemony; the core of Eastern civilization was the kingly way. Japan was once lost and pursued the quest for hegemony, but now it had awakened to the kingly way. In the Chinese context, agriculture and the kingly way were supposed to be consistent with one another.[46]

When his words were directed to a *Chinese* audience, Yamazaki Momoji's apologetic arguments for regionalism in science were different. They gestured toward a yearning for cosmopolitanism and were drawn from his own research specialty. In the spring of 1928, Yamazaki addressed an audience of about two hundred faculty and students at the College of Agriculture of the National Central University in Nanjing (the university was later reorganized into a number of institutions including today's Nanjing University), with the aid of a Chinese simultaneous translator who was an engineering graduate from Kyoto Imperial University. A summary of the talk appeared in the Kyoto-based popular science magazine *Warera no kagaku* (Our Chemistry) the following year.[47] In his address, Yamazaki began with an appeal to scientific internationalism, and then focused on its tension with national culture. He opened with the question of how exactly to contribute to global culture (*sekai bunka*), a question which he said had arisen like a chant in nations across the world since the Great War.[48]

A nation-state, Yamazaki said, was where each individual within the nation exercised their abilities—or talents, even—to improve and develop that nation's indigenous culture; if each nation of the world was so, then overall improvements and developments in global culture would appear. The various countries of Europe and America, compared to "our Asia" (*waga Ajia*), were more advanced in the natural sciences and their application, as was well known. But if "we Asian people" (*wareware Ajiajin*) merely followed European and American people, then in a hundred years or a thousand years, the difference would only increase and would not likely contract. An imitation could not be said to be a contribution to global culture. Thus how could Asia contribute to global culture? How could Asia *not* be an imitation of the West?

Yamazaki's answer was to use Asian history as scientific data. Initially he phrased it in the lecture as using Western knowledge, methods, and techniques as tools to conduct scientific research on Asia; but when he moved to the example of fermentation chemistry, it became clear what he meant. In the far East, like the ancient and large enterprise of the brewing industry, the perfect mastery of special kinds of fermentation fungi deserved admiration, he said. China (*Chūgoku*) was by no exaggeration a cornucopia of fermentation fungi, and it was an honor as well as a responsibility to the world that the same country bore to research those fungi.

As Yamazaki told his audience at the College of Agriculture, from 1914 to March 1927 he had resided in Shanghai, researching many Chinese classics in relation to the sciences. In his address he offered his impressions from exploring those books. Here he praised the people of the Chinese nation for their gift for observation of natural phenomena as well as their matchless flair for applying the results to everyday life. The only regrettable thing, he said, was that it was not written, "Why is it so?" The *Honzō kōmoku* (the Japanese reading of the sixteenth-century Chinese classic in materia medica), for example, mentioned the substance *shingiku* (神麴), which was still sold in drugstores and had been used as a digestive for about six hundred years.[49] When one conducted research on it, he continued, one found that it produced various kinds of enzymes and consisted of many fungi. Several decades ago, he explained, the digestive Takadiastase, a commercially popular enzyme preparation, had been invented and it, too, was made from fungi.[50] If six hundred years ago the question "Why is *shingiku* effective?" had been investigated, it would have represented advances in bacteriology and enzymology in China that would have been far ahead of Europe; for in the Yuan dynasty, China would already have had the discovery of diastase, he speculated. Similarly, the heat sterilization methods known as pasteurization had been carried out for wines in China six hundred years before — and in Japan two hundred years before — the time when they were discovered by the great French scientist Pasteur. If the question of why those methods were good had been researched, a Chinese bacteriologist would have preceded the pioneering Pasteur by more than five hundred years.[51]

Whether or not the reader agrees with Yamazaki Momoji's imagination of counterfactual timelines in the history of science is beside the point. The point is that Yamazaki's arguments extended from strategies of scientific inquiry that were already being used implicitly by Japanese chemists of the period in their work. For example, one of Yamazaki Momoji's mentors at Tokyo Imperial University, the agricultural chemist Suzuki Umetarō, claimed to have a research strategy that centered on a knowledge of local materials and that it eventually led to Suzuki's isolation and characterization of vitamin B from rice bran.[52] By Suzuki's account, his own mentor when he had studied in Germany — the chemist Emil Fischer — had advised him that upon his return to Japan he should focus his research not on proteins, as everyone in Europe was doing, but on problems distinctive to the East, such as rice. The reason, Fischer warned, was that if Suzuki followed the European fashion to investigate proteins, he would be unable to compete successfully for priority with chemists in Europe who were equipped with facilities far superior to those available in Japanese laboratories.[53]

The historian of chemistry Kaji Masanori makes a similar argument for organic chemist Majima Rikō, who discovered urushiol, the key ingredient in Japanese lacquer: "His research strategy involved *studying the structure of the components of Japan's local natural products using newly developed methods from Europe* to catch up to and compete with chemists in more advanced countries in the West. Majima's approach became the primary research method employed by organic chemists in Japan until the 1950s."[54]

The focus on local natural products and traditional industry within organic chemistry was part of a more general trend in Japanese chemistry as a whole, which extended across the first half of the twentieth century, as I have detailed for agricultural chemistry in chapter 1 of this book, and which began when modern science was institutionalized in Japan in the late nineteenth century.

As Kaji Masanori notes, and as I have argued in chapter 1, Japanese chemists were practical people engaged in experimentally oriented work, who were generally silent on the broader intellectual implications of their research for society. By explicitly encouraging scientists to use Western science as a tool to perform "Asian" (*Ajia no*) research on traditional industry and local products, and articulating its significance as a contribution to "global culture," Yamazaki Momoji took what Japanese scientists, particularly chemists, had consciously used as a national resource to achieve contributions within international science, and rhetorically turned it into one that was nationalistic. His narrative for the Chinese audience redrew the national unit of "Japan" to encompass instead the regional unit of "Asia," and appealed to a grander sense of history, referring to classical documents rather than industrial practices. At the same time, Yamazaki softened any emphasis on cultural sameness and temporal difference. Instead of sameness, he referred to China and Japan's responsibility to cooperate as "the two great independent countries in Asia." He took the trouble to defend himself against the accusation of promoting "cultural invasion," by arguing that his own name, which they could see written on his business card, was indebted to the Chinese cultural invasion of Japan, and that if not for China's and later the West's cultural invasions, Japan would be "as primitive as the South Seas" and not "one of the world's great powers."[55]

The formulation of an intellectual current in science that was self-consciously cosmopolitan-regional-nationalistic was Yamazaki's own expression of the motivation that eventually produced his *Study*. It was opposite to and in tension with the statist technocratic rhetoric of the Japanese role in "developing Asia" (*kōa*) that Yamazaki employed for a Japanese audience—a rhetoric that downplayed the historical existence of Asian scientific knowledge, prioritized provincialism over cosmopolitanism, and emphasized the appropriateness of Japanese leadership in the Western-style scientific modernization of Asia. The schizophrenia in Yamazaki's 1945 *Study*—driven by his interwar research in China using historical Chinese documents, but synthesized for a Japanese wartime readership—results from his hopeless attempt to reconcile the two contradictory ideological currents. The work remains today the most extensive definition of the identity of a "Japanese" fermentation tradition, and it is accomplished mainly through comparison with China and Korea, rather than with the West.

What exactly it is that makes *kōji* Japanese, rather than Chinese, Asian, or something else, is a question that has been answered directly by Yamazaki Momoji alone. H. T. Huang's volume *Fermentations and Food Science* in Joseph Needham's Science and Civilisation in China series, published in 2000 and comparable in length to (though differing in scope from) Yamazaki Momoji's *Study*, is the most extensive English-language account of Asian fermentation history to date and does not take up the issue.[56] The mold preparation *kōji*'s bearing on national identity has had a new resonance following the addition of "Japanese cuisine" (*washoku*) to the UNESCO Intangible Cultural Heritage list in 2013.[57] Although *kōji* is today the "national microbe" of Japan, the question does not necessarily arise; when it does, it is consistently Yamazaki's answer that is given, though direct reference to his work is left out as often as it is included. Yamazaki's answer, developed in the *Study*, is therefore definitive to the present day, and it is the following: Only Japanese culture uses *barakōji* (散麴) or *Aspergillus oryzae* to make wine (酒), whereas Chinese and other East and Southeast Asian cultures use *heikiku* (餅麴) or *Rhizopus* and *Mucor* to make wine from cereal grain (fig. 3.1).[58]

These are the two principal forms of mold preparation for brewing, also made from cereal grain. They share a common name in Chinese and Japanese, 麴 (Japanese: *kōji*; Chinese, Modified Wade-Giles as used by H. T. Huang: *chhü*), but *barakōji* is in loose granules, while *heikiku* is in cakes. Microbially, the fermentation ethnologist Ishige Naomichi puts the difference thus: "In China and Korea, *Rhizopus* and *Mucor* spores are placed on the outer surface of loaves of wheat flour that have been soaked in water before being kneaded. In Japan *Aspergillus* mold is cultivated on steamed rice, being placed on the surface of each individual grain."[59] H. T. Huang elaborates when describing mold cakes: "Conditions in the interior of the cake tend to favor the growth of *Rhizopus* species, while those on the surface, of *Aspergillus* species."[60] The difference has been explained in terms of color. In historical Chinese texts, for *barakōji* the appearance of mold on loose granules was described as a "yellow robe," whereas in *heikiku* making the appearance of a multicolor mold "coat" on the cakes was noted as a sign of the preparation's maturity.[61] The agricultural chemist Sakaguchi Kin'ichirō calls Japanese application of the yellow-green *barakōji* mold form in fer-

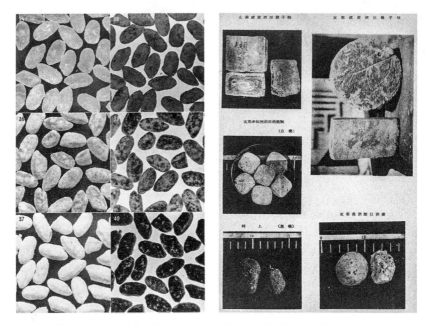

FIG. 3.1. The two main types of mold preparation used in brewing, both called 麴 (Japanese: *kōji*; Chinese, Modern Pinyin: *qū*, Modified Wade-Giles: *chhü*). Left: *barakōji* (散麴) or *Aspergillus oryzae* on individual grains, from Murakami Hideya, *Kōjigaku* (1986), front matter. Reproduced with permission of the Brewing Society of Japan. Right: *heikiku* (餅麴) or *Rhizopus* and *Mucor* in cakes or bricks, from Yamazaki Momoji, *Tōa hakkō kagaku ronkō* (1945). Reproduced with permission of Daiichi Shuppan.

mentation processes "monochromatic" across the entire product spectrum, from sake to soy sauce to miso, while Chinese and other Asian societies employ polychromatic *heikiku* preparations for wine, but yellow-green *barakōji* for condiments.[62]

Therefore, the so-called uniqueness of Japanese culture's use of *barakōji* applies to wine only. For the similar yellow-green *Aspergillus* mold preparations used in *other* fermentation processing, such as for shōyu, tofu, fermented fish products, and vegetable and meat pickling, the strength of historical regional commonalities and connections across Asia is evident and uncontested. The use of *Aspergillus* molds in miso and shōyu brewing in Japan, for example, is widely understood to have resulted directly from the transfer of Chinese fermentation processes.[63] The multiple meanings of the term *kōji* in Japanese also help to obscure the question: *kōji* is used by brewing specialists in a vernacular way to refer to brewing microbes of the *Aspergillus* genus, especially *Aspergillus oryzae*, as opposed to *kekabi* (*Mucor* genus) and *kumonosukabi* (*Rhizopus* genus) microbes, well as for the mold preparation as a whole. Unlike Huang's book, which spans a range of fermentation products in China, Yamazaki's *Study* focuses on wines alone. Covering East Asia (*Tōa*), it is a compilation of four sections of surveys on China, Japan, Korea, and "areas surrounding China," in that order and at successively decreasing length, tracing the evolution of *kōji* making and winemaking as they relate to their surrounding agricultural and food systems, and beginning with prehistory but based mostly on the historical texts of the respective society.

As to *why* the difference arose, the enzymologist Kitahara Kakuo as late as 1974 cites the "mutation theory" in Yamazaki's *Study* as the only response in existence.[64] The mutation theory appears three times in the work, and only in the Japan sections, in connection with *kōji* making. It serves to punctuate remarks on the independent creativity of the Japanese race, whereby accomplishments "prove the excellence of the *Tenson minzoku* [race descended from the gods] who make up the mainstream of the Japanese race."[65] Each time, Yamazaki finishes: "The author is convinced that the *Tenson minzoku* actually appeared by a mutation."[66] The race's "sensitive skill" is apparently shown by the route of its discovery of *kōji*: "That when awareness of the efficiency of molds [*kabi*] has heightened, they would begin to make deliberate effort toward methods to propagate molds on steamed rice [*han*] and dried cooked rice [*hoshii*], and then advance to making '*kamutachi*' [*kōji* mold preparation]—this is only natural." For a race who had "already gone so far as to succeed in selecting excellent rice," such achievements as "the manufacture of '*kamutachi*' and the experiential selection of mold species" would not take special exertion to realize, Yamazaki writes.[67]

The mutation theory should thus be understood in the context of the question of the independence of Japanese culture from the Asian continent. The question opens and closes the Japan section within the *Study*, but is not mentioned in the other three parts of the work. In the opening words of the Japan section, Yamazaki sets out to place the common assumption that "the brewing methods of *Nihonshu* [sake] were transferred from Korea and China" under the "necessary" scrutiny of "scientific investigation."[68] The conclusion of the Japan section reiterates: "China's and Korea's influence is not apparent beyond trivial details of tools and operations."[69] In the *Study*'s conclusion, the wrap-up of the Japan section adds to the list of Japanese achievements "the excellent *Nihonshu*, which has no equal in East Asia nor the world."[70] The mutation theory, which Yamazaki self-consciously states as a sheer assertion, serves therefore to accentuate the *Study*'s commitment to the absolute independence and uniqueness of Japanese culture, which is repeated throughout the Japan sections.

Yet the mutation theory is clearly not the primary driving force in the work, as betrayed by the *Study*'s ultimate conclusions. The real contradiction at the heart of the *Study* lies in the measures of scientific worth, or modernity, by which the mutation theory is articulated, and the ways in which the mutation theory relates to the other sections of the work, especially the China sections. In the *Study*'s overall conclusions, Yamazaki writes of how the "*Tōa minzoku* [East Asian race] with their sensitive intuition" have "serendipitously produced" mold preparations and cereal wines.

> In this way, only the Japanese and Chinese races have each independently accomplished these things, and it is not seen at all in other races and peoples. The Japanese race has in the aspect of "*getsu* 蘖" (*barakōji* 撒麹) achieved unique discoveries and inventions, and responded to the efficiency of indigenous fungi and skillfully used and mastered them. The Chinese race has regarding "*kōji* 麹" (*heikiku* 餅麹 and *shuyaku* 酒薬) achieved unique discoveries and inventions, been engrossed in their devices and improvements, responded to the efficiency of multiple genera of fungi, and ingeniously used and mastered them.[71]

He ends the conclusion with a statement on the global stage of science, namely, his hopes for the place of East Asia in the world and the relevance of East Asian fermentation to the modern chemical industries. Specifically, he makes a plea for how the problems of high energy consumption and waste production — in the synthetic processes that are typical of modern industrial chemistry — might be surmounted by fermentation processes with their ambient pressure, low temperature, and biological catalysts.[72] (This very idea would be picked up globally in the late twentieth century, and has recently become a key focus of contemporary materials research.[73]) In an effusively subjective afterword, after again hammering down his belief in the "mutational superiority" of the Japanese race, he writes: "I eagerly await the day when East Asian fermentation chemistry reigns over the world."[74]

The irreconcilable paradox of Japanese uniqueness lies in this fact: that it can only be articulated in a work that focuses on China, written by a scientist who has devoted the greater part of his career to studying Chinese fermentation processes. In the *Study*, Chinese achievements are the benchmark by which Japan's own modernity is measured. Chinese independent creativity is taken for granted and does not need to be proved, unlike that of Japan. In order to fit the study's framework to the ideology promoted by the wartime state, the main challenge that Yamazaki faces is not the development of the mutation theory, but rather the establishment of the steps necessary to essentialize Japanese *and Chinese* modernity as equal and opposite to each other. Just as the theory of Japanese uniqueness is inherently contradictory—after all, if there was no influence on Japanese identity from the Asian region, then why is it that Japanese identity can only be defined in intimate comparison against the region?—he is set to fail before he begins, if that is his primary goal. The argument for absolute Japanese cultural independence cannot be made any other way, but neither can it be made convincingly. Yamazaki's admiration for China is more apparent to the reader, as well as his unshakable awareness that Japan relies on the scientific knowledge of a vast region in Asia to find a leading place in a modern world dominated by a Western order—no matter how many times he repeats the mutation theory. However, the mutation theory is nonetheless the core of the *Japan* part of the *Study*, and it has been the work's main afterlife since the fundamental shift in the Japanese worldview later in 1945 made Yamazaki's East Asian frame unacceptable. As an explanation, it is tautological: Japan is unique because it is unique.

Such interpretations still have a place because the answers remain underdetermined by the evidence—whether textual, archaeological, or genomic—to the present day. Scientists who address these historical-mythological questions are wary of treading on the edge of a minefield.[75] A few things are agreed upon by scholars. First is the primacy of mold in East Asian grain fermentation, a technique unknown in premodern Europe, where an analogous function was performed by malt (cereal grain that has sprouted; in Europe, sprouted barley), since enzymes that break down proteins and turn starch into sugar are produced by brewing molds and malt alike. Second is that all the important developments in China date back to the Han dynasty (206 BCE–220 CE) or before, and in Japan they date at least back to the time of the tenth-century code the *Engishiki*.[76] Third is that the way Chinese, Korean, Japanese, and Southeast Asian developments relate to one another is not yet clear for *kōji*, whether it pertains to its application in winemaking or in fish fermentation.[77] If the world of origins far back in time is indeterminate, however, it is of interest because, in the words of one novelist describing those who construct their own life in the narration of it, "it tells you what they value, not what happened."[78] In debates on the origins of *kōji*, we see the changing weight given to awareness of intra-Asian historical connections in constructing Japanese modernity.

The historian Hiromi Mizuno has insightfully detailed intellectuals' moves to reconcile the wartime ideology of national uniqueness and superiority with scientific universalism, including attempts by "Japanist" scientists to claim that different races produced different sciences, as well as key efforts to canonize the "scientific" in the Japanese classics.[79] But according to her analysis, these intellectuals' struggles pivoted entirely on a Japan-West axis. Practitioners of fermentation science, too, who were far less articulate than many of the philosophers and humanists whom Mizuno considers, faced the problem of recovering a "scientific" past for Japan as they worked toward goals set out by the state. Yet their conflict between modern national identity and scientific universalism turned above all on a Japan-Asia axis, even while a number of the leading scientists worked in colonial locations. This unintuitive dynamic of ambivalence toward Asia in conceptions of scientific modernity is understudied in the historiography of modern Japan.[80]

The reticence itself on the question of origins is a postwar phenomenon, a burying of the transwar consciousness. In conversation in 2012, one retired agricultural chemist in his eighties chose an off-tape moment to speak to me of the ways in which microbial classification can trace historical cultural flows across southeast and east Asia. At a Japanese symposium titled "The Culture of Food in East Asia" in 1981, Sakaguchi Kin'ichirō began a talk on fermentation by jokingly apologizing for telling a "boastful . . . old-fashioned history."[81] Today, the highlights of fermentation history as told by Japanese scientists have shifted to other, more domestically focused questions: the medieval roots of the practice of saving kōji spores from the last brew to make the next brew (tanekōji making); the evidence for the taming and speciation of the kōji microbe Aspergillus oryzae from the wild, closely related, aflatoxin-producing microbe Aspergillus flavus; or simply the ways in which yeast is a much better model organism for genetic studies than kōji, and why not many microbiologists work on kōji anymore.[82] Chinese American H. T. Huang's definitive English-language study of fermentation history is drawn on a different map than Yamazaki's, and it encloses present-day China as a national unit. Yamazaki's map, by contrast, is bounded by Tōa (East Asia) and drawn specifically to make his studies in China relevant to the Japanese question of national origins. That map is no longer acceptable, and so the category of Tōa is at the center of the question's—and the Study's—problematic nature.

Both Huang (as part of the broader goals of the Needham series) and Yamazaki aimed to showcase coherent, culturally specific scientific traditions that were viable alternatives to Western civilization in China's and Tōa's fermentation histories respectively (table 3.1, row 1). With the Science and Civilisation in China series, Joseph Needham's goal to render, in comparativist and civilizational terms, the extent of China's "precedence

TABLE 3.1. Summary of the comparison between Yamazaki Momoji's *Tōa hakkō kagaku ronkō* (A Study of East Asian Fermentation Chemistry, 1945) and H. T. Huang's *Fermentations and Food Science* (2000)

	Yamazaki, *Tōa hakkō kagaku ronkō*	Huang, *Fermentations and Food Science*
1. Object of analysis	Essentialist modernity in *Tōa*	Essentialist modernity in China
2. Status of fermentation science	Globally weak (1945)	Globally strong (2000)
3. Primary measure of scientific progress	Saccharification method (depicts China and Japan as equal and opposite)	Use of step separation and pure culture (highlights obstacles to Western-style modernity in China)

and influence" in modern science also "functions as a critique of Western civilization," as Robert Finlay argues.[83] On the other hand, the categorization of *Tōa*, too, has a longer history, rooted in Japanese thinkers' desire to both present a civilization led by Japan that could compete with the West, and increase the weight of Japan vis-à-vis China within that region. As Jung Lee explains, the earliest use of the term was in the title of the Japanese translation of Ernest Fenollosa's *Epochs of Chinese and Japanese Art* in 1921—a work which, Benjamin Elman argues, drew its depiction of the aesthetic decline of Chinese art since the Song dynasty and the rise of Japanese artistic achievement in the modern period from Euro-American responses to Japanese victory in the Sino-Japanese War (1894–95). The term *Tōa* became widespread in parallel with the expansion of Japanese imperial aggression in China in the 1930s, and the vision of a Greater East Asia Co-Prosperity Sphere (*Dai Tōa kyōeiken*) in the 1940s during the Asia-Pacific War. Implicit in the term was "the claim that Japan had the unique ability to combine *Tōa* traditions with good things from the West."[84]

In fermentation histories of Asia today, we find a variety of maps that illustrate cultural zones (fig. 3.2). One can have a fish zone and a soy zone, in southeast and northeast Asia respectively; one can have a nation-microbe chart in which nations are colored according to their different predominant winemaking molds (*kumonosukabi* and *kekabi* for most of China, Korea, Nepal, Bhutan, Myanmar, Thailand, Laos, Cambodia, and Vietnam, as well as Malaysia and Indonesia; yellow *kōji* microbes for most of Japan; black *kōji* microbes for southern Kyūshū, Okinawa, and Hachijō; and red *kōji* microbes for Taiwan and Fujian). One can have a polycultural history of Japan pursued through ethnochemistry, focusing on chewing-method wines, malt-method wines, or regional traditions.[85] But *one cannot have an East*

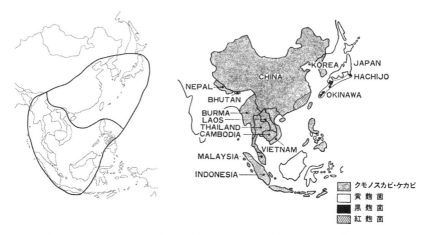

FIG. 3.2. Maps of traditions of fermentation in Asia. Left: fish fermentation zone in southeast Asia, and soy fermentation zone in northeast Asia. From H. T. Huang, *Fermentations and Food Science* (2000), fig. 88, after Ishige, "Fermented Fish Products" (1993), fig. 6. © Cambridge University Press 2000; reproduced with permission of Cambridge University Press through PLSclear. Right: nation-microbe chart. Microbially, the vernacular mold names translate as the following: *kumonosukabi* (*Rhizopus*) and *kekabi* (*Mucor*) for most of China, Korea, Nepal, Bhutan, Myanmar, Thailand, Laos, Cambodia, Vietnam, Malaysia, and Indonesia; yellow *kōji* microbes (*Aspergillus oryzae*) for most of Japan; black *kōji* microbes (*Aspergillus luchuensis*) for southern Kyūshū, Okinawa, and Hachijō; and red *kōji* microbes (*Monascus purpureus*) for Taiwan and Fujian. From Sakaguchi Kin'ichirō, "Hakkō" (1981), fig. 5.

Asia. This is why, fifty-five years later, H. T. Huang cites Yamazaki's *Study* cursorily and with puzzlement, misreading the author's name as "Yamasaki Hiyachi," and only because the *Study* continues to be cited by Fang Hsin-Fang, one of the preeminent Chinese scholars on whom Huang's synthesis relies.[86]

Huang narrates from a much stronger position than Yamazaki since, between 1945 and 2000, fermentation methods did become staple parts of global industrial chemistry (table 3.1, row 2). Where Yamazaki ends his book on his hopes for fermentation in the modern world, Huang finishes on the real accomplishments of mold fermentation in contemporary enzyme production, and places Japan's scientific role in its development alongside the West's.

> [A person who has] never heard of the word *chhü* or *koji* [still experiences its enzymes or others inspired by it] each time he/she consumes a piece of cake, drinks a glass of clarified fruit juice, gulps down a tankard of beer, gobbles a slice of bread, eats a bowl of fast-cooked oatmeal, inserts a piece of processed cheese in a sandwich, sprinkles grated Romano cheese on spaghetti, or pours corn syrup on a pancake or waffle.[87]

In Huang's account, modern Japan is a key node for the assimilation in Asia of Western scientific methods, such as the use of pure culture for mold preparations, *and* for facilitating the flow of fermentation methods in the opposite direction *from* Asia to the United States and Europe, "because [Japan] was the first country in East Asia to become industrialized."[88]

Yet both Huang's *Fermentations* and Yamazaki's *Study* share a primary concern to look back upon the past and find a modernity in Asia that measures favorably with that of the West. Huang's strategies, then, tell us something about the strategies at Yamazaki's disposal as well as his constraints, by way of contrast. Huang's and Yamazaki's measures of scientific progress (table 3.1, row 3) converge substantially, as both are trained chemists. Where their focus of comparison differs, it is due to a difference in audience: a wartime Japanese audience for Yamazaki, and an English-speaking audience at the turn of the twenty-first century for Huang. Whereas Huang puzzles over obstacles to Chinese modernity, Yamazaki aims to show that East Asia is a region superior to the West.

H. T. Huang's line of progress is similar to that of Needham's Science and Civilisation in China series as a whole.[89] It is the trajectory of European science, and he compares China directly to Europe. He demonstrates that Chinese civilization was precocious by tracing the origin of cereal wines in China to around the period when beer appeared in Mesopotamia.[90] Addressing why the two civilizations diverged in the saccharification method (the process for breaking down starch into sugar), he attributes the lack of mold fermentation in the West to a combination of nature—that is, microbial ecology—and the historical path of cooking technologies, whereby grains were ground into flour to make bread, instead of being dehusked and steamed as in China.[91] Huang asks why China nevertheless did not achieve the results of Western science. His measures of worth center on the separation of the saccharification and alcohol fermentation steps, and on the use of an inoculum from the last brew as a precursor to pure culture—both of which were implemented in premodern Europe, but not in China.[92]

Huang sees these operational changes as a reflection of a capacity for scientific analysis, and thus his categories have a tautology too. As the analytical components, he takes for granted the concepts that already existed in modern science when it was imported to East Asia from the West—saccharification and alcohol fermentation as individual chemical processes, and microbes as isolated pure-cultured strains. He does not consider them as simply magnifications of existent process divisions in premodern Europe. In European beer brewing, for example, malting occurs separately *before* alcohol fermentation, but in Chinese mold brewing the saccharification and alcohol fermentation steps are performed by the mold preparation simultaneously. So in the latter process only, there is no *operational* basis for a conceptual division between the two.[93]

Yamazaki's line of progress differs from Huang's. It runs from single-fermentation wines (such as grape wine and other fruit wines), in which only alcohol fermentation of the raw material is needed, to compound-fermentation wines, in which *both* saccharification and alcohol fermentation of the raw material are required (steps for the breakdown of starch into sugar *and* of sugar into alcohol, as are demanded by cereal wines).[94] Alcohol fermentation occurs spontaneously where sugar is present, as wild yeasts from the air settle on the raw material and carry out the conversion. This means that single-fermentation wines exist in nature without human intervention. In compound fermentation, on the other hand, yeasts likewise perform the alcohol fermentation step, but the saccharification step does take human invention, or what Yamazaki calls "intellect and sensitive intuition."[95] Yamazaki's main object of comparison is the saccharification method in compound-fermentation wines.[96]

In this way, in the first few lines of the *Study*, Yamazaki divides the world's civilizations into two, between the "Western (*Seihō*) cultural sphere," where malt (sprouted barley) is employed to perform saccharification and the exemplary product is barley wine (beer), and the "East Asian (*Tōa*) cultural sphere," where mold is employed for saccharification and the representative product is *chō* (醩; an ancient Chinese ritual wine, and to Yamazaki the prototype of all rice wines).[97] He emphasizes that while East Asia had known of malt methods since ancient times, the premodern West did not know of mold methods. His measures of scientific worth thus place East Asia at the pinnacle of progress in compound fermentation methods, superseding the West. They give him a rationale to consider East Asia only, and in the rest of the book he focuses on identifying the fundamental principles of winemaking methods in China, Japan, and Korea, and on describing how they developed in each nation.[98]

Nonetheless, Yamazaki and Huang's measures of worth converge in their commitment to finding a modernity in Asia. For them as chemists, it means uncovering a moment of innovation—that is, a route of serendipitous discovery—of the mold preparation, and then investigating the magnitude of its influence upon surrounding civilizations and the world. Both use elaborate flowcharts to think through the moments in the past at which the opportunities for serendipity would likely have arisen. Both argue that the technology of cooking pots for processing grain led to the invention of mold preparations in China, and that there was a significant change in the nature of the mold preparations with the shift from stone tools to clay pottery, which allowed grain to be steamed.[99] For both authors, Shaoxing wine exemplifies a tradition of core principles of Chinese winemaking that is continuous to the present.

Yamazaki describes Chinese mold preparation for wines as beginning with *getsu* 蘖 (meaning loose granular *barakōji* in his interpretation), which was displaced by *kōji* (meaning *heikiku* wheat-flour bricks in his interpretation) in the Han dynasty. In central China during the Song dynasty (960–1279 CE), these northern Chinese *heikiku* traditions were then synthesized with independent southern Chinese traditions of *sōkiku* 草麴 (mold preparations involving medicinal herb leaves) to create *shuyaku* 酒薬—an intermediate between *sōkiku* and *heikiku* that was closer to the latter, and which represented a compilation of methods from across northern and southern China.[100] Early wines in the Neolithic period, such as *chō*, were made by applying mold to what were probably congee preparations (*birei* 糜醴), but in the Northern Wei dynasty (386–535 CE), many wines were made by applying *heikiku* to steamed rice preparations. In Yamazaki's account, there were three major milestones in Chinese winemaking: the rise of the *shuyaku* form of mold preparation, the replacement of congee by steamed rice (*funrei* 饋醴) as the primary raw material for fermentation, and the implementation of heat sterilization in wine storage. These three developments culminated in the Song dynasty in the creation of Shaoxing wine, which for Yamazaki is the emperor of all Chinese wines.[101] Huang's milestones for Chinese wines generally match Yamazaki's, with the establishment of mold preparation making by the Zhou dynasty (1046–256 BCE), the emergence of the wheat-flour cake as its dominant form during the Han dynasty to be later modified by southern-inflected *shuyaku* technology, and the continuity represented by Shaoxing wine.

There are three key differences between Huang's narrative and Yamazaki's, besides the millennial endpoint. The first concerns the interpretation of the word *getsu*, which I discuss further below. Second, Huang places greater emphasis on the emergence of the highly distinctive "red *ferment*" (*benikōji* 紅麴, microbially dominated by *Monascus purpureus*), which is specific to Fujian province, in the Song-Yuan period (960–1368 CE). This is partly because his ancestral village is near Fuzhou and he first learned of the variety of Chinese fermentation processing techniques there, as he movingly describes in his author's note.[102] Third, Huang mentions the *transmission* of Chinese wine methods to Japan.[103] The application of sprouted rice (malt) to steamed rice to make wine, he suggests, was transmitted to Japan as *rei* 醴 and became today's *amazake* 甘酒 (a sweet wine with a low alcohol content, which is often drunk at festivals).

By contrast, for the *Japanese* origins of the mold preparation, in place of a route-of-discovery flowchart, Yamazaki Momoji has a personal anecdote from his youth. He recounts memories of himself and his siblings leaving *sekihan* rice as an offering at the Shintō altar at home and often forgetting to clear it. When under scolding from his family he would finally remember to clear it, he would see that mold had grown. He imagines Japan's Neolithic people cooking, storing, and carrying rice in pots, and forgetting to clear their offerings to the gods in the same way.[104] This is how Yamazaki argues that Japan discovered the mold preparation independently, unlike in Huang's narrative of transmission; and he makes the moment of discovery both a nationalistic and an emotive one by connecting it to Shintō rituals in his family home during his childhood.

Yamazaki traces Japanese winemaking from the spontaneous emergence of *amazake* in the sweetness of moldy congee, to the deliberate cultivation of mold on freshly steamed or dried cooked rice (*han* 飯 or *hoshii* 糒) to make *kamutachi* 加無太知—the mold preparation. The latter marks the actual moment of discovery, and it is the first place in the book where Yamazaki states the mutation theory. Applying *kamutachi* in turn to *amazake*, Japanese could also seek a product with a strong flavor or sweetness, leading to *ame* 飴 (a sugar product) and, along a different line of development, *seishu* 清酒 (sake) eventually. The improvement of *kamutachi* converged with the rise of rice cultivation in the Jōmon and Yayoi periods to result in the employing of *kōji*—that is, the indigenous *kamutachi*—to make a rice-based wine by the time of the Age of the Gods (the era preceding the reign of the legendary first emperor of Japan Jimmu [660–585 BCE] as chronicled in the *Nihon shoki* and the *Kojiki*), a wine which was used in Shintō ritual offerings.[105] For Yamazaki, this wine represents the fundamental principles of Japanese winemaking.

Yamazaki thus places all the important developments in the mythological or prehistoric age. He then follows the diversification of wines in the historical era. Considering heat sterilization for wine storage, he makes a point of the fact that the Japanese technique uses a lower temperature than the Chinese operation, and declares that Japan developed the technique independently.[106] He has a chapter directly addressing the influence of Korea and China on Japanese winemaking, including details in the *Kojiki* of a Korean man named Susukori who presented the court of the semi-legendary emperor Ōjin (third to fourth century CE) with wine, which the emperor was said to have enjoyed. Yamazaki concludes that, beyond such trifles of tools and operations as the form of the barrels, there was no continental influence upon the fundamental principles of Japanese winemaking.[107] These arguments of independence and lack of outside influence are all to be expected of a book that aims to conform to wartime clichés of absolute Japanese uniqueness and civilizational superiority based on race, and they are one of the most significant areas of deviation from Huang's presentation.

Far more telling, however, is a different major point of divergence: how Yamazaki interprets the term *getsu* within the history of Chinese mold preparation. Most specialists, including Huang, have read *getsu* clearly to mean malt (sprouted grain) in Chinese sources. According to this interpretation, malt-method wines—a kind of beer made with sprouted rice— were made in early China, and then disappeared in the Han dynasty when they were displaced by mold-method wines.[108] Yamazaki's interpretation of *getsu* in the early Chinese context to refer to mold instead of malt is idiosyncratic; for him, the meaning of *getsu* dissociates from mold and becomes malt only later, in the Northern Wei dynasty.[109] He specifically interprets *getsu* to mean the *barakōji* form of mold preparation that is now associated with Japanese winemaking and the yellow-green *Aspergillus* mold. He believes, then, that *barakōji* mold was displaced by *heikiku* wheat-flour mold cakes for winemaking during the Han dynasty.

In the Japanese context, the consensus among experts is that *getsu* is more likely to refer to mold there than in China, and that it can plausibly be interpreted in Japanese sources to mean either mold or malt.[110] In the Japan part of his *Study*, Yamazaki presents the "*Tenson minzoku*" strand of the "Japanese race" as having had a continuous *barakōji* tradition since Neolithic times. Later, he argues, Japanese used imported Chinese characters to describe the indigenous *kamutachi* mold preparation, and so designated to it both the characters *getsu* (to be used for the colored spores, or the equivalent of today's *tanekōji* 種麹) and *kōji* (to be used for the white growth without spores, the same as today's *kōji*). Yamazaki explains that the *getsu* character in the tenth-century *Engishiki* should be read *yonenomoyashi*, a term which is synonymous with the *tanekōji* of contemporary brewers.[111]

Reflecting later on the state of the field, the agricultural chemist Sakaguchi Kin'ichirō declares that the connection between *getsu* and *kōji* is certainly ambiguous in Japan; but he asks why Yamazaki would go so far as to assume that *getsu* has the same meaning in Chinese sources. Sakaguchi attributes it to Yamazaki's (over-) "confidence" from his long studies in the subject.[112] My own explanation is this: The reason for Yamazaki's eccentric interpretation of *getsu* in the Chinese context lies in the requirement to essentialize Chinese civilization as equal and opposite to Japan's, in order to make the *Tōa* concept work. If in China *getsu* means mold, then China and Japan are equal (East Asian, in using only mold for winemaking) and opposite (*heikiku* versus *barakōji*) in their respective continuous tradition, which is in line with the notion of *Tōa*, though it does imply that China possessed a precursor to Japanese sake. If in China *getsu* means malt, on the other hand, then China had beer (malt winemaking), and so China is superior to Japan because China also has elements of the West, thus over-turning the *Tōa* premise. Yamazaki's elaborate attempt to make his depiction of China and Japan conform to the map of *Tōa* is unmistakable. As he states in the conclusion of his section on Japanese *kōji* making, "In the East Asian cultural sphere . . . only Japan" uses "*getsu* (*barakōji*)" to make wine, while in China, by contrast, "*kōji* (*heikiku*)" was developed and *barakōji* discarded since ancient times.[113]

Nobody else but Yamazaki, publishing in 1945 before the surrender, has the need to make China and Japan equal and opposite, and the mold inter-pretation would be unlikely by other scholarly standards. As I have men-tioned above, H. T. Huang writes that a beerlike product in the form of *rei* was in fact transmitted from China to Japan, "probably through Korea," in the late Han period. In contrast to Yamazaki's earlier narrative of cul-tural independence, Huang's statements are based on the work of the agri-cultural chemist Ueda Seinosuke, who believes not only that chewing-method wines (see below) predominated in Japan prior to *rei* and were then eclipsed by the transmission from China of *rei*, but also that Japa-nese may have subsequently derived the *kōji* mold preparation from *rei*.[114] The fermentation scientist Katō Hyakuichi, who has written prolifically on sake history, does trace the mold preparation *kamutachi* in Japan back to the Yayoi period or earlier, on the basis of archaeological evidence. But his tentative explanations for Japan's distinctive use of *barakōji*—citing the his-torian of Chinese food Shinoda Osamu—are similarly premised on trans-mission from the continent. Perhaps Japan was more humid than the con-tinent, altering the molds that settled; or Korea's role was significant and remains little understood; or the Japanese mold technology of rice *barakōji* was transmitted from paddy-farming regions around the Lower Yangtze, rather than being influenced by the wheat-flour *heikiku* traditions of north-ern China or Korea.[115]

The Japan section of Yamazaki's *Study* includes its prewar formal empire—with the exception of Korea, which is given its own section (discussed further below)—as well as areas on the geographical and cultural periphery of the Japanese main islands. The Japanese perception of ethnic difference from both the Ainu and the people of Okinawa had already developed in the nineteenth century, and had coevolved with the marginalization and exploitation of those peoples in the Japanese economy.[116] In Japanese anthropologists' writings, as the historians Robert Tierney and Taylor Atkins have argued, the colonized were a foil to highlight temporal contrast with the metropole.[117]

Japanese anthropological research in the early twentieth century, as Tierney and Atkins emphasize, was motivated not only nor even primarily by the practical aims of managing colonial populations but by the search for the origins of Japan itself.[118] Japan incorporated both the Ainu of the northern island of Ezo and the people of the southwestern Ryūkyū island chain directly into the borders of the nation-state, as Hokkaidō prefecture in 1869 and Okinawa prefecture in 1879, respectively. In the wake of the colonization of Taiwan following victory in the Sino-Japanese War in 1895, the seizing of German-controlled Micronesia at the beginning of World War I (after which the islands were administered by Japan as a League of Nations mandate), and the invasion of European and US colonies in Southeast Asia in the name of a Greater East Asia Co-Prosperity Sphere in the 1940s, two key assumptions justified Japanese expansion and the rhetoric of a civilizing mission. One was the notion of prehistoric blood ties between Japan and its empire in the Asia-Pacific. Another relied on the cultural evolutionist premises of the era—the idea that colonized peoples, with their racial affinity to Japan, represented earlier stages in Japan's own development.

Unlike his presentation of China, Yamazaki's account of the areas under colonial control function conversely to depict the supposed characteristics of Japanese that were frequently featured in wartime propaganda—"belonging to a pure race and possessing a unique culture to distinguish them from their rivals and enemies in the West," as Tierney explains.[119] Yamazaki devotes two chapters to the Hayato and the Ainu respectively. In his description, the ancient Hayato 隼人 race (he uses the label to refer to various groups of peoples who would probably be called Austronesian today) encompassed tribes in the Ōsumi region of southern Kyūshū, the Ryūkyūs (present-day Okinawa), and Taiwan, and their contemporary descendants live in the "South Seas." He introduces the Hayato and Ainu as indigenous races of the Japanese archipelago, which fused with the *Tenson minzoku* eventually to form the Japanese race, and thus the Ainu had had contact with the Hayato before the *Tenson minzoku* pushed the Ainu north.

In the imperial hierarchy of culture, the indigenous peoples of Taiwan and Micronesia occupied what Tierney, quoting Michel-Rolph Trouillot, identifies as the "savage slot," and the Ainu were perceived to be "atavisms."[120] In Yamazaki Momoji's descriptions, indeed, their primitive quality is symbolized especially by chewing methods of saccharification to create *kuchikamizake* 口嚙酒. It is a technique of chewing plant material in the mouth and then using the chewed plant preparation to ferment cereals in pots as compound-fermentation wines. According to the *Study*, the saliva method still exists in Okinawa and Taiwan.[121]

For Yamazaki, *kuchikamizake* is representative of a primordial cultural sphere he calls the "Greater South Seas." Both the Hayato and the Ainu in the past apparently made *kuchikamizake* cereal wines, along with single-fermentation wines from plant juices. Indigenous mold-method wines among the Hayato and Ainu had also existed but were then, he argues, displaced by Japanese mold preparation methods. Yamazaki says it is uncertain whether *kuchikamizake* methods were invented by the Ainu or Hayato, or came from the continent. In Chinese historical documents, *kuchikamizake* has been recorded among the wines made by Tatars and Jurchens of Manchuria, and he argues for a connection with the Ainu. Yamazaki is adamant that unlike these minority groups, the *Tenson minzoku* themselves did not have chewing brewing methods.[122] However, as I have mentioned above, later experts do not agree with this strict separation.[123]

Other than saliva-method wines, Yamazaki notes two other kinds of compound-fermentation wines made by the Hayato. The Paaran tribe in Ōsumi made a *barakōji*-like mold preparation by wrapping millet in plant leaves, but Yamazaki writes that, after the tribe came into contact with the *Tenson minzoku*, these indigenous methods were probably lost via assimilation into *kamutachi*. Another is the distilled wine known as *awamori*, which is a famed tradition of the Ryūkyū islands. The corresponding *barakōji*-like mold preparation uses black *Aspergillus* microbes, which are strongly distinct from the yellow *Aspergillus* microbes used in other parts of Japan. Yamazaki speculates that it is likely that Hayato people assimilated the *Tenson minzoku*'s mold preparation methods, but the climate of the Ōsumi region led to the emergence of distinctive microbial species as well as cereals and plant materials used in brewing. He underlines similarities between the dumpling-like mold preparations made by some Taiwanese and Pacific island tribes—such as the Toroko and Taudaa—with southern Chinese herbal mold preparations—or *sōkiku*—and explains them in terms of a biological connection between these groups. Despite his acknowledgment of some of the cultural diversity of the region, then, Yamazaki merely uses that variety to present a pure, unique, *barakōji* mold tradition carried by the *Tenson minzoku*.

More than any other part of the work, it is in the Korea section where Yamazaki draws the temporal contrast between Japan and its "primitive selves"—to borrow Taylor Atkins's phrase—most directly.[124] Korea, which Japan annexed in 1910, was in a more ambiguous position culturally than other Japanese colonies, in that there was an especially strong scientific consensus among Japanese and Western ethnologists regarding Korea and Japan's common racial origins. Therefore, the Japanese portrayal of Koreans as being "mired in self-destructive stagnation, while their Japanese cousins progressed triumphantly into the modern age" was particularly acute.[125] Korea's supposed regression as a civilization is expressed explicitly in the *Study* in several ways. One is the emphasis on a lack of documentation. Yamazaki declares that he cannot find any evidence in the written record that Korea independently invented mold preparations (*kōji-getsu*), and that this fact makes Korea extremely different from the Chinese, the *Tenson minzoku*, and the Japanese race.[126]

Another is the characterization of Korea as a scientifically inferior copy of China. The *Study* details the process of Korean *kōji* making: wheat flour is wrapped in cloth, straw, and leaves and stamped upon to make a small, shallow, circular disk. The method is intermediate between China's *heikiku* and *shuyaku*, and somewhat closer to *shuyaku*.[127] When Yamazaki analyzes winemaking methods, he restricts himself to purely fermented wine, leaving out distilled wine. Overprocessed rice in congee form comprises a large proportion of the raw material for fermentation in the Korean winemaking method, he argues, in contrast to the use of steamed rice as the raw material base in the mainstream yellow-wine methods of China. Adding the observation that Koreans also make an indigenous distilled wine (soju) based on cooking sorghum into congee, he draws a connection with those congee methods that he says now predominate in northern China and Manchuria. Implied is these northern regions' cultural inferiority vis-à-vis southern China, which is resonant with the imperial hierarchy of his wartime present—with Manchuria having been under Japanese occupation, mostly as a puppet state, for almost fifteen years.

The underpar quality of Korean winemaking lies in its distinctive use of congee instead of steamed rice as the raw material for the wine, according to Yamazaki's analysis. To prove it, he replicates the winemaking method from Korean sources in his laboratory, and conducts quantitative chemical analysis on the product at different stages of the operation. He argues that a finely processed congee basis makes for an inferior tasting wine, since the bits and pieces do not gelatinize, and the microbes cannot stick directly to them but merely float viscously in between, leading to poor "amylo effect" (see chapter 4) in terms of mold propagation on and decomposition of the material.[128] The smell, too, is bad due to the fermentation of raw and unripe starch; here Yamazaki's description echoes other descriptions of wartime manufacturing, such as experiences of the smell of amino acid fluid.[129] He notes that Korean winemaking also lacks heat sterilization. Concluding, he writes that not only are Korea's mold preparation and winemaking methods mostly an imitation of Chinese *shuyaku* mold methods and *shuyaku* wines, but where indigenous Korean elements have intervened, they have only led to regress.

To show that Japanese history alone can stand as equal to China's, Yamazaki must demonstrate that it is so by setting it alongside the supposedly converse example of the history of Korea. Thus he is compelled, by the constraints of the framework of *Tōa*, to devote a section to a society about which he has spent little to none of his career acquiring in-depth knowledge. His final point—that his conclusions could be due to the shallow learning of the author, and that he is prepared to correct himself immediately upon the supply of evidence—could hardly make the Korea section more disheartening as a representation of his views. For our own historical purposes, there remains little scholarship today on Korean alcohol history available to an English-language readership; one important exception is Hyunhee Park's recent work on the transfer of distillation technology from the Mongols.[130]

The final and shortest section considers "areas surrounding China," using Chinese historical documents. Yamazaki's analysis ends by confirming that wines and their origins here, too, match his three cultural zones: the "Western cultural sphere" that has malt-method wines, the "East Asian cultural sphere" that has indigenous mold-method wines, and the saliva-method wines that constitute "Greater South Sea wines." Notably, there are historical records of saliva-method wines being made in Manchuria, Mongolia, and Primorsky Krai (the Siberian area on the Pacific coast that is nearest to Korea and Japan), and according to Yamazaki, the methods were transmitted from the Pacific rather than being indigenous in origin.[131] The claim of cultural and perhaps racial affinity between these regions resonates broadly with Japanese imperial ambitions of that era, and there is still an ongoing territorial dispute between Russia and Japan over the nearby southern Kuril Islands today. In this way, Yamazaki's book—which more than any other work defines the identity of a national "Japanese" fermentation science—begins and ends with Asia, not with the West.

CONCLUSION

The origins of a national "Japanese" fermentation tradition lie beneath the layers of silence that have fallen on the question of Japan's relationship to Asia since August 1945. The national microbial culture collections form a material record of that history and its deep entanglement with empire. For an explicitly articulated answer, however, the microbial strain collector Yamazaki Momoji in *Tōa hakkō kagaku ronkō* (A Study of East Asian Fermentation Chemistry) gives the first and only extensive response to the question of a national fermentation science, creating a key heuristic source for all subsequent scholarly histories of Japanese fermentation. To return to the problem of how the traditional nature of scientific knowledge and its debt to Asia could be reconciled with the imperial rhetoric of scientific modernization: it could not. The current of hope for a cosmopolitan-regional-nationalistic science that motivated Yamazaki's work on the *Study* was impossible to align in any coherent fashion with the clichés of wartime provincialism advocating Japanese uniqueness and superiority. Thus, the mutation theory that he repeats through the work is contradicted by the work's very structure, which takes China to be the benchmark against which Japanese scientific modernity is measured. Not only that; he is unable to complete the task without addressing the achievements of other Asian nations, particularly Korea, in a comparative light—whatever the quality of the analysis or the nature of the conclusions that he draws.

Yamazaki decided in the *Study* to present Asian history as an immense resource of Western-style scientific data, building on trends already existent in prewar Japanese chemical research to focus on local materials and traditional industry. In so doing, he turned strategies that had been national into rhetoric that was both nationalistic and, in principle at least, applicable across Asia. The emphasis on finding innovation in human cultures' "response to the efficiency of indigenous molds, and the skillful use and mastery of them" was an answer to the question of Japan's historical relations to Asia that would set the contours for all histories of Japanese fermentation to follow. But the question itself would fall into obscurity and remain unanswered as the problematic nature of the *Tōa* category became recognized after August 1945.

What defined the nature of the ideological tension between state ideals and Yamazaki's work was the map of *Tōa*—the wartime Japanese conceptualization of Japan's relations to Asia. The vision of *Tōa* captured in the *Study* in March 1945, as well as physically and somewhat silently in the national microbial culture collections as a whole, reveals how Japanese scientists tackled the question of modern national identity using intra-Asian comparisons rather than comparisons to Europe. Essentialism is inherent in the question, but what quickly drops out of the frame as secondary issues are the dichotomies of technology/science, traditional/modern, and even East/West that tend to dominate our historical perspective on science in East Asia and which, for example, emerge so strongly from Joseph Needham's seminal project.[132] In their place, what emerges above all is the *awareness* of the debt that Japanese modernity owed to Asia, which we can recognize most clearly as ambivalence. Japanese scientists and technicians, including Yamazaki as well as all those prominent strain collectors who, like him, had carried out much of their work in Japan's informal or formal empire, perceptibly struggled to see themselves as harbingers of modernity in Asia, counter to wartime and colonial apology.

Contrary to the misconception that the construction of modern Japanese nationalism has been simply an exercise of "leaving Asia" to join the ranks of civilized nations of the West—a misconception that is especially strong for science—the assertion of Japanese identity in the modern period necessarily involved resolving Asia conceptually.[133] The question of the modern Japanese nation's and nationalism's relationship to Asia goes far beyond science. For example, the twenty-first-century reign name Reiwa—a character compound derived from the native poetry compilation the *Man'yōshū*—was officially chosen to break with precedent by being the first reign name to be drawn from a Japanese rather than a Chinese classic. Yet, as was also televised at the time of the name's announcement, the relevant lines were recorded in an era that was one of Japan's most cosmopolitan, the Nara period (710–94 CE), and were themselves inspired by Chinese literature. To return to the work of the microbial strain collector Saitō Kendō, who had worked for sixteen years in Manchuria, in a memoir in 1949 he described *kōji* as an important and special microbe, *which had been discovered* not only in Japan but across Asia, from Korea, Manchuria, and China to India and the Pacific (emphasis mine, in contrast to Yamazaki's conclusions).[134] Similarly, I argue that if we misinterpret the self-conscious ambivalence of scientists to be mere incoherence while they worked to construct modern Japanese nationalism, we consequently miss the central place of Asian contributions to Japanese scientific modernity itself.

This fact becomes more obvious when we trace the history of the microbial culture collections further down the time-specific layers of visions of Asia to the work of fermentation scientists in the formal colonies, who never wrote or philosophized on such abstract issues as definitions of science or of national tradition. Thus Yamazaki Momoji's singular articulation of the importance of Asian regionality at the historical-mythological level of modern national identity, only months before the surrender, must be considered as only part of the background for understanding the practical uses of regional knowledge across the Japanese empire since the turn of the twentieth century. As the next chapter will show, it was scientists and technicians at the colonial frontier who led the work of industrial alcohol production, a sector which was absolutely vital to the military expansion of the Japanese empire. They not only acknowledged the existence of fermentation traditions in other parts of Asia outside Japan and beyond those of the West; they found themselves compelled to use knowledge of regional scientific traditions in practice, especially in wartime.

4

Alcohol

EMPIRE IN PRACTICE

Theoretical research on the amylo-*kōji* eclectic method, which focused on
the contrast between the enzyme systems of amylo microbes and traditional
Japanese *kōji* microbes, was not simply a major contribution to advances
in alcohol fermentation but the foundation for the development of enzyme
chemistry in Japan.
—Kinoshita Toshiaki, "Arukōru seizō gijutsu no shinpo" (Progress of Alcohol
Manufacturing Technology), 300–301

Mass production technologies for industrial alcohol in imperial Japan drew
on heritage cultures. Japan's microbial type culture collections record that
fact and, as a result, many of the collections' most famous strains were gathered at the frontier. To take just one example, black-colored *kōji* strains
used to make the distilled spirit *awamori* in Okinawa were introduced to
southern Kyūshū's *shōchū* (another type of distilled wine) industry in the
first decade of the twentieth century, and there they displaced the former
yellow strains, because the black strains formed citric acid and allowed
more stable brewing in warm regions. These black strains then replaced
the previous yellow *kōji* strains in the mainstream *fusuma-kōji* method of
industrial alcohol production, as they were more powerful.[1] During World
War II, the heavy losses of life from the fighting in Okinawa included the
islands' *awamori* microbes. In Tokyo, firebombing entirely destroyed
the national Brewing Experiment Station's culture collections. However,
the microbial collections of the Faculty of Agriculture of Tokyo Imperial
University survived, having been sent to Morioka in Iwate Prefecture for
temporary refuge.[2] Recently, Okinawa brewers have been attempting to
recover some of the prewar black *awamori* microbes from the University
of Tokyo.[3]

The adaptation of European technology to fit Japanese domestic conditions of industrialization and modernization has been extensively addressed in the historiography on Meiji-era technology transfer. Much more
rarely explored, however, is the adaptation of regional Asian knowledge

to modern institutions of science and technology in Japan, especially after the Meiji period.[4] As discussed in chapter 3, the precise origins of the three most prominent fermentation methods of malt, amylo (derived from "Chinese yeast," and explained further in this chapter), and *kōji* were underdetermined; but in the eyes of Japanese scientists and technicians, these methods nonetheless each had a distinctive national and civilizational character. By accounting for the development of industrial alcohol production in the Japanese formal empire, this chapter sheds light on how fermentation scientists studied and combined European, Japanese, and other Asian microbial technologies in practice. Alcoholic fermentation upheld Japanese expansion in Asia in the early twentieth century, demonstrating that, contrary to the imperialist claim to bring Western-style modernity to the region, scientists relied on knowledge from the Asian region to build a Japanese modernity.[5]

This chapter traces the significance of the colonial frontier at the leading edge of innovation. I depart from the focus on heavy industry (especially in Manchuria from the 1930s) that historians of Japanese technology and science have tended to emphasize, and I show that the importance of colonial-era institutions in Taiwan as cutting-edge centers of scientific production and in particular agriculture is underappreciated.[6] As an experimental science base, Taiwan far preceded Manchuria. The semitropical island was the primary center of sugar production in the Japanese empire and a hub for fermentation research, with a focus on cultivating yeast microbes upon waste molasses from the sugar industry. There and across the Japanese empire, alcohol fermentation technology developed at a nexus of military arsenals, private enterprises, colonial government research stations, and semigovernmental companies.

The geographical extension of the scope of various Asian fermentation traditions followed a pattern that was the reverse of the one envisioned by the Japanese government's discourse of *nanshin* ("southern advance"), whereby the Japanese military and enterprises would use institutions developed in the metropole, as well as in colonies such as Taiwan, as bases to expand control in a southward direction: toward south China, European and American colonies in Southeast Asia, and the Pacific. While Japanese military aggression did take that route in the 1940s, the flow of fermentation technologies in imperial Japan went in the opposite direction.[7] The successful implementation of alcohol production technologies required special attention to the local conditions of agricultural raw materials, including rice, sweet potato, and corn, and to how the material resource landscape could be transformed into new chemical industries using microbes.

The values of industrial alcohol fermentation differed in several significant ways from drinking or medicinal alcohol, since in the latter, issues of
taste and cultural preservation took priority (chapter 3). Industrial alcohol
was directly linked to the nation's ability to defend itself in times of war.
Quantity was imperative. Scientists and technicians prioritized absolute
production values at low cost, aiming for manufactures to replace imports.
In the context of resource constraints and the goal of maximizing resource
utilization, yield—or the ratio of product to raw material—was a key measure in assessing processes. Quality, moreover, meant product purity, not
taste, and was essential for the modern chemical industries that used alcohol as a raw material. The measure of technological capability in the Japanese alcohol industries, then, was a question of how much alcohol could
be made from the raw materials at what cost, and to what degree of purity.
This was how industry practitioners evaluated the methods of malt, amylo,
and *kōji* with respect to their local material environments.

FROM MALT TO *KŌJI*

At the end of the Meiji period, there were two chief manufacturers of alcohol in Japan: the state-run Uji Gunpowder Plant in the city of Uji, Kyoto
Prefecture, and the private Kamiya Shuzō in Asahikawa, Hokkaidō, run by
the entrepreneur Kamiya Denbei (1856–1922; fig. 4.1). A handful of private
enterprises had begun to appear across the country around the time of the
Sino-Japanese (1894–95) and Russo-Japanese (1904–5) Wars, but Uji followed by Kamiya were the largest and best-equipped manufacturers of the
time.[8] Built in 1896, the Uji Gunpowder Plant made the new smokeless gunpowder and dynamite, for which alcohol and ether were needed to make
nitrocellulose, the explosive component in gunpowder. Uji's in-house alcohol factory was one sign of the growing military uses of alcohol that were
driving up alcohol demand. From the Tokugawa period up until the early
Meiji period, the primary uses of higher-purity alcohol had been for traditional pharmaceutical preparations of plant extracts (similar to galenical
preparations in Europe), and it was pharmaceutical companies that had
imported alcohol, or sometimes made it using distillation apparatuses to
further distill *shōchū* or spoiled sake.

The quantity of domestic production of industrial-grade alcohol in Japan
was very small, and most of the alcohol that was used for military or pharmaceutical purposes, or in new manufacturing sectors such as manufacture of celluloid (an ingredient in plastic products and photographic film),
was imported. The low quantity of domestic production was exacerbated

FIG. 4.1. Above: remains of the Uji Gunpowder Plant, from Katō Benzaburō, *Nihon no arukōru no rekishi* (1974), front matter. Below: Kamiya Shuzō factory in Asahikawa, July 1902, from Gōdō shusei shashi hensan iinkai, *Gōdō shusei shashi* (1970), 56. Reproduced with permission of Oenon Holdings.

by the Japanese government's lack of tariff autonomy under the unequal treaties, as well as its heavy tax policies on alcohol into the first decade of the twentieth century.[9] In 1904 Ishidō Toyota, director of the Uji Gunpowder Plant, complained that it was regrettable that the country had to rely on foreign imports, especially as the bulk of industrial-grade alcohol now went to making gunpowder as well as ether. The country did not by tradition have a single specialized manufacturer of industrial-grade alcohol, and although there was now Kamiya's enterprise in Hokkaidō, Ishidō felt that it was strange that there were no other private ventures, such that there was no other choice but to make alcohol inside the gunpowder plant. He pleaded that the country develop the alcohol industry regardless of cost, in order to compete with foreign imports and not lose to them.[10]

Kamiya Denbei had learned Western brewing in Yokohama as a young man while working for three years at Furetsure and Company's wine factory, run by a Frenchman in the foreign settlement of the city. He had gone on to start a business in making grape wine domestically, apparently to combat the rampant presence of imported liquor, before beginning his alcohol project in Hokkaidō. In a personal petition to the Imperial Diet in 1906 against the repeal of tax rebates on medicinal alcohol, he pointed out the vast gap between the amounts of imported and domestic alcohol, and that demand was likely to increase after victory in the Russo-Japanese War. He stressed that, as an agricultural nation, Japan should strive to meet demand by using agricultural products with a high starch content as raw material in order to prevent foreign imports, adding that this would help to promote agriculture.[11]

At that time, Japanese alcohol was mostly produced either by sake brewers as a side industry or by pharmaceutical makers, using sake lees or spoiled sake as the raw material, respectively, and employing the *ranbiki* distillation apparatus—a still that had first been introduced to Kyūshū in the sixteenth century and which had become widespread in Japan during the Tokugawa period (fig. 4.2).[12] The product had an alcohol component of 60 to 70 percent at best.[13] Foreign imports, mainly German, came from fermenting wheat, rice, or potato and had an alcohol component of 95 to 96 percent. In Europe, the nineteenth century was a time of great development for the alcohol industry, with the introduction of continuous distillation apparatuses beginning with the English Coffey still in 1831. In 1871 a new process was introduced in Germany for treating the main raw material, potato, by cooking under high pressure, which allowed the starch to be almost completely gelatinized and so made accessible for fermentation.

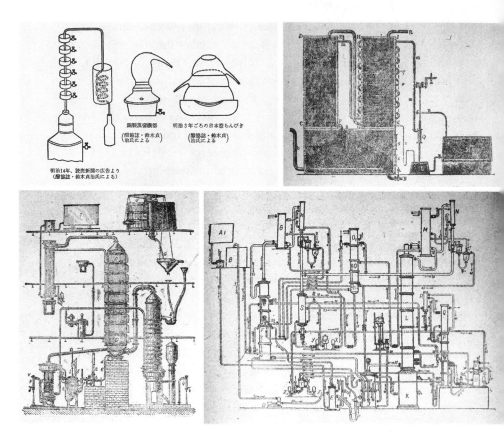

FIG. 4.2. Top left: diagrams of early Meiji-period stills, including a *Yomiuri Shimbun* advertisement (1881), a copper still, and a Japanese *ranbiki* (1870), from Suzuki Sadaharu, "Nihon no arukōru kōgyōshi (sono 2)" (1972), 322. Reproduced with permission of the Japan Bioindustry Association. Top right and bottom: diagrams of continuous distillation columns, namely the Coffey (top right), Ilges (bottom left), and Guillaume (bottom right) stills. From Nakamura Shizuka and Ichino Kazuma, *Saishin arukōru kōgyō* (1948), figs. 94, 98, 101. Reproduced with permission of Sangyō Tosho.

Frederick III's policy of encouraging national industry, moreover, aided the German alcohol industry and supported work to improve potato breeds.[14]

Both the Uji Gunpowder Plant and Kamiya Shuzō relied on a German malt method, the most advanced at the time.[15] Uji hired a German expert to help plan construction, and gathered the newest apparatus in a three-story building. Using sweet potato cooked under high pressure, they were able to get 94 percent alcohol out of the process.[16] Kamiya likewise hired two German experts and bought equipment through a prominent Yokohama dealer, Shimon Ebaasu. By 1907, using potatoes grown in Hokkaidō, the plant was able to make alcohol at a standard comparable to that of Western imports, and began receiving orders from arsenals.[17]

However, Ishidō believed that when it came to explaining the dearth of private ventures in domestic industrial alcohol, acquiring or operating the equipment was not more difficult than for other industries, and that the chief obstacle was, rather, the unavailability of cheap agricultural raw materials in Japan. After five years of testing, he had not yet found any material that would bring the cost of the process below that of purchasing foreign alcohol. Rice nowhere in East Asia was cheap enough. Potato, primarily grown in the Kantō plain, was prohibitively expensive. Dried strips of sweet potato from Kyūshū were the best choice by cost, but spoiled quickly. Molasses in Japan was widely used for low-grade sweets, and this brought its price far above that in foreign countries. Sake lees likewise was in demand for making fertilizer and animal feed, thus driving up the cost. Even in Hokkaidō, the starch content in Japanese potatoes was lower than in foreign potatoes.[18]

Thus Japanese manufacturers first attempted to employ the malt method dominant in Europe and the United States; but they soon turned to adapting the indigenous *kōji* method instead, since it turned out to be a more suitable one for the domestic material environment. At the Uji Gunpowder Plant during the Russo-Japanese War, Iwai Kiichirō, in charge of manufacturing technology at the alcohol factory, developed the *fusuma* (wheat bran)–*kōji* method as an alternative to malt for the saccharification of raw materials (breaking down carbohydrate into sugar).

Kōji was already well known among European and American scientists to have powerful diastatic (starch-to-sugar conversion) properties, thanks to the Japanese chemist-entrepreneur Takamine Jōkichi's invention of the enzyme preparation Takadiastase in 1894. The son of samurai physicians in wealthy Kaga domain, and later the discoverer of the hormone adrenaline and the donor of cherry trees to Washington, DC, Takamine had attempted to revolutionize the American whiskey business by replacing malt with *kōji*. He was part of the first graduating class of the Imperial College of Engineering (later part of Tokyo Imperial University), and had worked for a time at the Ministry of Agriculture and Commerce, where he had pursued a personal interest in applying science to study indigenous industries. He had carried this interest with him as he moved to the United States with his American wife—whom he had met at the New Orleans World's Fair while traveling there as a Japanese commissioner—and worked from a distillery in Peoria, Illinois. There was based the headquarters of the Distillers' and Cattle Feeders' Company (that is, the Whiskey Trust), which had hired him. Their hope was that *kōji*, compared to malt, would prove able to increase whiskey yield and thereby decrease the cost per bushel of corn.

The plan failed, but Takamine's preparation of the enzymes extracted from *kōji*, Takadiastase, soon gained dramatic popularity not only as a stomach digestive in Japan, Europe, and the United States but also as a key laboratory material for enzymologists in Germany.[19]

For Takadiastase, Takamine had cultured the *kōji* mold on waste wheat bran (*fusuma*) instead of rice to save costs, and the preparation turned out to have a stronger diastatic power than malt.[20] At the Uji Gunpowder Plant, the *fusuma-kōji* preparation was first applied to industrial alcohol manufacture. The primary material for alcohol manufacture at Uji was sweet potato, rather than cereals as was typical in Europe or the United States. As a result, using malt for saccharifying the sweet potatoes produced a mash that was too viscous, thus lowering yield. Acquiring barley was expensive, the malting equipment was cumbersome, and malt production took many days; all in all making the malt method "extremely uneconomical" at the Uji plant.[21] In the *fusuma-kōji* method on the other hand, saccharification (converting carbohydrate into sugar) and fermentation (turning sugar into alcohol) took place at the same time, as in traditional brewing, and the enzymes from *kōji* in contrast to malt had strong liquefying power, giving results that were much more yield- and cost-efficient than in the malt method.

Unlike malt, the *fusuma-kōji* method could be used with existent small-scale technology, and so it became widespread among alcohol and *shōchū* factories across Japan around the end of World War I, displacing the malt method.[22] More industrial alcohol and "new-style *shōchū*" companies were finally emerging on the home islands at this time, encouraged by tax protection for the industry after the Russo-Japanese War, as well as munitions demand during World War I. The two types of manufacturers were more or less interconvertible with one another, deeply interlinking the industrial alcohol and drinking alcohol sectors. "New-style *shōchū*" was simply alcohol diluted with water, sometimes with *shōchū* added. Originating in the 1890s boom years of imitation liquor, its production had eclipsed that of conventional *shōchū* (made from sake lees, or sweet potato or other starchy crops, by pot distillation).[23] These new alcohol makers possessed continuous distillation columns imported from Europe, the use of which had spread out from the Uji Gunpowder Plant to private companies through a combination of movements of skilled personnel, personal contacts, and contracts.[24]

Along with the person-to-person flow of specialized technology moved microbial knowledge. Whether they preferred an Ilges column or a Guillaume column for distillation (fig. 4.2), the factories all used the *fusuma-kōji* method, along with another technique first introduced at the Uji plant of adding lactic acid bacteria to combat contamination. The lactic acid bacteria method, which created an acidic environment to deter harmful microbes, was resonant with existent techniques in both sake making—which involved open vats, and so depended on such microbial ecological techniques to prevent spoilage, rather than on the solution of engineering closed containers—and German beer brewing.[25]

ON THE FRONTIER

The growth of an industry, industrial alcohol, that was dependent on fermentation technology came hand in hand with the transformation of land use and the agricultural political economy. Kamiya Denbei had bought the factory in Asahikawa, Hokkaidō, in 1899 for the reason that the climate and space on the colonial frontier were suitable for growing new foods, like potato or corn, that were appropriate for the malt method. In fact, the Japanese state had modeled agricultural development in Hokkaidō along American lines, complete with a new American-style agricultural college in Sapporo and the promotion of cattle ranching—thus initiating a series of unforeseen environmental repercussions, including the extermination of the native Hokkaidō wolf by the first decade of the twentieth century.[26] Kamiya encouraged farmers to grow corn, wheat, and oats and helped them to select strains. In 1910 he brought back two new breeds of potato from Germany for cultivation on the island. In 1913 he traveled across the island to encourage farmers to make starch, with the result that starch production more than doubled by the following year. When starch producers struggled because alcohol firms on the home islands were declining in competition with Taiwanese-made alcohol, Kamiya developed *shōchū* made from potato and starch lees—which became newly recognized in the *shōchū* category in tax regulations in 1918—to aid them.[27]

The reshaping of social relations in tandem with the development of the alcohol fermentation industry was equally salient in Taiwan, where sugarcane plantations along with rice fields came to fill the landscape in a two-crop economy. With the help of the government-general's policies, sugar companies were among the largest landowners in Taiwan. For example, after a number of sugar companies were merged in a move to centralize the economy for wartime production in the 1940s, one sugar corporation alone owned factories that covered more than one-eighth of the cultivated land in Taiwan; another, more than one-fourth.[28] The vast majority of the sugar was exported, sustaining Japanese self-sufficiency. In the late 1930s, Taiwan supplied more than 90 percent of the sugar imported by the Japanese home

FIG. 4.3. The Taiwan Government-General Central Research Institute, from Taiwan sōtokufu, *Taiwan sake senbaishi* (1941).

islands.[29] A production structure consisting of monopsonistic sugar companies with close ties to the government-general—which enabled them to cultivate sugar on large land tracts, and buy all the sugarcane in their designated region for refining—as well as small owner-operators and tenants kept sugar prices low while achieving the implementation of technological advances in cultivation.[30] It also led to an oppressive economic structure for farmers, since land ownership was concentrated in the hands of a small number of sugar companies.[31]

The sugar refineries sought a way to dispose of the waste molasses. Fermenting molasses, skipping the saccharification step, was the cheapest way to produce alcohol. Alcohol producers on the home islands, who relied on starchy crops for raw material, were uncompetitive against Taiwanese manufacturers, particularly after 1911, when the newly imported Guillaume distillation apparatus was able to remove the poor smell of the molasses from the resulting product. The alcohol industry in Taiwan came to dominate absolutely over the home islands' alcohol industry from the 1910s until the beginning of World War II, due to the better raw materials available in Taiwan—namely, the molasses that was a by-product of the sugar industry.[32]

At the Taiwan Government-General Central Research Institute (Taiwan sōtokufu chūō kenkyūjo; fig. 4.3), scientists had an intense consciousness of the island's agricultural distinctiveness, seeing it as a "treasure trove" of resources to be summed up in the abbreviations ABC and PQRST: alcohol, banana, camphor, pineapple, quinine, rice, sugar, and timber.[33] Fermentation scientists at the Central Research Institute carried out microbial classification and engineering research. For the development of alcohol fermentation from molasses, they extensively investigated yeasts suited to tropical conditions, since the type of yeast had a dramatic effect on the yield.[34] The background to this expertise was the university disciplines of agricultural science and especially agricultural chemistry, not only on the home islands but also in Taiwan, including at Taihoku Imperial University.

On the Japanese home islands, meanwhile, under increasing pressure from Taiwanese manufacturers and the post–World War I depression, the alcohol industry converted to making new-style *shōchū*. They still faced a crisis, as new-style *shōchū* would not soon replace conventionally brewed products in people's tastes and the market. To create new demand for alcohol, producers on the home islands began to look toward developing it as a source of liquid biofuel, one that could offer an "unlimited" and "clean" alternative to gasoline and which, they noted, was already under research in a number of Western countries. They stressed that liquid fuel was essential for national defense, to move planes, tanks, submarines, and automobiles.[35] Yet there was the issue not only of the still rising prices of rice— which had triggered the rice riots of 1918—but the prices also of potato, sweet potato, corn, and molasses on the home islands, as the technical experts' perception of national agricultural resource scarcity mounted to a full-blown sense of crisis.[36]

Experts advocated theoretical research on new manufacturing processes, to reach higher yields with the same amount of starch.[37] In other views, such as Suzuki Umetarō's (chapter 2), fermentation was perhaps a bad idea altogether for Japan's resource and security problems, since it meant that food and industrial production extracted from the same pool of constricted agricultural resources—a debate that continues to rage globally in the twenty-first century around corn-based renewable biofuel production, especially in Brazil, the United States, the European Union, and China.[38]

AMYLO

As a new national demand for fuel security emerged in the 1930s following the Japanese invasion of Manchuria, alcohol research concentrated on the southern frontier of empire, with the goal of scraping to the bottom of whatever existed of agricultural resources in the land under Japanese administration, and exploiting those resources to their limits in terms of industrial yield. It was a continuation of the pattern of extraction from the agricultural sector for Japanese industrialization and militarization, which saw the siphoning of capital out of a distressed countryside through the early twentieth century, and the export of locally produced food out of colonial Korea and Taiwan to benefit the home islands.[39] But the fermentation community accomplished the extraction in new ways by applying Asian microbial technology harnessed from across the scope of Japan's territorial control and beyond. This was especially so in the "amylo method," which was developed in the tropics of French Indochina and in Japanese-ruled Taiwan, and which was understood to have Chinese origins.

By the 1920s, European scientists had come to notice mold fermentation technologies in both Japan and other parts of Asia. In 1892, Albert Calmette (1863–1933), director of what would become the Pasteur Institute branch in Saigon—and of later fame as the "C" in the name of the BCG vaccine against tuberculosis—observed that in China and Indochina, diverse rice liquors were made using the mold *Amylomyces*, the "Chinese yeast."[40] He wanted to offer French producers a way to master the "Chinese ferment" (mold starter cakes) that had long been manufactured and supplied by the ethnic Chinese minority. In this way he hoped to break the broader commercial dominance of ethnic Chinese in the liquor brewing economy of Cochinchina and Cambodia.[41] *Amylomyces* had strong diastatic powers, and the distinctive property of functioning well at high temperatures, as well as being able to saccharify and perform alcohol fermentation at the same time. Calmette characterized a strain of this microbe, later also classified as *Mucor*. At the Pasteur Institute in Lille from 1894, he developed an "amylo method" to make alcohol with Auguste Boidin, who in turn partnered with the Belgian distilling expert August Collette to research its industrialization.

Amylo-method factories were built shortly afterward in Saigon, and in France around Lille and Rouen. The large-scale Saigon factory, along with others modeled on it, would become a symbol of colonial exploitation in Indochina, producing alcohol on behalf of the French monopoly regime.[42] By this time, scientists had noticed that the "amylo microbes" of Chinese wine starters consisted of two different genera, and that *Rhizopus* was more powerful than *Mucor*. This was unlike the *kōji* microbes found in Japanese *moyashi*, which were of a single genus, *Aspergillus* (see chapter 3). They eventually employed *R. delemar* in the amylo method, which came to be used in Belgium, Italy, Hungary, Brazil, Argentina, Mexico, and parts of the French empire, particularly where warm climes made malt manufacture difficult.[43]

The differences between the use of amylo molds (*Rhizopus* and *Mucor*) and *kōji* molds (*Aspergillus*) were not devoid of cultural resonance in the eyes of Japanese scientists. The use of amylo microbes in liquor brewing was clearly seen as a Chinese and Southeast Asian practice, while only Japan brewed alcohol with *kōji* (chapter 3). In French Indochina, as Gerard Sasges shows, colonial officials' claim to produce a native alcoholic beverage via the industrial amylo process that was equivalent to the indigenous drink was met with widespread skepticism.[44] Elsewhere in Asia in the same period, however, not only European but also Asian scientists—especially those operating outside the conditions of formal colonialism—desired the industrialization of alcohol from mold fermentation techniques, and perceived the mass production methods thus derived to be native in origin. In China during the Nanjing Decade (1928–37), as Tristan Revells argues, Chinese scientists investigated the amylo method in efforts to produce a national science.[45]

In Taiwan, the Japanese colonial government imposed an alcohol monopoly early on, in 1922. This meant that the state could implement technological changes without concern for manufacturer or consumer resistance on the island. As the work of Fan Yajiun has elucidated, a variety of drinks existed in the local liquor landscape, including aboriginal chewing-method liquor, aboriginal grass-ferment liquor, Han white-ferment liquor (which was based on the same techniques of mold preparation as aboriginal grass-ferment liquor), and Han red-ferment liquor (which was imported from China due to the highly specialized nature of red-ferment manufacture), as well as imported liquors from Japan, including those from Okinawa.[46]

The Monopoly Bureau oversaw the manufacture of all kinds of liquor, such as the Chinese distilled rice liquor *biichū* (米酒), which accounted for about half of Taiwan's total liquor production. By 1924, Nakazawa Ryōji, Takeda Yoshito, and other scientists at the Taiwan Government-General Central Research Institute had isolated the main mold involved in *biichū* brewing, *Rhizopus peka* (named for the Taiwanese mold starter *peeka* 白糀; fig. 4.4), as well as the yeast.[47]

In 1928 the Central Research Institute sent the brewing technician Kamiya Shun'ichi to investigate rice varieties for *biichū* in Southeast Asia, and there Kamiya unexpectedly encountered an amylo-method alcohol factory at the Société des distilleries de l'Indochine, known as the best equipped rice refinery in Saigon. His guide, "Mr. M" of the leading Japa-

FIG. 4.4. *Peeka*-making room at the Taiwan Government-General Central Research Institute. From Nakazawa Ryōji and Takeda Yoshito, *Taiwan hakkō kōgyō* (1940), fig. 27. Reproduced with permission of Kōseisha Kōseikaku.

nese foreign trade firm Mitsui Bussan, was initially hesitant but was able to help him gain permission to tour the refinery, on the basis that Mitsui Bussan was one of the merchant enterprises that purchased rice from the refinery. Kamiya found that half of the refinery was used as an alcohol factory, and he noted its enormous size. What surprised him in particular was that the primary product of the alcohol factory was not alcohol but the alcoholic beverage *sim-sim* or *shamsyu*, prepared using the amylo method, distilled, and then diluted to make the final product. Upon tasting two or three different samples of it, he experienced a shock of recognition as he noted that it was similar to Taiwanese *biichū* in being very light. He felt that the way the factory packaged the liquor in Western-style casks and sent it away to the countryside, and the way *biichū* was distributed in Taiwan, were "exactly alike." On his return to Taiwan he decided to implement the French-Indochinese amylo method in Taiwan, but for making *biichū*.[48]

The amylo method altered the traditional technique of making *biichū* by using pure cultures of both mold and yeast, and carrying out the saccharification and fermentation of rice within a closed, sterile environment. To facilitate the process, the rice was first crushed and mixed with hydrochloric acid. As in the traditional method, saccharification and fermentation happened simultaneously, but the amylo method skipped the step of making a yeast starter—which mixed mold, yeast, and water together and allowed them to propagate on the appropriate raw material, after which the yeast starter was applied to the rice to perform alcohol fermentation. Instead, the new method inoculated the rice with separate pure cultures of mold and yeast directly, though this lengthened the fermentation time.

There were some difficulties involved in transferring the technology to Taiwan. The engineering of the closed sterile tank environment was the most difficult part of implementing the amylo method. The team attempted to apply alternative anticontamination techniques from the *kōji* process—such as the lactic acid bacteria approach as well as the gradual adding of sake yeast—but did not succeed, and instead built a completely sterile tank after much trial and error.[49] The lack of specific engineering data in scientific publications was a challenge. Delbrück's *Handbuch der Spiritusfabrikation* served as the only manual, and the team estimated tank and equipment dimensions by looking at the ratio of their size to that of the people in the book's photographs (fig. 4.5).[50]

Outside Taiwan, on the home islands, the amylo method first received Japanese attention as the only "European" alcohol method to employ molds. A report in 1914 explained the amylo method's roots in Chinese technology, how there was a kind of *kōji* used to make *shōchū* in China and the Cochin region, and it described how this Chinese *kōji* or yeast was prepared and sold in markets. Research on the technology had involved some exchange with Japan, as the French embassy had sent samples of Chinese *kōji*, made in Tokyo and other parts of Japan, to Collette and Boidin in France for investigation.[51] In 1925 the fermentation engineer Katō Benzaburō compared the enzyme systems of the commonly employed *kōji-* and amylo-method microbes, *Aspergillus oryzae* and *Rhizopus delemar*.

Katō found that the saccharifying power of *Rhizopus* enzymes was superior, but that they were inferior to *Aspergillus* enzymes in liquefying power. This hindered *Rhizopus*'s efficiency for solid, starchy materials such as sweet potato. *Rhizopus* microbes also consumed sugar, which lowered fermentation yield. The amylo method could potentially give higher yields of alcohol due to the closed environment (fig. 4.6). But, Katō argued, with the raw materials used in the Japanese home islands, such as sweet potato, and even more with dried sweet potato and sorghum, it would be difficult to liquefy the starch completely before subjecting it to the action of the amylo microbes, and therefore the *kōji* method remained more suitable. Contamination by other microbes brought amylo fermentation to a standstill, and the sterile environment and complexity of control demanded by the amylo method thus made it impractical.[52]

FIG. 4.5. Photographs from Maercker and Delbrück, *Handbuch der Spiritusfabrikation* (1908), showing amylo-method factories in Seclin (top) and Rouen (bottom), as reproduced in Nakano Masahiro, "Taiwan ni okeru arukōru sangyō" (1974), 211.

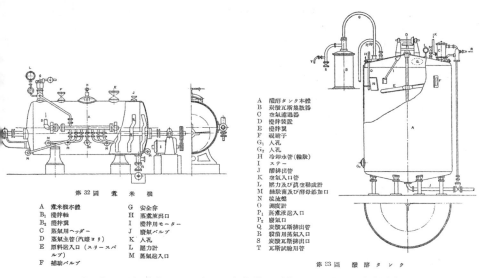

A 醸酵タンク本體
B 炭酸瓦斯集散器
C 空氣遮過器
D 攪拌裝置
E 攪拌翼
F 視硝子
G₁ 人孔
G₂ 人孔
H 冷却水管(輪狀)
I ステー
J 醪排出管
K 空氣入口管
L 壓力及び眞空聯成計
M 麯肤菌及び酵母添加口
N 拡流盤
O 測液計
P₁ 蒸煮液送入口
P₂ 廢氣口
Q 炭酸瓦斯排出管
R 殺菌用蒸氣入口
S 炭酸瓦斯排出口
T 瓦斯試験用管

第 32 図　　煮　米　機

A 煮米機本體　　　　　G 安全弁
B₁ 攪拌軸　　　　　　 H 蒸煮液出口
B₂ 攪拌翼　　　　　　 I 攪拌用モーター
C 蒸氣用ヘッダー　　　J 廢氣バルブ
D 蒸氣主管(汽罐ヨリ)　K 人孔
E 原料送入口(スリースベ L 壓力計
　ルブ)　　　　　　　 M 蒸氣送入口
F 補助バルブ

第 33 図　醸酵タンク

FIG. 4.6. Amylo-method rice steaming tank (left) and fermentation tank (right). From Nakazawa Ryōji and Takeda Yoshito, *Taiwan hakkō kōgyō* (1940), figs. 32 and 33. Reproduced with permission of Kōseisha Kōseikaku.

The development of techniques was supported by intensive work on classifying the microbes of brewing processes in both the Japanese empire and other parts of East and Southeast Asia. Fermentation scientists' research on microbial classification functioned to capture cultural diversity, and therefore microbial biodiversity. There was great variety; in the case of Japan alone, it was said that there were a "hundred schools of sake making," and that the specialist *tanekōji* makers who made the characteristic *kōji* for each of these schools themselves consisted of over forty companies.[53] Although *Rhizopus delemar* was commonly used in the amylo method for industrial-grade alcohol, in Taiwan Takeda Yoshito's extensive investigations of microbial strains turned up a strain isolated from Javanese *ragi* starter cake in 1935. Named *R. javanicus*, the strain was superior to *R. delemar* in that it consumed less sugar and therefore gave higher yields of alcohol from sweet potato (since sweet potato contained more sugar than other raw materials).[54] The yeast he isolated from Taiwanese starter, *Saccharomyces peka*, would similarly become widespread later in the amylo process in the Japanese empire.[55]

THE AMYLO-*KŌJI* "ECLECTIC" METHOD IN WARTIME

It was during the Asia-Pacific War, when yield and resource economization became the top priority, that the amylo method expanded beyond Taiwan to other parts of the Japanese empire. Government bureaucrats had begun to think about the problem of fuel security when imports were disrupted during and after World War I, and the Home Ministry carried out oil surveys, which revealed its scarcity. After the Manchurian Incident in 1931, government policies encouraged the development of domestic oil production, including in Karafuto (present-day Sakhalin), and research into the economical manufacture of alcohol as an alternative fuel. Moreover, industrial demand for alcohol had risen because the chemical industry itself expanded rapidly in the 1930s.[56] In the Japanese home islands and in Korea, alcohol companies unable to compete with producers in Taiwan had largely converted to *shōchū* making, but the developments of the 1930s spurred research into absolute alcohol production methods for fuels. Alcohol fermentation research competed with other lines of alternative fuel development, such as coal liquefaction in Manchuria. Proponents of fermentation highlighted the difficulties of chemical methods of obtaining benzine—that is, petroleum distillates—from coal derivatives due to coal shortages, as well as the fact that adding alcohol to benzine increased its antiknock properties.[57] Osaka Imperial University's Faculty of Engineering undertook research on new distillation apparatus for making alcohol more cheaply; but industrial practitioners encountered difficulties. The *kōji* method was not suited to mass production, because the fermentation yield was low.[58]

In this context, companies on the home islands began research on making absolute alcohol using the amylo method which, in the Japanese empire until then, had only been used in Taiwan. As early as the 1920s, Japanese experts had observed that closed-tank brewing gave higher yields, and that the domestic alcohol industry was already lower in terms of theoretical (maximum) yield, at 40 to 45 percent, than other industries.[59] In 1932, Miyazaki Shizu—who had first studied alcohol technologies as an employee in the alcohol factory of the Uji Gunpowder Plant—filed a patent for a process to produce *shōchū* by the amylo-*kōji* eclectic method (or combined method), at the factory of the liquor distilling company Takara Shuzō in Hachiōji, Tokyo.[60]

Japanese scientists perceived the amylo-*kōji* eclectic method to be a cutting-edge blend of contemporary Euro-Chinese amylo technology with traditional Japanese brewing techniques. It removed the disadvantage of the amylo method's long fermentation time by reintroducing the yeast starter from traditional brewing. It also solved the problem of the viscosity of the fermentation broth by first adding *fusuma-kōji* to the main sweet potato mash and allowing the starch-liquefying *kōji* enzymes to work, and only subsequently adding the yeast starter made with amylo mold to ferment the mash. It retained the environment of the closed tank that was featured in the amylo method, which allowed a high yield, but the addition of *kōji* made the manual supervision of such aspects as ventilation, stirring, and temperature easier, since the amylo yeast starter lessened the effects of contamination. If there were contaminating yeasts from the *kōji* step, the good yeast microbes of the starter still overwhelmed them in numbers and effect. The main aspect that resisted further scale-up was *kōji* making, which was manual and therefore labor-intensive.[61] The yields from the "amylo-*kōji* eclectic method" were reported to be extremely good, and the *shōchū* cheap.[62]

In a political economy of war, margins make the difference, and thus the questions of resource scarcity and conservation are deeply military issues as well as environmental ones. Due to the importance of yield, the amylo method became favored over the *kōji* method and was widely used in the Japanese empire. In 1937, after the outbreak of full-scale war with China, and following conferral with the private National New-Style *Shōchū* Association, the government implemented an Alcohol Monopoly Law and a system whereby alcohol would be mixed at 20 percent with benzine for liquid fuel, the primary raw material would be sweet potato, and the manufacturing method would be amylo fermentation.[63]

By 1939, on the home islands there were eleven new government factories manufacturing absolute alcohol, and eight private companies licensed to manufacture it.[64] Government factories used a "pure" amylo method, combating the problem of viscosity by first pulverizing the sweet potato mash and cooking it rapidly under a low pressure.[65] They achieved a fermentation yield of 85 percent, markedly higher than for *kōji*.[66] Some private factories used the amylo-*kōji* combined method invented by Miyazaki, which gave fermentation yields of more than 90 percent.[67]

The amylo method expanded geographically into the northeast Asian parts of the Japanese empire because of the need for greater quantities of agricultural raw material. Daisen Hakkō, an alcohol company based in Busan, Korea, had noticed the Fukuoka-based parent company Nippon Shurui's success in the amylo method with making *shōchū*, and tested growing sweet potatoes on Jeju island. This drew attention from alcohol manufacturers to Jeju as a possible sweet potato site. In 1938, the Government-General of Korea contracted with the semigovernmental colonization corporation Tōyō Takushoku to construct an amylo-method factory in Jeju and grow sweet potatoes there. The company not only made alcohol but supplied food from the sweet potatoes. Factories across the empire, as far away as Karafuto, fermented absolute alcohol under government contracts.[68] As food became scarce, alcohol raw materials shifted to molasses from Southeast Asia and corn from Manchuria.[69]

Fermentation was only one way of creating new industries from the land—land that was short of the oil that the leadership wanted for liquid fuel, especially as Allied blockades in the Pacific tightened and air raids destroyed many industrial facilities in 1944–45. Synthetic chemical processes were also pursued, but microbes remained crucial for shoring up the liquid fuel supply. One industry expert estimates that ethanol comprised 26.7 percent of total liquid fuel in 1945.[70] By another estimate, ethanol constituted 21 percent of the Imperial Japanese Army Air Force's and 40 percent of the Navy's consumption of liquid fuel in July 1945.[71] Other programs for synthesizing liquid fuel from coal also produced fuel on a scale similar to that of the fermentation efforts.[72] The large-scale use of ethanol as fuel (alone or blended with gasoline) due to fuel shortages significantly impacted pilot training programs in the Imperial Japanese Navy and Army, and scholars have argued that fuel shortages were therefore important in shaping the strategies adopted by the Navy and Army late in the war.[73]

In the final two years of the war, the state shifted emphasis to butanol production for the high-octane fuel needed to fly planes. Acetone, too, was indispensable as an industrial solvent. Acetone-butanol fermentation developed separately from the Asian methods of amylo and *kōji* that were used for ethanol, as it relied on imported microbial technology from Europe. However, its trajectory demonstrates the centrality of agricultural chemistry in the military-academic-industrial complex as early as the 1920s. At that time, the director of the gunpowder section of the naval arsenal had obtained Chaim Weizmann's microbe for acetone and butanol production (which had played a prominent role in Britain during World War I), and passed the strain to Tokyo Imperial University's Department of Agricultural Chemistry for research.[74] In the 1940s, and especially from 1944, the fermentation community turned its attention to the problem at sites including Riken, the South Manchuria Railway Central Laboratory, the Hokkaidō Industrial Experiment Station, the Department of Agricultural Chemistry of Kyūshū Imperial University, the Taiwan Government-General Central Research Institute, and numerous private corporations—including what eventually became Kyōwa Hakkō Kōgyō on the home islands and the Tōtaku sugar company in Taiwan, to name just two.[75]

As the war effort put increasing pressure on agricultural raw materials, fermentation research came to center on what different types of raw materials could be used. Viewed decades later, the technologies of alcohol fermentation would appear deceptively simple compared to antibiotic fermentation, but the meaning of alcohol fermentation in the wartime period was particular: the raw materials were unconventional and difficult to use.[76] Great effort went into seeing whether there was a way to ferment alcohol effectively from wood. Researchers also investigated coconut, *kikuimo* (Jerusalem artichoke), Manchurian sorghum, and urban trash as raw materials for fermenting alcohol.[77] If much of the butanol produced late in the war was destroyed in the air raids and never came to be used in airplanes, it was the experience of paying close attention to raw materials—their nature, availability, and cost of processing—that would remain when the fermentation community turned its efforts to penicillin production.

After World War II, the majority of alcohol factories in Japan adopted the amylo-*kōji* combined method, which experts there described as the summit of alcohol manufacturing technology; and there was consistent research on its improvement, such as the automation of *kōji* making.[78] In the late 1960s fermentation was supplemented by the introduction of oil-based chemical synthesis, which was far cheaper for alcohol products not destined for human consumption. In Japan, however, the engineering of the closed-tank environment for the amylo method was directly transferred to penicillin production and submerged-culture fermentation, having provided experience with some of the technology for sterile liquid culture.[79] More generally, the fermentation expertise of scientists in the alcohol and butanol industry, most visible in personnel and institutional connections, contributed to early biotechnology and the mass manufacture of amino acids in the 1950s and 60s, where Japan played a leading role globally.[80]

CONCLUSION

The case of alcohol manufacture in Japan sheds light on how fermentation scientists studied and combined European, Japanese, and other Asian microbial technologies when they developed the alcohol industry. In order to upgrade alcohol production to the standards needed for modern military and chemical industries and to replace imports, scientists first attempted to transfer the malt method dominant in Europe and the United States. They then turned to adapting the indigenous *kōji* method used in sake, miso, and soy sauce brewing as more suitable for domestically available raw materials and technology. With the changing demands brought about by the need for fuel substitution in the 1930s, they added the amylo method, a Chinese-based technique developed by French scientists in Saigon and Japanese scientists in Taiwan in the context of the Japanese colonial state's heavy investment in agricultural development in the tropics. They modified it to suit materials in the home islands. Through industrial alcohol production, fermentation scientists supported modern Japanese military expansion and shored up the Imperial Japanese Army and Navy's liquid fuel supply during World War II, under the Allied oil embargo and blockade in the Pacific.

In alcoholic fermentation, Japanese scientists used the Asian ambit that became *Tōa* (East Asia under the Japanese wartime empire) as a resource not only for raw materials but for scientific knowledge. It is as clear in their practical work on the industrialization of alcohol manufacture as it is in the articulated historical exposition by Yamazaki Momoji, analyzed in the preceding chapter of this book. Historians have emphasized how many Europeans in the late nineteenth and early twentieth centuries perceived the scientific potential of East and Southeast Asia to be mostly that of a sick place, where microbiologists could make their name by discovering the cause of exotic bacterial diseases.[81] Yet Japanese fermentation scientists and technicians were invested in the brewing microbes of Asia in a broader Pasteurian sense that was grounded in agricultural work, where microbial activity went beyond putrefaction and disease and was "correlative to life."[82] From classification to engineering, their work attempted to capture the microbial biodiversity of heritage cultures and apply it to other materials. At the core of their work was the idea that the environment could be transformed by microbes, which could then make new industries.

Consequently, Japanese scientists engaged closely not only with European and American traditions but also with technological traditions in Asia, compelled by broader political and economic forces. Within the context of agricultural chemistry, we see how the indigenous fermentation tradition was strongly rooted in modern academic institutions since the early Meiji period, aided by government policies that stressed applied science as the most urgent for Japan both economically and in terms of national security. Against this background, alcohol fermentation as a "national" science came to deal with key questions of self-sufficiency and the scarcity of agricultural resources in Japan, in the process incorporating the Japanese and Asian tradition of viewing microbes and especially molds as both a beneficent natural tool to be manipulated and a cultural emblem within Asian science. The strength in fermentation science allowed important continuities in terms of technologies, personnel, institutions, and understanding of raw materials that were critical in postwar Japan's ability to emerge as a global leader in antibiotic innovation, and subsequently in early biotechnology.

5

Antibiotics

DOMESTICATING PENICILLIN

> Their habit in Japan, it seems to me, is to consider screening not as a
> streamlined mechanical operation, but as research problems in techniques,
> microorganisms, and chemistry. Screening is undertaken comprehensively,
> and fans out in the directions of interest to the chief investigator.
> —J. W. Foster, "A View of Microbiological Science in Japan," 446[1]

On August 15, 1946, the presidents of thirty-nine Japanese companies gathered with Ministry of Health and Welfare officials at the opening meeting of the Japan Penicillin Manufacturing Association (Shadan hōjin
Nihon penishirin kyōkai) at the Seiyōken Hall in Ueno, Tokyo. The attend-
ees included the presidents of the largest permitted manufacturers at the
time—pharmaceutical companies Banyū Seiyaku, Morinaga Yakuhin, and
Wakamoto Seiyaku—as well as other pharmaceutical and chemical manufacturers such as Dainippon Seiyaku, Yaesu Kagaku, and Wakōdō, and the
dairy company Meiji Nyūgyō. Iwadare Tōru, then president of Banyū Seiyaku, later recollected that "it seems strange to think of it now, but at the
time both the government and firms were not very enthusiastic about developing penicillin."[2] In fact, at the time almost no factories in Tokyo were
in operation at all, except small workshops turning out goods for the black
market. It was the first anniversary of the surrender that had ended World
War II.

In the difficult conditions of sheer material scarcity that marked the
years immediately after the war, officials, academic scientists, and industry
leaders met to begin discussions on the domestication of penicillin mass
manufacture. To address the problem of raising the quantity and quality
of domestic production, the Ministry of Health and Welfare had proposed
two new associations on behalf of the Allied occupation government: one
a corporate body to encourage exchange between penicillin manufacturing
firms, the Japan Penicillin Manufacturing Association as described above;
and a second, separate, *academic* body to coordinate laboratory research

on production problems, the Japan Penicillin Research Association (Nihon penishirin gakujutsu kyōgikai; hereafter JPRA). It is the work of the second body, the academic association known as the JPRA, that is the focus of this chapter.

The mass production of penicillin was originally a triumphant legacy of the World War II biomedical research complex in the United States. Penicillin was not difficult to produce in small quantities at the laboratory bench, but the challenge was in making cheap large-scale manufacture possible so that penicillin could be widely available for clinical use. During wartime, the British team of scientists who had attained small amounts of penicillin from the *Penicillium* mold at the laboratory bench took it to the United States to seek manufacturers willing to scale up production. At the US Department of Agriculture's Northern Regional Research Laboratory in Peoria, Illinois, scientists made commercial-scale fermentation possible using a submerged culture (also called deep fermentation) tank, where strains were grown throughout the culture medium, rather than only on the surface, as had been previously done at the bench. Unlike surface culture, submerged culture for aerobic processes such as penicillin fermentation was complex to engineer: it required a supply of air into the liquid culture medium in the tank, and stirring to disperse the air bubbles to the strains, as well as temperature control and sterile conditions. A major innovation that allowed inexpensive manufacture was the use of corn steep liquor as a culture medium, which was as effective as it was cheap and abundant in the region.[3]

Because of the specific narrative of the American achievement in mass production, historians who have considered Japanese penicillin as a case of technology transfer within the pharmaceutical industry often identify submerged-culture fermentation (deep fermentation) to be the heart of expertise in penicillin and antibiotic production technology.[4] Yet, although the transfer of submerged-culture technology from the United States was indeed new and pivotal to Japanese fermentation expertise, a story of the technology transfer of submerged culture alone does not sufficiently account either for *how* Japanese scientists and manufacturers so rapidly assimilated antibiotic technology in domestic material conditions, or for *why* antibiotic science in Japan subsequently developed as it did. Penicillin production during the Allied occupation period built upon as well as transformed Japanese microbial science and industry. The country successfully mass-produced penicillin domestically and achieved self-sufficiency as early as 1948, the third country to do so after the United States and Britain. In the subsequent decades, Japan emerged as a leading center of antibiotic research and innovation—including critical work in stabilizing fermentation methods, elucidating the genetic mechanisms of antibiotic resistance in bacteria, and developing new mold-based drugs including statins and avermectin. (The anticholesterol drugs known as statins are among the best-selling drugs in pharmaceutical history, and work on the antibiotic avermectin earned Ōmura Satoshi of the Kitasato Institute the 2015 Nobel Prize in Physiology or Medicine.)[5]

Technology transfer, especially from the United States, is a persistent theme in accounts of Japan's postwar high-speed economic growth from the 1950s to the early 1970s. In particular, accounts often emphasize the guiding hand of the Ministry of International Trade and Industry (MITI) in promoting technology transfer for strategic sectors and industry winners, especially in the electronics and automobile industries. The story of penicillin production, which took place *before* Japan's economic miracle, casts a different light on the high-growth era.[6] An examination of early antibiotic science in the occupation years reveals the indigenous contribution of both *institutions* and *expertise* to the postwar development of Japanese science and technology. In his classic study of MITI, Chalmers Johnson highlights the transwar origins of industrial policy itself, tracing MITI's continuity with wartime and prewar bureaucratic organizations.[7] More recently, historians of Japanese engineering as well as biomedicine have argued similarly that post–World War II achievements relied on experts' wartime and prewar experiences, rather than emphasizing post–World War II knowledge transfer.[8]

In place of technology transfer, this chapter focuses on the phenomenon of "domestication" (*kokusanka*) in order to explore the creativity that is necessary in import substitution. How did scientists try to make things work?[9] Experts in Japan faced a quite different set of material constraints immediately after World War II than they would in the following decades. Both Japanese and American perspectives from the period stress the starkness of material scarcity. The term "domestication" was the main term used by government officials, scientists, and manufacturers to describe the goals for penicillin production in Japan at the time. In that context, the term specifically referred to achieving the capacity to manufacture penicillin—in mass quantities and to an adequate quality—using raw materials available locally.

But the word "domestication" had another, broader meaning, which would have been equally resonant to Japanese technical experts in the period. The historian Daqing Yang describes how the connotations of "domestication" shifted from merely indigenous manufacturing to reduce imports of specialized equipment in the 1920s, to a movement that sought to promote the completely independent development of innovative technologies that would use raw materials from Japan's Asian empire in the 1930s and 1940s.[10] Thus, the word *kokusanka* carried wartime associations with both autarky and imperialism; while it is translated by Yang as "domestic production," it was also allied with the military-linked idea of self-sufficiency. Here "domestication" is chosen as the translation for *kokusanka* because the term conveys the attempt to achieve change (*ka*) along a continuum from technology transfer to import substitution, rather than positing a sharp distinction between the two in actors' categories. It is worth remembering, however, that the related notion of Japan as a resource-poor country was used to justify its imperial expansion.[11]

Contrary to wartime rhetoric, Yang emphasizes that the shift in the meaning of "domestication" was motivated by a combination of "material, ideological, and personal" demands, and not solely by material need arising from geopolitical circumstances.[12] Material scarcity was most apparent only in the final years of the war and after the surrender.[13] Japanese fermentation scientists working immediately after World War II drew on similar experiences from the wartime period and applied them to the problem of penicillin domestication. For a variety of historical reasons, then, which were partly but not entirely material, fermentation scientists' knowledge and institutions were organized around the salience of resource scarcity in motivating experimentation. More specifically, they saw microbes as tools of abundance that could transform an environment in the midst of resource scarcity.

Such a perception of microbes as alchemists of the environment, I argue, organized the existent knowledge by which penicillin scientists made the environment work, in a material as well as sociopolitical sense. This chapter explores the specific dimensions of Japanese fermentation expertise in the occupation era in two ways. First, I examine processes of the domestication of penicillin production through the JPRA records, which have not hitherto been analyzed historically.[14] Second, I compare Japanese developments with a number of other national cases of penicillin domestication, drawing on a strong secondary literature on Europe in the wake of World War II. The comparisons are both institutional and conceptual; since skill is embodied in personnel, attention to institutions is required to fully understand the nature of knowledge.[15]

In this way I contribute to a growing literature on biological research in the chemical industry. Production-related questions for penicillin—for example, problems of screening (How does one select the microbial strains that can best perform the task of penicillin fermentation?) or contamination (How does one ensure that the stray presence of other microbial strains in the fermentation tank will not impede penicillin fermentation?)—point to a distinctive history of *biological materials* within chemical manufacturing. Since the early twentieth century, medicines such as salvarsan, aspirin, and the sulfonamides have represented the ideal of science-based drug development: results of chemists' efforts to purify, structurally characterize, and then synthetically manufacture organic compounds. Yet preparations from biological materials, especially plants, likely made up the majority of drugs on the market.[16] Penicillin domestication at midcentury was precisely the moment at which the overwhelming dominance of synthetic organic chemistry as the ideal model of pharmaceutical research began to shift. After World War II, governments across the world invested in microbial expertise on an unprecedented scale, in order to produce penicillin locally.[17]

I follow developments from the establishment of the JPRA in 1946 to when the focus of the JPRA shifted from penicillin to other antibiotics, as signified by its name change to the Japan Antibiotics Research Association (Nihon kōseibusshitsu gakujutsu kyōgikai; hereafter JARA), in 1951. I concentrate especially on the first half of this period (up to mid-1948) during which most of the basic problems of domestication were worked out by the JPRA's Central Laboratory. Production-related questions were academic research questions at the scale of the laboratory bench; therefore, how scientists approached them is revealing of the contours of Japanese fermentation expertise.[18] Beyond the transfer of submerged-culture fermentation technology for antibiotic mass production, a distinctive engagement with agricultural chemistry's long-standing perception of microbes—as alchemists of the environment, with the ability to transform resource scarcity into productive abundance—organized the knowledge by which penicillin scientists made the domestic environment work, and deeply shaped antibiotic research in the subsequent decades in Japan.

PENICILLIN PRODUCTION IMMEDIATELY
AFTER WORLD WAR II

The Japan Penicillin Manufacturing Association (Shadan hōjin Nihon penishirin kyōkai; hereafter JPMA) was formed in response to a meeting held in July 1946 at the Ministry of Health and Welfare under the directive of GHQ (General Headquarters of the occupation government).[19] There, the ministry offered clarification concerning GHQ's decision in February to ban the sale of penicillin before issuing manufacturing permits again in May. It was a necessary step toward raising the standards of domestic production, which were unacceptably uneven. It was to this end that the ministry proposed the formation of two new associations on behalf of GHQ, the JPMA and the JPRA. In turn, the ministry promised that GHQ would do what it could to aid the transfer of American technology and invite foreign experts to Japan, as well as offer microbial strains and allow penicillin manufacturers special access to essential materials such as electricity and coal. The occupation authorities thus presented penicillin, as they would also do for the insecticide DDT, as a gift from the United States to Japan.[20]

When American troops, riding in jeeps, entered Japan to occupy the country after Japan's unconditional surrender in 1945, they encountered a people in exhaustion.[21] Japan had been at war for fifteen years, as Japan's Kwantung Army had invaded Chinese territory in Manchuria in 1931 before full-scale war broke out in China in 1937. Cities had been flattened by firebombing; Hiroshima and Nagasaki, by atomic bombs. The occupation government was headed by General Douglas MacArthur, the supreme commander for the Allied Powers (whose administration was often referred to as SCAP, or GHQ for "General Headquarters"). SCAP arrived with an agenda to implement sweeping reforms and democratize Japan—or, in a common phrase of the time, to enforce a "revolution from above."[22] Meanwhile, Japan was cut off from the former empire that had supplied much of its food, and starvation and disease were rife: reports counted 146,241 deaths from tuberculosis in 1947 and 99,654 deaths from other infectious diseases between 1945 and 1948.[23] Trains between Tokyo and the countryside overflowed with crowds in search of food for which they could barter their clothes. As a part of public health policy, the Japanese government set up a series of licensed brothels for American troops to contain the spread of venereal disease (which SCAP at first condemned but eventually allowed), while SCAP had fields sprayed with DDT to kill ticks.[24]

Following the opening meeting of the JPMA on August 15, 1946, the JPRA held its own opening meeting soon after, on August 26. In his introductory remarks, Katsumata Minoru, chief of the Public Health Bureau of the Ministry of Health and Welfare, assured the scientists in attendance that the domestication of penicillin production was a matter for which GHQ too held "great concern."[25] The reasons for this concern remained unspoken, but they would have been obvious to those present. Of the first authorized batch of penicillin released by the pharmaceutical firm Banyū Seiyaku in May, a total of 167 bottles all subject to distribution controls, 50 bottles had gone to the Recreation and Amusement Association (the network of brothels set up by the Japanese government in preparation for the arrival of US troops), and 27 bottles had gone to Yoshiwara Hospital in Tokyo's red-light district. As Robert L. Eichelberger, commanding general of the Eighth US Army, remarked, there was more to fear from venereal disease than from the atomic bomb.[26] Japan was not alone in this situation; in Europe, Allied forces were prioritizing penicillin for countering syphilis in occupied West Germany.[27] A SCAP pamphlet published by the Public Health and Welfare Section in 1949 explained:

> In planning to provide adequate medical supplies and equipment to meet the needs of the civilian population, the problem of utmost importance that confronted SCAP was (1) should all needed supplies be imported at the expense of the American taxpayer, or (2) should every effort be made to increase and stimulate indigenous Japanese production and import only those materials, preferably in raw form, which would not be available in Japanese supply. It was decided that the latter course would be followed and immediate steps were taken to rehabilitate the Japanese medical supply and equipment industry.[28]

At the time that the JPMA and JPRA were established under GHQ's directive, penicillin was already being produced domestically—a remainder from the Japanese wartime project. During the war, based on information in journals delivered from German submarines, scientists at the Army Medical School in Tokyo formed the Hekiso Committee ("blue-essence," or penicillin committee) with the aim to industrialize penicillin manufacture by surface culture. At the same time, the eclipse in scientific communication had made researchers hungry for information about new advances abroad. The young researcher Umezawa Hamao (1914–86) later recalled seeing the foreign periodical that would introduce him to penicillin on a desk in 1943, and feeling "like a starving man coming across food."[29] The wartime committee was a large-scale coordination of efforts by prominent scientists, including agricultural chemists with expertise in both microbiology and microbial chemistry, plant physiologists and plant chemists, medical bacteriologists, a synthetic organic chemist, and physicians.[30] The committee developed strains, culture methods, and refinement and assay methods, and approached the confectioners Morinaga and Meiji Seika (the latter then part of Yamagata Gōdō) as well as the pharmaceutical firm Banyū Seiyaku, to begin manufacturing penicillin by surface culture in dairy bottles. Firebombing destroyed factories, but immediately before the surrender Banyū produced the first batch of thirty grams.

Under occupation, not only was the military disbanded and the large business conglomerates known as the *zaibatsu* targeted for dismantling; all research deemed relevant to military application was banned, surviving facilities were suspended and allocated for reparations, and undertaking any research project required GHQ's permission.[31] In November 1945 the Institute of Physical and Chemical Research's cyclotrons were torched to pieces and dumped in Tokyo Bay.[32] Drug stocks, which had been military goods, were confiscated from October 1945, and the Ministry of Health and Welfare took over distribution controls.[33] Penicillin manufacturers slowly repaired facilities, and more firms joined in production. The penicillin produced averaged 29 units per milliliter (u/mL) in 1946, and still under 100 u/mL in 1947. It was so impure that it made patients jump up in pain when injected.[34] The standard unit for penicillin is the Oxford unit, which is defined by a fixed zone of inhibition of bacterial growth in a standard assay. Pure penicillin, for example, contains 1,650 units per milligram (u/mg). GHQ's overall goal was to have penicillin manufactured to the same standard as the US product as quickly as possible, though in the meantime the "working standard" was more relaxed than that of the US Food and Drug Administration.[35] The working standard that domestic manufacturers aimed to meet was 152 u/mg in December 1946.[36]

In promoting the domestication of penicillin production in Japan, SCAP's primary concern was to reduce the cost of the occupation to the American taxpayer, rather than give priority to protecting intellectual property. Future competition from Japanese industries seemed anything but a likely prospect at the time, and a GHQ Public Health and Welfare Section pamphlet noted that, "due to the lack of raw materials and the deterioration of equipment, the remaining factories were producing only 20% of prewar requirements."[37] The issue of intellectual property goes entirely unmentioned in the regular SCAP publications that summarized the occupation's activities and accomplishments. GHQ's focus was first and foremost on bringing down prices by increasing the quantity of penicillin production and, once mass-quantity production was achieved, on increasing the quality of the penicillin produced to a satisfactory standard. Public Health and Welfare Section publications summarizing achievements for the years 1949 and 1950 celebrate progress in domestic manufacture in terms of the outcome in price reductions as well as the shift in emphasis from quantity to quality improvement, and they delineate the limits of domestic manufacture in terms of continued importation of supplies.[38] The figures included clearly show the dramatic increase in the quantity of domestic penicillin manufacture (fig. 5.1) and the decrease in prices (fig. 5.2).

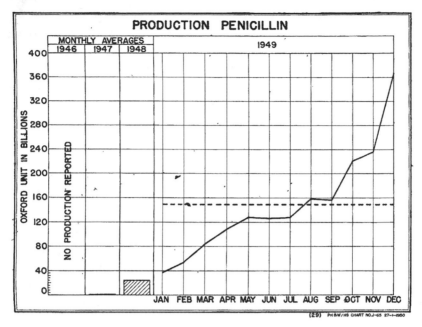

FIG. 5.1. Graph showing monthly production amounts for penicillin. A dotted line at 152 billion units serves as a reference point for reading production amounts against the working standard, which was 152 units per milligram. From General Headquarters, Supreme Commander for the Allied Powers, Public Health and Welfare Section, *Public Health and Welfare in Japan, Annual Summary 1949, Volume I* (1949), 124.

PENICILLIN PRODUCTION

CY	Units	Units/100,000	Av Price per 100,000 units	Total Value Million/Yen
1946	negligible	--	--	negligible
1947	13,821,390,000	138,214	¥1,333 (offi-	184
1948	297,029,810,000	2,970,298	500 cial)	1,485
1949	1,798,300,177,000	17,983,002	140 "	2,518
1950	7,495,530,385,000	74,955,304	45 (est av)	3,373

FIG. 5.2. Table showing the dramatic decrease in the cost of penicillin. "The value in 1947, 1948, and 1949 is based on official prices established by the Japanese Price Board. In 1950 the price control was removed. Value in 1950 is based on an estimated average price of ¥45 per 100,000 units." From General Headquarters, Supreme Commander for the Allied Powers, Public Health and Welfare Section, *Public Health and Welfare in Japan, Annual Summary 1950, Volume I* (1950), 83.

Patent rights became an issue only from September 1, 1949, onward, when the Afterwar Remedy Order of the United Nations' Industrial Property came into effect. The law recognized patents registered by United Nations members within the period dating back to one year before the start of the war, and it meant that a number of penicillin producers in Japan who had been manufacturing without licenses would now have to procure licenses in order to continue. (Penicillin itself was not patented for humanitarian reasons, though some of the manufacturing processes were patented.)[39] But until that moment—as the JPMA's institutional history explains—productivity and cooperation between firms in penicillin manufacture had been emphasized over the enforcement of patent rights, in the name of widespread dissemination and application of the drug to patients. This all changed in late 1949, despite an unsuccessful attempt on the part of GHQ's Public Health and Welfare Section to negotiate an exception for penicillin with the US Department of State.[40]

The son of the president of Banyū Seiyaku helped negotiate an agreement with Bristol, the American pharmaceutical enterprise, for manufacturing the penicillin derivative G procaine in 1953 so that all twenty Japanese producers were able to continue manufacturing penicillin G procaine without conflict over rights.[41] However, as the business historian Julia Yongue argues, that would be the last instance of open cooperation between firms in pharmaceutical manufacturing, just as the penicillin boom was ending. A new commercial era began in the 1950s and continued into the 1960s, in which Japanese pharmaceutical firms individually negotiated their own licenses with foreign businesses for technology transfer, and competed with each other in litigation over patents.[42] But by then, the period of the most crucial developments in the domestication of penicillin production was already over, as the country reached self-sufficiency in penicillin well before the 1949 law came into effect. For the very first antibiotic, penicillin, it was not firm-to-firm licensing agreements that served as the vehicle for technology domestication. Rather, it was the activities of the JPRA.

THE ROLE OF THE JAPAN PENICILLIN
RESEARCH ASSOCIATION

The JPRA, as an association of academic researchers, clearly had a role to play as a designated intermediary between government and industry. This made it distinct from the JPMA, which was a private body of firms. Donations from the JPMA and grants from the Ministry of Education funded JPRA research, and a Ministry of Health and Welfare official was appointed to sit in JPRA meetings. On November 1, 1946, at GHQ's Public Health and Welfare Section, with Ministry of Health and Welfare officials present, GHQ officials introduced JPRA scientists to Jackson W. Foster from the University of Texas at Austin.[43] Foster was a student of Selman Waksman and had worked at the New Jersey–based pharmaceutical company and penicillin manufacturer Merck during the war. His role as a foreign consultant would be to embody the six years (and $25 million) of American experience in the field, which he said his government had asked him to bring for Japan's "peacetime battle" against disease.[44]

Afterward, GHQ officials issued an outline of objectives to the JPRA.[45] The JPRA's tasks included establishing a Central Laboratory in order to expand basic research, which would use existing facilities in universities; constructing a submerged-culture pilot tank, which would need to be built anew at a university or research institute; and assessing factories and choosing the most promising ones to support in order to use limited resources effectively. The Ministry of Health and Welfare, with GHQ's approval, would appoint assistants to direct research in consultation with the ministry, and those assistants would in turn appoint the heads of each research section in the JPRA. A central assay laboratory would be constructed under the domain of the Ministry of Health and Welfare. The JPRA would consult with GHQ on how to break through bottlenecks and strive toward the increase in production that GHQ requested. Twice a month the Central Laboratory would present detailed research reports to GHQ, the Ministry of Health and Welfare, and each laboratory and factory; and twice a month manufacturers would report on the production situation to GHQ.

The Technical Committee was the core of the JPRA's Central Laboratory, and the scientists whom with Foster would work most closely in the following months. It included medical researchers from the University of Tokyo's Institute of Infectious Diseases, such as Umezawa Hamao and Hosoya Seigo, but most of the members were senior researchers from the Department of Agricultural Chemistry in the University of Tokyo's Faculty of Agriculture, such as Yabuta Teijirō, Sakaguchi Kin'ichirō, Asai Takenobu, and Sumiki Yusuke.[46] The committee was more or less the same as that of the wartime project. Sakaguchi, a fermentation expert, would go on to set up the Institute of Applied Microbiology at the University of Tokyo in 1953, while Yabuta, a leading expert on molds, had been the scientist to isolate the first plant hormone gibberellin. A Clinical Committee was also established to collate clinical experiences of penicillin treatment; its members were limited to researchers in state hospitals, since these were the only hospitals receiving penicillin supplies under the distribution controls.

Foster gave a three-day series of lectures in Tokyo, attended by 120 scientists, 6 bureaucrats from the Ministry of Health and Welfare and the Ministry of Education, and 201 representatives of 47 companies from the 51 members of the JPMA, in order to help bring Japan up to date on technical knowledge about penicillin.[47] Later, Foster served as a consultant on submerged-culture plant construction, for manufacturers as well as the

JPRA. In addition, some of the raw materials required—those which were new, or simply difficult to obtain in late-1940s Japan—were flown over from the United States, put in a jeep, and delivered to the JPRA Central Laboratory's Culture Section (then the laboratory of the agricultural chemist Sakaguchi Kin'ichirō at the University of Tokyo), where Foster handed them over to scientists on November 19, 1946. These materials included various strains for surface culture and the Q176 strain for submerged culture, two liters of corn steep liquor, and lactose and phenyl acetate for culture media.[48] The Q176 strain was four to five times more powerful than the Japanese strains under investigation at that point.[49] The tool of induced mutation for creating more strain varieties was also new, and Central Laboratory scientists quickly adopted the technique.[50] GHQ reported that the "latest American scientific literature has been made available and procurement and allocation programs for certain critical raw materials such as phenyl acetic acid, lactose and amyl acetate have been set up."[51]

Section divisions within the Central Laboratory reflected the main research problems involved in penicillin production. The Strains Section focused on screening, or selecting microbial strains most suitable to the task of penicillin manufacture. The Culture Section developed media for mass production that relied on domestic raw materials as much as possible, for *both* surface culture and submerged culture—aiming ultimately for a transition to submerged-culture production, but using surface culture to bridge the production gap that would otherwise be caused by the transition. The Refinement Section similarly researched refinement methods. The Central Laboratory was also tasked with building a submerged-culture pilot tank, where contamination—the infiltration of miscellaneous microbes that might decrease yield—was an especially challenging problem to solve.

The Assay Section assessed the quality of penicillin produced by manufacturers, and officially authorized them. On GHQ's decision, the Assay Section was relocated along with other antibiotic facilities from the University of Tokyo's Institute of Infectious Diseases to the new National Institute of Health (Kokuritsu yobō eisei kenkyūjo, or NIH; this Japanese institution had been established early in 1947 and was attached to the Ministry of Health and Welfare).[52] In his lectures Foster had stressed the importance of upgrading the assay method from the dilution method, which was resulting in large errors, to the internationally adopted cup method.[53] But overcoming the limitations of local resources was not a small challenge. One of the main problems was that the cup was supposed to be made of aluminum. The economic conditions meant that scientists had to use instead a cut-glass tube, but it was impossible to make the cut part flat, and Assay

Section scientists were anxious about this problem even in March 1947, as they were finalizing the draft of an assay method proposal to be sent out to physicians and factory technicians.[54] In December 1946, when the chemical company Yaesu Kagaku managed to produce penicillin at 152 u/mg, the Assay Section noted that it met the working standard.[55]

The JPRA facilitated exchange between academic scientists and experts in the industrial and clinical spheres. The academic scientists in the wartime Hekiso Committee had not included engineering specialists. However, submerged-culture production required a new kind of large-scale apparatus—the sterile aerobic fermentation tank—and thus demanded participation from industry, in particular from chemical engineering and heavy chemical firms.[56] Thus it was the JPMA and not the JPRA that was responsible for preparing two sections to develop industrial culturing and refinement equipment.[57] In 1948 the Central Laboratory added two chemical engineers from the Tokyo Institute of Technology to oversee the construction of the JPRA's submerged-culture pilot tank and refinement equipment, which in turn would be made by Mitsui and Hitachi respectively.[58] Commercial firms were faster than the JPRA to build submerged-culture pilot plants, with the first opening at Tōyō Rayon on March 11, 1947, and others quickly following.[59] JPRA machinery association meetings in Tokyo and Osaka allowed academic scientists and factory technicians to exchange designs and data.[60] In the meantime, JPRA representatives, including a Ministry of Health and Welfare bureaucrat, visited the Acetone Industrial Association in February 1947 to explain their need for solvents for the refinement process.[61] Even by September 1947, however, butanol factories were still idle; the solvent industry would not revive until about the end of the decade.[62] It was as late as June 1948 when the Central Laboratory's full-sized pilot plant came into operation at the Japanese NIH, and the refinement methods were upgraded with the latest high-performance machines in the early 1950s.[63]

The JPRA's Clinical Section allowed information from clinical trials to be conveyed back to penicillin manufacturers by way of the Central Laboratory, which was crucial in effecting product standardization, especially after an adequate production quantity of penicillin had been achieved. Physicians conveyed their views on product quality, pricing, and development back to the Central Laboratory's Assay Section via the Penicillin Standards Investigation Committee, with Ministry of Health and Welfare officials involved as intermediaries.[64] In a November 1947 meeting, for example, physicians' concerns included increasing product potency to decrease side effects; limiting penicillin prices to facilitate physicians' turning to penicillin as the first line of treatment; and requesting the development of new forms of penicillin that would maintain the concentration of penicillin in the blood for longer periods after injection. During a December 1947 visit to the factory of one supplier, Meiji Seika, Central Laboratory scientists assured physicians that although previously it had been necessary to focus on quantity over quality, scientists would now be working to solve the problem of side effects, which were correlated with refinement methods.[65] Side effects differed with the manufacturer due to varying refinement procedures, and also seemed to depend on the microbial strain used in production.[66]

The occupation state's success in coordinating *academic* research on the mass production process, rather than leaving the research initiative to firms, is notable. The fact that the postwar penicillin project followed fifteen years of war helped this particular organizational configuration to function effectively.[67] Not only were the key researchers largely the same as in the wartime committee; the centralized, state-led coordination of the project, the devotion of prominent university scientists exclusively to one production problem, and state policies that confined industrial possibility to this sector by rationing raw materials and providing other economic incentives were all important parallels between technical projects in Japan before and after 1945.

By contrast, in postwar Italy for example, the director of the Istituto Superiore di Sanità (ISS) in Rome, Domenico Marotta, as well as the visiting British penicillin scientist, Ernst Chain, held visions for the ISS's penicillin factory that were similar to the function of the JPRA: to function as a public research establishment, a center for both biochemical and biotechnological innovation, and a service to industry players through its fermentation pilot plant linking laboratory science to manufacturing improvements. It was an important center for a time, with results such as the discovery of 6-APA (the basis of semisynthetic penicillins). However, after Chain left in 1961, the liberal protectionist climate of postwar Italy changed. The ISS's production component fared badly, and Marotta was prosecuted and attacked for corrupting the ISS's public health mission.[68]

Reasons for the flourishing of the JPRA (and then the JARA) also lie in the longer history of Japanese fermentation research. Functionally, as an academic intermediary between government objectives and industrial production, it was comparable to the national and regional experiment stations (*shikenjo*) that were set up by Japanese government ministries from the end of the nineteenth century to aid small and medium-sized enterprises. Like the JPRA, this network of institutions was a state-supported information mechanism to facilitate novel technology domestication and raise the competitiveness of domestic businesses via laboratory research on manufacturing processes and industrial surveys.[69] The experiment stations employed scientists from the universities and technical colleges. For both academic and industrial scientists in fermentation-related fields, such institutions were familiar precedents for the kind of state-backed research coordination on commercial production problems that the JPRA represented.

APPROACHES TO PRODUCTION PROBLEMS
AT THE CENTRAL LABORATORY

All aspects of production that were under research in the Central Laboratory—strains, culture media, refinement methods, and even assaying procedures—required domestication. Apart from the chemical engineering efforts that went into building the physical components of mass production plants (which were largely overseen by the JPMA instead), the intellectual skills for domestication were to be found in the fermentation knowledge that was already existent from wartime, and which carried over directly into the postwar JPRA's Central Laboratory because of the continuity in personnel.

From the beginning, JPRA scientists carrying out "general and basic research on penicillin" indicated that along with achieving the penicillin production objectives, they wanted to do research of their own free direction.[70] At the first meeting of the Strains and Culture Sections, assignments included not only penicillin-related topics such as submerged culture, surface culture, and increasing the power of strains, but also looking for strains outside of the blue mold that would produce antibiotics, and investigating strains that would produce antitoxins.[71] At the same time, the JPRA scientists held a pragmatic view of the local industrial conditions. Both they and Foster knew that material limitations would compel Japanese firms to continue surface-culture production for many months, even though submerged culture would ultimately achieve the necessary step-up in production quantities. Because of this, JPRA scientists developed strains and culture media for both production methods in parallel.

The French wartime penicillin project offers an illuminating contrast because of the existence of comparable microbiological skill at the same time as there were differences in the precise nature of that microbiological knowledge. As in Japan, penicillin development in France was led by academic research rather than firms—namely, by medical microbiologists at the Pasteur Institute under a military administration. Scientists in the

French project possessed a configuration of expertise similar to that of the scientists in the Japanese project (though the Japanese team had more chemical expertise), with a biological emphasis on strains, culturing, and assays. Moreover, since the Pasteur Institute was also a vaccine and serum factory, microbiologists were keen to extend the technological possibilities of biological production. However, the French microbiologists' excitement about scientifically advanced biotechnology meant that they pushed for taking the many more months required to build a submerged-culture plant, whereas the military engineers disagreed about time and built a surface-culture plant without the microbiologists' support. In the end, production failed to materialize before the end of the war, and penicillin production was simply undertaken by the private sector after the war through licensing agreements.[72] Japanese microbiologists in the discipline of agricultural chemistry, on the other hand, had had the wartime experience of developing production technologies for resource-intensive goods such as fuel alcohols, which was one reason behind their sensitivity to economic constraints in industry when undertaking the postwar project.[73]

Moreover, in the pre–World War II period, Japanese agricultural chemists had developed ways of approaching microorganisms that would become significant in both the theoretical and the applied spheres. For the historian Robert Bud, the accumulation of know-how in applied science through early research on organic acid fermentations, at the German University in Prague and the New York firm Pfizer, was a key factor in the success of the Anglo-American penicillin program.[74] Interwar Japanese microbiological research within agricultural chemistry at Tokyo Imperial University (later the University of Tokyo) is a revealing comparison because this work, too, focused heavily on organic acid fermentations of *Aspergillus* and other molds, having expanded from studies of the molds used in traditional sake and soy sauce brewing. From the 1920s, however, the research was deliberately theoretical rather than practically oriented— aimed at understanding the biochemistry of the mold, and without links to breweries or other industrial spaces.[75]

Whereas Bud characterizes the work at Prague as being part of a "low status but industrially well-connected network," the Japanese interwar work in organic acid fermentations differs in being moderately high in status and distant from industry.[76] Its distance from industry may help explain why the Japanese work did not produce the innovations in submerged-culture fermentation that the German work produced, which would later prove critical to penicillin manufacture. But the subsequent rapid domestication of penicillin and antibiotic research in Japan indicates that there were aspects other than submerged culture at the heart of antibiotic production and innovation—namely, a biological approach to microbes, and a sense for what microbes could do. The implications of the Japanese fermentation scientists' approach can be seen especially in screening work—the task of finding and selecting microbial strains with specific desired properties, especially those that made the strains suitable for use in the mass production of a metabolite (a substance formed as a result of biochemical processes in a cell).[77]

Scientists indicated that they did not see screening work as entirely routine. At meetings there were steady reports of work on new antibiotics, though they were often not given priority, and came after reports on penicillin work. There was research on antibiotics produced by actinomycetes, *Penicillium*, and *Aspergillus candidus*, for example, as well as gramicidin from a *B. brevis* soil microbe, and streptomycin-lookalike compounds from actinomycetes strains.[78] In order to select strains, one researcher in the Strains Section reported, it was necessary not only to be systematic but also to see the physiological characteristics as important, and to use culture media that would make the physiological differences easy to see.[79]

Such consciousness of the variability and diversity of microbes as biological organisms, each with their own biochemical and physiological capacities, suggests that scientists drew on prior practices in the discipline of agricultural chemistry.[80] It helps explain the vibrancy and rapid outcomes of JPRA research on antibiotic-producing strains, whether directed toward applied goals for penicillin production or toward gaining knowledge of microbial physiology and ecology more broadly through antibiotic research. On March 20, 1948, Central Laboratory scientists announced that from then on, they would not prepare particular shared topics of research, and instead the laboratories would simply communicate with each other while doing their own research individually; this marked a point when laboratory research on penicillin was mostly complete.[81] Yet the JPRA's Culture Section—with diligent adherence to JPMA firms' requests—continued to give a long report on strain research for penicillin, and on a new mutant strain of Q176 that would produce colorless as opposed to yellow penicillin, which interested the industry side.[82] In June 1948, research on bacterial acquired resistance to penicillin and streptomycin as a laboratory (not clinical) phenomenon came first in the list of research reports.[83] Thus, interest in antibiotic resistance in the laboratory context *preceded* the wide occurrence of antibiotic resistance in the clinical context, which was to emerge in the next decade.

JPRA scientists possessed a strong sense of what materials were available or not in Japan, making painstaking comparisons of US and Japanese products, and not only because GHQ had instructed them to do so in their list of directives. During the war, agricultural chemists had done similar work to reconcile manufacturing technologies and natural resources when developing alcohol production for fuels.[84] Investigations of the culture medium for surface culture to produce a higher-potency broth took up much of the Central Laboratory's energies until late in 1947. As the University of Tokyo agricultural chemist Sumiki Yusuke (1901–74) later described, developing the best culture medium was a messy craft that could be accomplished only by trial and error, since it was impossible to grasp the conditions of every strain that grew upon every culture medium and was affected by many factors.[85] Importing materials to use in the culture medium was not appealing, so scientists attempted to investigate nitrogen sources other than corn steep liquor and peptone, and carbon sources other than lactose.

The hunt for a substitute for the corn steep liquor as a nitrogen source included tests of pupae, rice lees, the side products of Japanese brewing industries, and many other chemicals.[86] From early on, soybean was tested as a medium alongside the other standard media.[87] All kinds of ingredients for testing appear in the records of the Culture Section for the years of 1946 and 1947; burdock, rabbit bone, *gomame*, whole dried sardines, potatoes, taro, onion, and *nattō* are only some of them, and this was for surface culture, which was only a temporary means of penicillin production.[88] At a meeting of the Culture Section on June 20, 1947, the group announced that experiments concerning surface culture were largely complete, and that they would proceed to research on submerged culture.[89] Even in a new political environment where autarky was not a necessity, JPRA scientists' approaches to the problem of penicillin production drew on autarkic experiences from the wartime period.

In a manner comparable to that of the centralized, interdisciplinary institutions of the wartime era, the JPRA facilitated exchange of results among many scientists, which was especially useful for problems as highly specific as the culture broth. In one meeting, for example, the explanation for a particularly good culture result ran as follows:

> Of the three types of waste fluid produced by the textiles factory, the secondary product of waste hot water is the best, and it is good to add starch saccharifier (glucose conversion 1%) to it. To the waste hot water culture broth do not undertake high-pressure sterilization; in the climate, especially in summer, carry out low-temperature drying. If one adds the P substance

donated by Foster then the potency increases and it is maintained for a long time.[90]

Along these lines, the best culture media were trial-and-error outcomes for which there were no systematic or rational formulas, and which were thus an area where information exchange was particularly valuable. That such information was openly shared among laboratories was striking, since manufacturing data for commercial products would normally be closely guarded.[91]

The refinement process for penicillin was a similarly messy procedure to improve.[92] Scientists tested the two methods of carbon adsorption and solvent extraction for the products of each company. Most of all, they were concerned about the limited supply of the solvents needed for the extraction method. They sought substitutions for the ammonium sulfate required in the butanol extraction method, and tried butyl acetate as a replacement for amyl acetate.[93] In June 1947, scientists were still worrying about local resources. If the acetone supply was insufficient, they needed a method without acetone, and if butyl acetate was hard to obtain, they needed a substitute; it was necessary to investigate alternatives systematically.[94]

One of the most punishing problems in submerged penicillin fermentation was contamination, which could be addressed by keeping the tank environment sterile with the utmost care. The degree to which it affected yield was new to the fermentation industries worldwide.[95] On March 15, 1947, before he left Japan, Foster reiterated to the JPRA the importance of solving the contamination problem whatever the cost in money and time.[96] As elsewhere, this would eventually be addressed as an engineering problem of sterilizing the tank components and air supply—but in their early studies, Central Laboratory researchers also tried to draw on the knowledge that the agricultural chemists possessed on traditional brewing. At one point, the medical bacteriologist Hosoya Seigo of the University of Tokyo's Institute of Infectious Diseases investigated substances, such as monoiodoacetic acid, that might prevent the action of penicillin-decomposing enzymes coming from contaminating bacteria in the air.[97] Counteracting contamination within the culture medium, instead of preventing contact with contaminating microbes entirely, had resonance with brewing practices of sake and soy sauce in which lactic acid bacteria were deliberately allowed to acidify the broth to make it a less favorable environment for the growth of other microbes.

As in Germany, pharmaceutical companies in Japan had historically concentrated on chemical synthesis as the methodological path to novel drug innovation.[98] But even if some large Japanese pharmaceuticals might have hesitated to invest in fermentation and hoped instead to create a competitive niche for themselves in penicillin synthesis, they would have been marginalized in penicillin development, due to GHQ's institutional organization of the domestication project under the JPRA.[99] For chemical firms across a whole range of sectors from textiles to steel, penicillin offered a means to revive at a time when raw materials were scarce, military procurements had vanished, and GHQ rationing policies encouraged development exclusively in penicillin.[100] The scope of incentives for penicillin production went beyond inexpensive bottle (surface-culture) fermentation and the more technologically demanding submerged-culture fermentation, to the manufacturing of solvents for the refinement process and machinery components. Academic scientists on behalf of the state directed research and issued advice to companies—initially under the Hekiso Committee in wartime, and then under the JPRA in the occupation era.

The prominent role played by agricultural chemists in the JPRA ensured the involvement of *both* microbiological and chemical expertise, facilitating rapid assimilation of the new antibiotic fermentation technologies. This was unlike the situation in Germany, where penicillin research was similarly coordinated by state managers, and yet the leading initiative was left to chemists in pharmaceutical firms as well as powerful academic chemists in a consultancy relationship to them. In the case of the pharmaceutical firm Schering and the Kaiser Wilhelm Institute–based biochemist Adolf Butenandt, both favored the strategy of building upon prior expertise to develop a commercial niche in chemical synthesis, given submerged-culture fermentation's technical difficulties, and other factors such as the division of Berlin (where Schering and Butenandt were located), which impeded the transfer of information regarding new technologies. The synthetic venture failed, and in the end German pharmaceutical firms simply imported American submerged-culture technology through patent licensing agreements.[101]

ANTIBIOTIC SCIENCE AFTER PENICILLIN

In October 1948, the Clinical Section revised its penicillin user manual to reflect the changes in product supply and quality: from under ten thousand units per bottle (which set the dose) in November 1947 to one hundred thousand units per dose. And whereas previously physicians could use penicillin instead of sulfa drugs only for the most serious cases, now it was possible to use penicillin more generally.[102] In June 1948, JPRA physicians had noted *Staphylococcus aureus* resistance to penicillin in a patient for the first time.[103] The problem of antibiotic resistance would only become more serious. Nonetheless, by 1950 production was so ample that physicians began to discuss using penicillin for the prevention rather than treatment of human disease. A series of clinical trials were conducted, focusing on prostitutes as testing subjects and potential users; it lasted for three years in several urban centers across Japan.[104] At the time, physicians dismissed worries about provoking "unconfirmed" antibiotic resistance phenomena in favor of practical need. Central Laboratory microbiologists had already encountered antibiotic resistance as a laboratory phenomenon, and understood microbes to be part of a wider ecology; but, due to their need to deal with immediate problems of illness on a day-to-day basis, physicians in this period were more likely to take a militaristic approach that aimed to eradicate infection in patients. This was similar to the British hospital context of the 1950s.[105]

Leaving behind the focus on penicillin, in October 1948 the JPRA's *Journal of Penicillin* became the *Journal of Antibiotics*, and in January 1951, the Japan Penicillin Research Association changed its name to the Japan Antibiotics Research Association (JARA).[106] The 1949 Afterwar Remedy Order of the United Nations' Industrial Property prompted a shift to more restricted producer participation in which only firms that procured licenses could undertake antibiotic manufacturing. As the historian Julia Yongue argues, this marked a shift from a business atmosphere of cooperation to one of competition and patent litigation between antibiotic-producing firms.[107] For the JPRA, on the other hand, the change meant that the academic association stepped back from the front-seat role it had previously taken in directing developments in the antibiotic industry. In 1949, Japan imported the new antibiotic streptomycin through mechanisms similar to those used for penicillin—that is, through GHQ coordination and JPRA research on domestication—but unlike for penicillin, only a handful of firms obtained licenses to manufacture streptomycin.[108]

In the subsequent decades, beginning with a procurement boost from the Korean War and continuing far beyond it, the antibiotic industry flourished in Japan. A diverse array of antibiotics—including both new drugs discovered domestically and imitations of foreign products developed under the process-based patent system—came to market and were prescribed frequently. The high consumption of a variety of antibiotics created the widespread emergence of resistant strains of bacteria, which were often resistant to multiple antibiotics at once.[109] With each appearance of strains resistant to an antibiotic came further therapeutic and commercial incentives to search for new antibiotics. Research on antibiotics took place both in company laboratories and in academic medical institutions including the antibiotics section at the NIH and the Kitasato Institute.[110] Although Japanese pharmaceutical companies did acquire numerous licenses for antibiotic production from foreign companies, many new drugs were also produced by Japanese companies following their discovery and development in Japanese academic laboratories.[111]

In the context of the country's economic growth in the 1950s and 1960s, screening work in laboratories could be seen as exploiting low-cost intensive scientific labor.[112] Yet, to Japanese microbiologists, screening was not routine. Rather, designing a suitable screening system required synthesizing chemical and microbiological knowledge across fields, and appreciating the variety of microbial physiology and ecology.[113] The historian María Jesús Santesmases's account of 1950s antibiotic screening in the Spanish firm Compañia Española de Penicilinas y Antibióticos (CEPA) offers a number of reasons for why screening is sometimes portrayed as a routine task.[114] One is the hierarchical international division of labor: the American firm Merck outsourced the screening program to CEPA, which took instructions, training, and equipment from Merck. Another is the scale of the program, involving tens of thousands of strain samples each year, and its systematic nature, whereby the order of individual test procedures and even the target output rate could be standardized like a "testing assembly line."[115] The original aspects of the work tended to be kept hidden or at a low profile as commercial secrets; and moreover, CEPA antibiotic researchers tended to be bound to that firm's laboratory over their careers.

All of these reasons mean that it is easy to overlook the contribution of such elements as knowing which microbes to screen, and the design of the screening system. In Japan, the intellectual and integrative dimensions of antibiotic screening would become institutionally recognized in the decades after the domestication of penicillin production. Unlike in the United States, antibiotic screening was a problem whereby a researcher could earn a PhD degree.[116]

Medical microbiologists and agricultural chemists had come to share common intellectual approaches—such as to antibiotic screening—as a result of working together in the Hekiso Committee and then the JPRA. The disciplinary lines were also sometimes blurred for particular individuals (for example, Sumiki Yusuke was both an agricultural chemist and an antibiotic scientist). Even when research activities in corporate laboratories took more prominence from the 1950s and 1960s, the occupation-era JPRA served as a precedent for later interdisciplinary academic institutions, allowing scientists to engage simultaneously with practical problems and broader microbiological research questions. The establishment of the University of Tokyo's Institute of Applied Microbiology in 1953, for example, created a central space that brought together researchers from across the disciplines of agriculture, engineering, science, and medicine, including in the antibiotic field.[117] Such institutions propagated intellectual approaches to antibiotic research that were legacies of the domestication of penicillin production in the JPRA—and which were, in turn, rooted in fermentation expertise from the wartime and prewar periods.

CONCLUSION

Chemical engineers would often remark with tongue in cheek that penicillin was "a medicine for companies, not a medicine for people."[118] Yet companies themselves did not take the lead in directing the Japanese domestication of penicillin production, and therefore a focus on firm-to-firm technology transfer alone would miss the scientific aspects of the process. This was especially the case at a time when government authorities prioritized raising the country's overall production capacity over the enforcement of intellectual property rights. Exploring the critical research role of the interdisciplinary academic association that was the JPRA reveals the specific challenges of domestication, and how academic traditions mattered—both institutionally and intellectually—in providing a source of creativity to address those challenges. The fact that penicillin domestication occurred relatively rapidly and smoothly, by coordinating academic research on production problems through an interdisciplinary association that mediated between the demands of government policy and those of industrial practitioners, suggests that the JPRA's distinctive functioning relied on organizational precedents in pre–World War II as well as wartime fermentation research. Indeed, the personnel in the wartime Hekiso Committee were largely the same as those in the Technical Committee of the JPRA's Central Laboratory. This chapter's examination of the occupation period preceding Japan's high-growth era and its concomitant expansion of corporate laboratories, then, highlights the ways in which indigenous expertise shaped the postwar development of antibiotic science in Japan.

With the growth of antibiotic mass production, fermentation approaches to microbial research became prominent in the medical and industrial fields of antibiotic science, whether in company or in academic laboratories. During penicillin domestication, academic scientists in the

Central Laboratory of the JPRA simultaneously pursued immediate practical problems for industry and broader, longer-term research questions. It was not only the existence of microbiological expertise itself but the *kind* of microbiological expertise that was specific to Japan, a kind that blurred the lines between "pure" and "applied." The configuration of expertise contrasts with that in the French wartime penicillin project, where the microbiologists involved did not share the military's sensitivity to economic limitations in manufacturing, as well as with the prewar Prague-centered "applied science" fermentation knowledge, which later became important in Anglo-American penicillin production and which was less engaged with theoretical questions. The microbial engineering expertise showcased in the Prague example became significant globally for antibiotic science after World War II, but what was also important in antibiotic science and often overlooked was a biological approach to microbes and a sense for what microbes could do. In addressing the intellectual problems of penicillin domestication—from strain development for both surface and submerged culture to the investigation of culture media, refinement methods, and contamination countermeasures—Central Laboratory scientists drew on fermentation approaches from the discipline of agricultural chemistry, seeing microbes as a way to manufacture essential chemicals locally in conditions of resource scarcity. The bounty of their scientific toolbox resulted from a historic *perception* of the salience of resource scarcity in motivating experimentation, and they made the domestic environment work by concentrating on microbes' various physiological and ecological capacities to transform it.

Thus microbiologists would later speak of antibiotics as gifts not from the occupation state, but from the microbes themselves.[119] Between 1947 and 1983, the average life expectancy at birth in Japan rose from 50.05 years for men and 53.96 for women to the highest in the world at 74.20 years for men and 79.78 for women, while the infant mortality rate dropped from 76.7 per 1,000 births to the lowest in the world at 6.2 per 1,000 births.[120] The wide availability and consumption of a multiplicity of antibiotics contributed to this transformation. It also provoked the pervasive incidence of resistant strains, to which scientists responded with more antibiotics even as they studied the mechanisms of resistance. Once attention to antibiotic discovery had receded in Europe and North America, Japan became one of the main centers that continued to produce advances in this field. The industrial view of microbes as an abundant source of new antibiotics, and the clinical view of microbes as resistant pathogens to be fought, fed upon each other. The historian Edmund Russell's observation on pesticides is equally apt for antibacterials: "War and control of nature coevolved: the control of nature expanded the scale of war, and war expanded the scale on which people controlled nature."[121] In the aim to preserve human life by pitching microbes against other microbes, the interactions between agricultural science and medicine in World War II and occupation-era Japan created a simultaneous vision of militaristic control and eradication, and beneficent variety and innovation.

6

Flavor

TO SCREEN FOR GIFTS

In amino acid and nucleotide fermentation, the fermentation community pursued the microbial production of novelty in the same way as Suntory pursued its blue roses and blue carnations.
—Ōshima Yasuji, professor emeritus, Faculty of Engineering, Osaka University, interview by author[1]

In the 1950s, the Japanese economy gradually recovered from the devastation of the Asia-Pacific War, following procurements from the Korean War. The government—retaining powers over foreign exchange, though it no longer had the degree of command over the economy that it had during the war and occupation—continued to promote technology transfer and industrial recovery. In 1953 the founders of the infant company Sony purchased a license to make transistors. Slowly, ordinary households were returning to the comforts of everyday life that had been unaffordable during the war, such as consumption of miso. The first instant ramen product from Nisshin Shokuhin appeared on the market in 1958 on the very cusp of Japan's high-growth era. In this period, early in 1956, scientists discovered that the flavor product monosodium glutamate (MSG) could be made using a microbe, making the process vastly cheaper than the prior method of extraction from hydrolyzed vegetable protein. The innovation did not come from Ajinomoto, the company that had pioneered MSG production and given the flavoring its trade name in Asia for almost half a century, but from the much smaller company of Kyōwa Hakkō Kōgyō, a newcomer from the alcohol industry. The technology to ferment glutamic acid (the amino acid that forms the basis of MSG) was the first case of technology transfer out of Japan in the post–World War II period, as the American pharmaceutical company Merck bought the license in 1958. The discovery pushed the possibilities for fermentation far beyond antibiotics in the pharmaceutical industry and created a significantly expanded food flavorings market worldwide.

In the new field of amino and nucleic acid fermentation—which went further than the technologies of the antibiotic industry by using microbes to anabolically produce not only secondary but also primary metabolites—research laboratories in corporations took the lead, as the Japanese industrial economy transitioned from wartime and occupation planning to market-driven competition.[2] As international scientific communications resumed and improved, the period saw a greatly increased flow of scientific exchange. Moreover, government funding for industrial associations and research groups supported domestic information exchange between laboratories in private companies, semipublic research centers, and universities.[3] As Tessa Morris-Suzuki explains, most of the information did not directly concern technology transfer, or the replication of the entire body of specific know-how required to manufacture a product. Rather, information forums mainly allowed laboratories—whether academic, government, or industrial—to share "views on the general direction of technological change" as well as market conditions.[4] Details of scientific developments were one major component of such information, yet the role of *scientific* information in Japan's high-speed growth remains little examined, even in the biomedical sector for which scholars have acknowledged the special importance of basic research.[5] This chapter traces microbial mass production technologies in commercial amino acid and nucleotide manufacturing during the high-growth decades, centering on the corporate laboratories of such companies as Kyōwa Hakkō Kōgyō (alcohols and solvents), Ajinomoto (MSG), Takeda Yakuhin (pharmaceuticals), and Yamasa Shōyu (soy sauce). It shows how, through dense information exchange, a now globally oriented science shaped local product development and the market itself.[6]

Information exchange was mediated by the nature of the central technologies involved: microbes, which were at once scientific research objects and closely guarded technologies.[7] Looking only slightly beneath the surface of an internationally homogenous technical language reveals the resilience of local expertise. Instead of analyzing how Japanese scientists adapted imported technologies to local environments, this chapter shifts the frame from the dynamic of global pressure and local response to exploring how local skill, embodied in the community of fermentation scientists, was reshaped to the specific circumstances of 1950s to 1970s Japan. At the same time, it charts how Japanese fermentation scientists made contributions to global scientific and technological fields. To do so, this chapter presents a portrait of the fermentation community as it follows the thread of technical and institution innovation in the wake of glutamic acid fermentation.[8]

Elsewhere, these very materials—amino acids and nucleotides—were under study for their cellular roles in protein synthesis and gene replication, and were key materials in the emerging discipline of molecular biology; but biologists did not usually think of them as *tasting* like anything. Instead, biochemists and molecular biologists investigated amino acids—the building blocks of proteins—and nucleotides—the building blocks of the nucleic acids DNA and RNA—as answers to the "secret of life."[9] Japanese fermentation scientists studied fundamental cellular components for their industrial uses, and manipulated the microbial sphere as a means to intervene in the nutritional and flavoring world, rather than engaging in "pure" areas of life sciences research alone. More than reading the book of life to puzzle out what chemicals made living things, the idea that tied Japanese fermentation scientists together across different fields of experimental biology was "metabolic engineering," to figure out how living things made chemicals.[10] This was because from the 1950s, they worked within an institutional structure closely linked to high-tech manufacturing.

The emergence of state-of-the-art scientific laboratories in the private sector, equipped with facilities superior to those in the government's laboratories, recalled the origins of an economy dominated by *zaibatsu* (industrial monopolies) in the 1880s. Although occupation authorities had moved to break up the *zaibatsu* immediately after World War II, from the perspective of scientific activity there was little to signal the decline of large corporations either in the 1950s or subsequently.[11] In the fermentation field, it was a period when competition in mass manufacturing relied on cutting-edge laboratory research—research that was led by industrial firms, informed by the academy, and supported by government funding.[12]

When it came to the role of science, agricultural chemistry's disciplinary identity had long seen its scientists as responsible for both environmental and economic management. Fermentation scientists continued to rely on local traditions of handling and thinking about microbes, even amid the massification of production and consumer culture due to the country's rapidly growing prosperity and expanding domestic markets.[13] Between the 1950s and 1970s, while Japan's GDP grew at an unprecedented pace to become the third largest after those of the United States and the Soviet Union, the work of fermentation scientists in corporate laboratories still built on pre–World War II knowledge of microbes as methods of national resource management. As I will argue in this chapter, they used what had previously been austerity techniques—with their resource-saving mentalities, labor-intensive setups, tight-knit information coordination, and familiarity with working with microbes precisely as sources of entrepreneurial creativity—to develop new biotechnological goods made for efficient national production and consumption.[14]

Partly as a result, in the high-growth era the leading edge of life sciences research in Japan took the form of high-tech, yet resource-conserving, biological innovation within large, pyramidal, corporate-linked information structures. It contrasted with the setting of biochemistry and molecular biology in the United States, which was dominated by capital-intensive academic research based on high-end instrumentation.[15] Considering scientists' work casts a different light on the political economy of Japan's "miracle" growth, toward which historians have tended to take one of two approaches. The first approach is top-down, focusing on a developmental state whose policies facilitated industrialization and growth.[16] Another is bottom-up, taking the viewpoint of consumers who produced a mass market, or pollution victims campaigning for redress.[17] Centering the scientific ideas that underlay technological developments adds nuance to narratives polarized between the developmental state and social response, and emphasizes that the perspectives of middle-level technical experts are central to understanding the interplay between industrialization and environment in the modern period.[18] Amino acid and nucleotide flavorings shifted in consumers' perceptions from the 1950s through the 1970s, whether as goods of desirable luxury or morally deplorable convenience, or as "chemical" or "natural" substances. Yet fermentation scientists understood Japan's ecologies of production not only by thinking about commercial profit, but also about a kind of resource management. It was clear to fermentation scientists that microbes were able to redistribute elements in the landscape, and they could see their interdependence with human economies.

COLLECTING HISTORY

The microbial type culture collections themselves trace a history of the forces shaping the Japanese fermentation community, which worked to gather and keep alive those organisms thought of as having special value. The existence of a collection signals massive inputs of labor, as it demands the contribution, maintenance, and documentation of every individual object of which it is constituted.[19] Hasegawa Takeji (1914–2006), who was appointed director of Japan's largest microbial type culture collection in 1946, has documented this history. His reconstruction was the result of painstaking scientific effort, and was necessary to complete his task of systematizing the collection and cleaning up the taxonomic designation of strains.[20] The culture collections reflected the coherence of the community of fermentation scientists, despite the numerous disciplinary labels associated with them, and as the microbial type culture collections were elevated to a national-scale organization in the 1950s, fermentation scientists developed a strong identification with their field.[21] In particular, the strains preserved in the collections testify to the institutional continuity between craft industries and the high-tech life sciences sector in Japan, and to the distinctive value attributed to productive microorganisms there.

Individual brewing houses such as *moyashi* makers had long preserved their own collections, but the first national collections began at the national Brewing Experiment Station in Tokyo from the first decade of the twentieth century. Since the government experiment stations sought to aid small and medium-sized industries across Japan by doing research on their behalf, the collections concentrated on varieties of microbes that would be used in a brewery, such as *kōji* microbes, yeasts, and lactic acid bacteria. Microbial cultures traveled between all of these collections and much smaller collections for teaching at the Agricultural Chemistry Department in the Faculty of Agriculture at Tokyo Imperial University, where researchers tinkered with organic acid fermentations, and the Brewing Department at the Osaka Higher Technical School.[22] International exchange of strains was limited, but there was correspondence in the form of both letters and cultures with laboratories maintaining significant collections of industrial microorganisms in Europe, including the Centraalbureau voor Schimmelcultures in the Netherlands and the Institute of Brewing in Britain.[23]

From the 1910s, strains from the Brewing Experiment Station followed their collectors to Japan's informal and formal empire. Collections of Asian brewing microbes grew, especially at the Central Laboratory of the South Manchuria Railway Company (CLMR) and the Taiwan Government-General Central Research Institute (the Government Research Institute of Formosa, or GRIF), where scientists worked on producing alcohol from the sugar industry that had developed in the semitropical colony. Alcohol research culminated in an all-encompassing drive to ferment fuels during the Asia-Pacific War, involving chemical companies across the Japanese empire and research at university laboratories.[24] Consequently, by the time Japanese settlers and soldiers returned to the home islands after the war ended, many microbial cultures collected in the empire had been transferred to the home islands as well. Along with their curators, copies of CLMR strains went to the brewing departments of the Osaka Higher Technical School in 1927 and the Hiroshima Higher Technical School in 1930.[25] When the Aeronautic Fermentation Institute (Kōkū hakkō kenkyūjo)—jointly supported by the Cabinet Technology Office (Naikaku gijutsuin) and the pharmaceutical giant Takeda Yakuhin to undertake fuel research—opened in 1944, its microbial collections were copies of the GRIF cultures.[26]

When the Technology Office broke up after the surrender, both Takeda Yakuhin and the Ministry of Education supported the continuation of the Aeronautic Fermentation Institute, with its name changed to the Institute for Fermentation, Osaka (Hakkō kenkyūjo; IFO).[27] The national Brewing Experiment Station's collections had been lost in the war. Immediately after the war, the occupation authorities prioritized penicillin production, and university laboratories sent out American strains to companies.[28] In the 1950s and 1960s, major companies built their own central research laboratories, and chemical firms with fermentation capability began to accumulate collections.[29] Now the largest national microbial type culture collections held at the Institute for Fermentation, Osaka, no longer resembled the microbes found in local breweries (which were decreasing in number), and they served large food, chemical, and pharmaceutical companies. Yet the continuity in the culture collections—still containing earlier brewing microbes from Japan and the former Japanese empire, and still carrying an overwhelming emphasis on useful molds over disease-causing bacteria—was a manifestation of the continuity in community and skills that underlay the shifting aims and sites of microbial technology.[30]

Collecting work emphasizes the epistemic (and commercial) value of biodiversity. From the 1950s to the 1970s, while biochemists and molecular biologists elsewhere were collecting protein and nucleic acid sequences, fermentation scientists in Japan were collecting microorganisms. These were labor-intensive experimental practices, conducted at the laboratory bench.[31] Collecting microbes required isolating them and culturing them, and to *use* the biodiversity of the microbial world, above all, meant screening organisms. Isolating, culturing, screening: these were the classic practices of twentieth-century Japanese fermentation science, honed to distill the power of microbial variety. The historian Bruno Strasser argues that collecting ran counter to the mode of much of twentieth-century experimental biology, in which a few exemplary model organisms tended to stand for the rest. Collecting, by contrast, enabled comparison among the variety of living things. Only later in the twenty-first century would biology bring "biodiversity back into the laboratory" on a large scale via genome databases, Strasser contends, and force a reconciliation between the two contradictory ways of working.[32] Japanese fermentation science diverges from this pattern, as is detailed further in this chapter.[33] Since collectors were experimenters and curators were scientists, configurations of the moral economy of collections were associated less with a conflict between experimenters and collectors, and rather with the tensions between collections' academic and commercial uses.

RESERVOIRS OF POTENTIALITY

Kyōwa Hakkō Kōgyō's turn to the microbial world to produce glutamic acid was unorthodox in the worldwide biochemical community, where common knowledge held that it would not be possible. Amino acids were primary metabolites necessary for building cell constituents and were metabolized quickly, and there were control mechanisms in the cell to stop them from accumulating in large quantities.[34] Yet the idea itself of making glutamic acid by fermentation built on long-term local concerns. Kyōwa Hakkō was a relatively new company, emerging from the dissolution of a larger company as part of the occupation government's efforts to break up the *zaibatsu*. That company, in turn, had origins in a cooperative association of alcohol companies in western Japan, which was first set up in 1936. Thus, Kyōwa Hakkō's main product line was in alcohols and solvents. The company was a recent entrant into the pharmaceutical market due to the state support it gained from participating in the penicillin efforts (making solvents using cheap imported molasses as raw material) and the subsequent license it obtained from Merck for streptomycin manufacture.[35]

Udaka Shigezō (1930–2015), the researcher who discovered the glutamic acid-producing microbe, entered the Kyōwa Hakkō laboratory in late 1955. Prior to that, he had studied as an undergraduate at Sakaguchi Kin'ichirō's laboratory in the Agricultural Chemistry Department of the Faculty of Agriculture, the University of Tokyo (Tōdai), and then had undertaken a two-year master's degree in biochemistry at the University of Chicago on private funds.[36] A handful of his peers in agricultural chemistry had joined Ajinomoto's laboratory, but it was Kyōwa Hakkō that accepted Udaka when he sought a job.[37] He was the only one among his Tōdai peers to go to the company, which at that point had a less established reputation in scientific research than did Ajinomoto. Kinoshita Shukuo (1915–2011), the director of the laboratory and Udaka's boss (also a graduate of Sakaguchi Kin'ichirō's lab at Tōdai), oversaw efforts toward the three research goals handed down by company president Katō Benzaburō back in 1946. The first two goals — making penicillin and streptomycin — had been accomplished; the third, making protein, remained open.[38]

After obtaining permission from Kinoshita to work on glutamic acid production by fermentation, Udaka designed a fast, reliable screening method that allowed him in less than three months to detect a glutamic

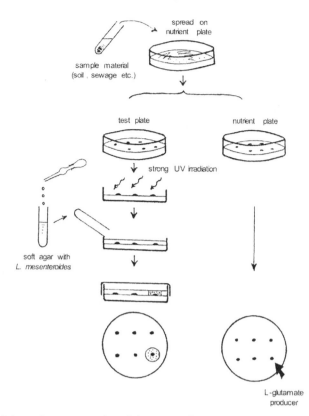

spread on
nutrient plate

sample material
(soil, sewage etc.)

test plate   nutrient plate

strong UV irradiation

soft agar with
L. mesenteroides

L-glutamate
producer

FIG. 6.1. Schematic representation of the successful screening procedure leading
to isolate no. 534, the first identified colony of *Corynebacterium glutamicum*. Diverse
samples from different environments were spread on agar plates. Colonies appearing
on these plates were killed by strong ultraviolet irradiation. Then the plates were
overlaid with a basal agar medium for glutamic acid bioassay containing *Leuconostoc
mesenteroides*, an L-glutamic acid auxotrophic bacterium. Upon further incubation, the
bacteria that produced a halo (growth of the assay organism) on the surrounding areas
of their colonies were selected from a nutrient plate inoculated in parallel. Image and
caption from Udaka, "Discovery of *Corynebacterium glutamicum*" (2008), fig. 1.1.

acid-producing microbe (fig. 6.1). He overlaid agar plates of test strains
killed by ultraviolet radiation with a basal agar medium containing *Leuconostoc mesenteroides* strain P-60, which required glutamic acid for growth,
and left them overnight. The next day, he would look for a halo around the
sample that would show the growth of the assay organism. Around 15 percent of the strains showed faint halos, which convinced him that this was
a viable line of research. After the screening of about five hundred strains,
sample no. 534 gave a clear halo. Udaka confirmed glutamic acid production with paper chromatography. The strain produced glutamic acid yields
as high as 21 percent from glucose.[39]

Udaka's screening method relied on the new techniques of the bioassay and paper chromatography. But he insisted that he had learned the bioassay and "pay-cro" at home, from Tōdai and Kyōwa Hakkō researchers, and not at Chicago.[40] His approach was uncommon in also searching for capabilities in unknown microbes, rather than screening only known microbes, as researchers tended to do. He reasoned that perhaps glutamic acid might be produced in large quantities if there existed a very "unusual" (*kawatta*) microbe.[41] Though his approach that emphasized reliance on the diversity of the microbial world was distinctive at the time, it would soon be powerfully institutionalized in fermentation science. In his experiments he also drew on an integrative approach to microbial methods, in the tradition of agricultural chemistry.[42] For example, he used his knowledge of the biosynthetic pathway to infer the most appropriate culture medium for glutamic acid production, which was then unknown.[43]

As novel as the discovery was, Udaka later presented his very idea to find a cheaper way of making MSG in terms of national needs, just as the fermentation community had done previously for penicillin, fuel, and synthetic sake, even though in this case the product sought was an everyday consumer commodity. Cheap production of monosodium glutamate would contribute to national health by making it easier to eat bland yet somewhat nutritious food, at a time when the memory of eating boiled acorn gruel in the wartime and immediate postwar period was still vivid. Acorn gruel was tasteless, he remembered, but at least it put something in the stomach. The worst were the rare times when one decided to leave the house and go outside, and one felt one's body swaying from hunger.[44]

The wartime economy restricted industries' use of such valuable agricultural produce as cereal protein for making glutamic acid, and in homes there was a shortage of sugar to make miso. The pretty crystals of MSG were an expensive and extravagant luxury.[45] And yet, Udaka recalled, if one could only add a little MSG to the food, it would become tasty, all the more so when one was hungry.[46] In the mid-1950s, the economy was slowly recovering after the stimulus of military procurements from the Korean War, and sales of MSG were beginning to revive.[47] Still, these were precisely some of the same questions that had preoccupied prewar fermentation researchers, who had begun to expand out of the traditional food and alcohol industry with amino acids and vitamins; and amid the latest literature, Udaka was especially fascinated by vitamin fermentation.[48] In seeking microbes as a solution to problems of nutritional and industrial resource efficiency, in this case by using a microbe to turn the abundant raw material of waste molasses from Southeast Asia into glutamic acid in one step, the design of the process resembled prewar and wartime work.

Kyōwa Hakkō Kōgyō president Katō Benzaburō announced to the public in a press conference on September 20, 1956, that the company had developed a new method of fermentation for making glutamic acid. Production would begin in October with one ton, and was expected to reach one hundred tons per month by January 1957.[49] Amid the sensation and competition that ensued, the most heated response predictably came from Ajinomoto. A complicated rivalry followed. In the wake of the Tōdai physical chemist Ikeda Kikunae's discovery in 1908 that MSG was responsible for an "umami" taste in kombu (kelp), Ajinomoto had begun commercial production of MSG in 1910 with large amounts of capital, decomposing vegetable protein with hydrochloric acid and using wheat gluten as the starting material. Over the years, Ajinomoto had employed scientific research both inside and outside the company to solve complex problems related to chemical manufacture.[50]

Ajinomoto first disputed Kyōwa Hakkō's priority with a patent for a similar method for which it had applied in 1954, though the microbe itself (which had been discovered by the Saitama University professor Tada Seiji) had turned out to be ineffective, and its patent was eventually annulled by a court decision in 1962 for problems having to do with the clarity of the details in the application.[51] In the meantime, Kyōwa Hakkō struggled with the refinement process; it eventually took the company three years to reach the production target that Katō Benzaburō at the press conference had estimated would take six months. Kinoshita Shukuo regretted that this gave "the other company" time to rival its microbial process, and Udaka lamented that for underestimating the refinement process, Kyōwa Hakkō wound up "under Ajinomoto's thumb."[52] As a much newer and smaller company, Kyōwa Hakkō was hesitant to venture into a new market, MSG, that Ajinomoto held so absolutely.[53]

In 1959 a deal was reached with Yamamoto Tamesaburō, president of the beer company Asahi Beer, as mediator. Kyōwa Hakkō would sell its entire yield of crude glutamic acid to Ajinomoto, and Ajinomoto would refine the product into finished commercial MSG. However, Kyōwa Hakkō canceled the deal in 1961. Eventually, in 1962, Ajinomoto paid Kyōwa Hakkō a large settlement of fifteen *oku* yen, and the two companies agreed to cross-license all their patents with each other and not to try to block each other's advancement into overseas markets with patents, as they had been doing in the Asia market up to that point.[54]

Aside from Ajinomoto, however, many other chemical companies saw the potential in the flavor and pharmaceutical market of Kyōwa Hakkō's discovery. Glumatic acid production expanded the possibilities of fermentation to a horizon wider than only secondary metabolites such as antibiotics; and when it came to making primary metabolic end products, it extended the prospects beyond catabolic processes (of molecular breakdown) in making a limited range of products including alcohol and organic acids, to anabolic processes (of biosynthesis of molecules) as well. The controlled economy had left the legacy of large-capacity fermentation tanks for penicillin and antibiotic manufacturing with the companies, which were now seeking to create new marketable products with the same tanks.[55] When Udaka Shigezō presented his work to the microbiology community in a large room in the Faculty of Agriculture at Tōdai in early 1957, he heard many camera shutters snapping to pick up his slides.[56]

In 1957 Japan held its first international academic conference in the postwar period, the International Symposium on Enzyme Chemistry in Tokyo and Kyoto.[57] In a lecture at the conference, the biochemist Tamiya Hiroshi referred to *kōji* as the "national enzyme" of Japan.[58] Kinoshita Shukuo received permission from company president Katō Benzaburō to report there at the Gakushi Kaikan in Kanda on Kyōwa Hakkō's research.[59] Among the audience were the USSR Academy of Sciences members A. I. Oparin, V. N. Bukin, W. L. Kretovich, and R. B. Feniksova, as well as the American scientists E. E. Snell and L. A. Underkofler (who asked a question).[60] Following the conference, Soviet scientists invited Kinoshita Shukuo to speak at an international microbiology meeting in Moscow. Merck sent its research director, Max Tishler, to Japan to discuss technology transfer. The victor nation paid a license fee to the defeated nation; this was somewhat marvelous, Kinoshita later wrote of the 1958 transaction.[61]

It was rumored that Merck had also considered a method developed at the University of Texas at Austin by Jackson Foster, the American consultant who had aided penicillin transfer to Japan during the occupation, but that it had chosen Kyōwa Hakkō's.[62] Arnold Demain, a research microbiologist at Merck, visited Kyōwa Hakkō Kōgyō's glutamic acid factory in Hōfū in Yamaguchi Prefecture. Too late, Pfizer and IMC both expressed interest. The impact of the discovery transcended the local in several senses: surpassing Kyōwa Hakkō's existent alcohol market to advance into the MSG market that Ajinomoto dominated; breaking out of the one major high-tech industry, antibiotics, in which microbes were thought to be useful; and pushing beyond Japan, as Merck first paid Kyōwa Hakkō to receive a license for the technology.

Flavoring production now tapped straight into the expertise of the fermentation community. Company disputes were mediated by microbes as a technology, because the strain itself held the key to effective manu-

月 日	Source of bacteria	No.	on urea medium 注1)			on test medium 注2)		
			growth	ring felt	colonial morph.	growth	ring forms	colonial morph.
	雑食及草食獣糞	402	+	+ᵍᵛ±		+	±	
	"	406	"		.	"	±	
	"	407	"	"		"	±	
	"	419	"	"	.	"	±	
	"	447	"	"		"	+	
		448	"			"	−	
	鳥類糞	534	+	+₀.₄	irregular small	+	−	
	"	525	±	−		"	+₀.₃	
	"	539	±	−		"	+₀.₄	yellow pellicle
	"	544	+	−		"	±	light brown
	"	549	±	−	s	"	+₀.₄	yellow
	soil & putrefaction	601	+	±	small	±	−	
	"	602	+	.	"	±		
	"	604	┤	＼		+	−	
	"	605	┤	.		┤	±	
	"	606	+	"		+	±	
	"	611	┤	"		±	±	
	"	612	+	.		+	±	
	"	622	+	.	white circular	┤	±	
	"	623	┤	+	white thin	+	±	

FIG. 6.2. Udaka Shigezō's laboratory notebook at Kyōwa Hakkō, 1956. In the row of notes for isolate no. 534, under the column for "source of bacteria," is recorded "bird poop." From Udaka Shigezō, "Aminosan hakkō no reimeiki" (2009), 213. Reproduced with permission of the Japan Society for Bioscience, Biotechnology, and Agrochemistry.

facture. Hearing that Udaka Shigezō's microbe had come from bird poop in Ueno Zoo (fig. 6.2), academic and industrial laboratories shifted from their usual practice of looking only for known antibiotic producers such as actinomycetes, and sent people out into the wild to collect samples. Nobody knew the exact constitution of the sample from which Udaka had isolated his microbe, and many efforts met with frustration, since no one could reproduce the exact conditions of the time and place at which the microbe had been present. Nor could it be found in the poop of the birds kept at the Tokyo University of Agriculture; perhaps the poop there was too pure, rather than being mixed with dirt and soil.[63] When American researchers found strong glutamic acid-producing microbes, it turned out that they too had come from bird poop.[64] Why bird poop? For Udaka, the microbial world behaved in a mysterious way.[65] Rumors circulated of companies stealing each other's microbes from the factory waste water, which was possible at the time because the waste water was poorly treated. In a dispute surrounding patents on the fermentation of the amino acid lysine, Kyōwa Hakkō researchers showed that a microbial strain they collected from the waste water coming out of Ajinomoto's Kyūshū factory was the same as the strain specified in Kyōwa Hakkō's patent; and on this basis, a court ruling in 1972 ordered Ajinomoto to stop manufacturing lysine and to pay compensation to Kyōwa Hakkō.[66]

The species of the microbe was at the center of the process-based patents, and so finding one's way around a patent meant finding a different species of microbe that could make the same product, pinning company stakes to microbial classification. This was more than a decade earlier than the 1980 *Diamond v. Chakrabarty* case in the United States, which marked the first time a living organism, a microbe, could be patented.[67] In order to be able to submit the patent application for glutamic acid fermentation, Kinoshita asked researchers to begin taxonomic work immediately. Referring to *Bergey's Manual of Determinative Bacteriology*, they identified strain no. 534 as a new species, *Micrococcus glutamicus*. Drafting the patent application was tricky because, while it required specifying the productive strain's properties, a patent based on the strain alone would make the patent a narrow one. In the submitted application, the company chose to emphasize three particular features of the process: the strain's characteristics (which they put at the forefront of the application), the fact that it was a biotin-requiring strain, and the manufacture of glutamic acid using a microbe.

Kyōwa Hakkō Kōgyō researchers later expressed great regret that, unlike in American companies such as Merck, company laboratories in Japan at the time generally did not have patent specialists to address these challenges thoroughly, and instead wrote patent applications with a "that will do" approach, which could mean that all their efforts in research might in the end be defied.[68] Other researchers could bypass the restrictions of existing patents if they claimed to use a different species in the manufacturing process. This could involve taking advantage of the persistent ambiguity in microbial classification, a subject dealing with organisms in which tiny variations in experimental conditions would give rise to subtle differences in morphology and physiology.[69] The Tōdai and wider Kantō academic communities were split, as some microbiologists in the early patent wars sided with the claims of Kyōwa Hakkō and others with Ajinomoto.[70] As the field of amino and nucleic acid fermentation blossomed, so did the number of new names for microbes.[71]

The high technical requirements of making MSG much more cheaply by fermentation pushed out the numerous local small and medium-sized makers that had made MSG by chemical methods of protein decomposition, and which disappeared in the 1960s. At that point the only players producing for the MSG market were Ajinomoto, Kyōwa Hakkō Kōgyō, the chemical company Asahi Kasei, and the pharmaceutical company Takeda Yakuhin; Ajinomoto was supplying about half the total demand.[72] The work of fermentation experts in the large company laboratories trod the line between pure science and applied design as they helped develop new products. Science itself functioned as an information medium, and as a result, while scientific trends did not determine the work of company laboratories, companies expanded their flavoring product lines in highly similar directions.

In the years after World War II, a prominent line of biochemical research developed in North America and Europe that focused on making auxotrophic bacterial mutants—whose growth and reproduction required the additional supply of specific nutrients—using, for example, ultraviolet rays or nitrogen mustard as mutation-inducing agents. The bacterial mutants would accumulate particular intermediates in the biosynthetic pathway as a result of genetic blocks, and they could thus be used to map biosynthetic pathways, drawing on the expertise of both biochemists and geneticists.[73] Shortly after the discovery of strain no. 534, researchers at Kyōwa Hakkō Kōgyō hoped to use mutants of *Micrococcus glutamicus* to create microbes that would produce amino acids other than glutamic acid in large quantities. Udaka Shigezō's experiments had already found microbes that produced alanine and valine. In particular, the laboratory aimed for lysine, which might be sold as a nutritional supplement in food and in animal feed. Kinoshita Shukuo put Udaka in charge of characterizing the initial mutant strains produced in the project, and put another researcher in charge of finding a strong lysine-producing microbe.

As part of this work, which was largely carried out in 1957, Udaka investigated the enzymes involved in the biosynthetic pathway of ornithine, using an ornithine-producing mutant of *M. glutamicus*, and elucidated the biochemical regulatory mechanism for cellular control of ornithine production. At the time, he believed that he was the first in the world to demonstrate the metabolic regulatory mechanism for amino acid production in cells, though he later found out that Edwin Umbarger at Harvard University had already shown negative feedback inhibition in *Escherichia coli* in 1956 in a different pathway, isoleucine synthesis.[74]

As researchers worked out regulatory mechanisms in *E. coli* through the 1950s, knowledge of these mechanisms helped Japanese researchers to select mutant strains that would accumulate specific amino acids for the purpose of industrial mass production. As Kinoshita Shukuo put it, these techniques for selecting mutants using the latest biochemical knowledge were very different from the ways of traditional microbe smiths (*jūrai no biseibutsuya*). To Kinoshita and other Japanese fermentation scientists, these microbes were created, not found, and were engineered for a part of the metabolic pathway to be genetically blocked for human needs.[75] Research on regulatory mechanisms in cells was a central part of the emerging discipline of molecular biology, and the work of Udaka and others at Kyōwa Hakkō Kōgyō on the metabolic regulation of amino acid production earned them invitations to attend the 1961 Cold Spring Harbor symposium "Cellular Regulatory Mechanisms," seen as a milestone conference in the history of the discipline.[76] The historian Lily Kay characterizes the period as one in which cells became conceptualized as self-contained informational programs—or, in other words, as "reservoirs of genetic potentialities."[77] Japanese fermentation scientists exploited these genetic potentialities for industrial production while, through the 1960s, biochemical investigation of mechanisms of feedback inhibition continued in conversation with work in molecular genetics.[78]

Similarly, work to open up products and a market for nucleotide flavorings such as IMP and GMP—which, unlike MSG, were as yet undeveloped in the flavorings market—was linked to Japanese researchers' interest in global academic developments in biochemistry and molecular biology. In 1951, two years before the publication of James Watson and Francis Crick's 1953 paper on the double helical structure of DNA, Sakaguchi Kin'ichirō told graduate student Kuninaka Akira, who was working in Sakaguchi's fermentation science laboratory in the Agricultural Chemistry Department at Tōdai, that nucleic acids—that is, DNA and RNA—would be interesting substances to bring into research on microbial decomposition. There were countless reports on using microbes to decompose proteins, but what about, for example, using *kōji* microbes to decompose the nucleic acids that were so abundant in living matter, and characterizing the biochemical routes of decomposition? Since fermentation science was also an applied field, Sakaguchi encouraged Kuninaka to bear in mind how in 1913 Kodama Shintarō had found that the umami flavor of *katsuobushi* (dried skipjack tuna shavings) was largely due to the histidine salt of the nucleotide inosinic acid, and yet inosinic acid's umami properties had generally not been mentioned again since Kodama's publication.[79]

In 1953, Kuninaka Akira entered the research laboratory of the soy sauce company Yamasa Shōyu and worked to mass-produce inosinic acid as an umami substance. Collaborating with colleagues at the Tōdai fermentation science lab and the Institute of Applied Microbiology, he explored two different methods. One method worked via the decomposition of yeast RNA, which required screening for a microbe to produce an enzyme that would cut RNA into 5'-nucleotides (the 3'-inosinic acid produced through decomposition by *kōji* microbes had no umami taste). The other method created an adenosine-requiring mutant strain that would accumulate 5'-inosinic acid (IMP) directly. The researchers first publicized their success in the former method through a patent application by Yamasa Shōyu in 1957, and their success in the latter method by a report in the *Nippon nōgei kagaku kaishi* (Journal of the Agricultural Chemical Society of Japan) in 1961.[80] Upon investigating the other nucleotides produced through cleaving yeast RNA at the 5' position, researchers found that 5'- guanylic acid (GMP) also showed strong umami properties, this time associated with shiitake mushrooms. It soon became clear that when MSG (kombu umami), IMP (*katsuobushi* umami), and GMP (shiitake umami) were mixed together, they had a synergistic effect and the umami taste was terrifically strong.[81]

On learning of these developments, Ajinomoto developed IMP manufacture by first extracting inosinic acid from *niboshi* (dried baby sardines), and later pursuing nucleotide fermentation as well. In 1960, the company began to sell MSG crystals coated with IMP as Ajinomoto Purasu ("Ajinomoto plus"), the first in a new line of flavoring products.[82] The pharmaceutical giant Takeda Yakuhin, too, built up the capability and plants to produce nucleotide flavorings by RNA decomposition, and began to create a sales strategy for its product, I no Ichiban, using its links to rice businesses.[83] In 1961, Takeda Yakuhin and Kyōwa Hakkō jointly set up a new flavorings company, Nihon Chōmiryō, with the aim of combating Ajinomoto's dominance. Kyōwa Hakkō would supply the MSG, while Takeda Yakuhin would supply the IMP and GMP. Nihon Chōmiryō targeted instant ramen manufacturing, soy sauce companies, and manufacturers of processed fish products, such as *kamaboko* (boiled fish paste), as potential users.[84]

When Takeda Yakuhin developed its own ability to ferment glutamic acid, and canceled the partnership in 1966—accompanied by chagrin on Kyōwa Hakkō's part, further patent wars, and the loss for Kyōwa Hakkō of its new sales network in flavorings—Kyōwa Hakkō allied instead with Yamasa Shōyu. Throughout this time, Yamasa Shōyu had been selling a mixture of the sodium salts of IMP and GMP, named Yamasa IG, to food processing businesses. In the new partnership with Kyōwa Hakkō, the soy sauce company began to sell MSG crystals coated with IG as Yamasa Fureebu ("Yamasa flave") to homes.[85] It was also a period of rapid expansion in miso consumption, and demand for MSG rose along with miso production.[86] Having been "ambushed" by both Ajinomoto and Takeda Yakuhin in glutamic acid fermentation, Kyōwa Hakkō finally succeeded in nucleotide manufacture in 1966 by direct fermentation; but, being behind other companies, it could not hope to gain a good share of the flavorings market. Although Kyōwa Hakkō continued to lead in amino acid fermentation even as rival companies including Ajinomoto acquired the same technology, amino acids other than glutamic acid were not flavor products and had a smaller market than flavorings.[87]

One of the major users to benefit from the new flavoring products was the food company Nisshin Shokuhin, which developed instant ramen in that period. The company became a big purchaser of MSG and nucleotide flavorings. The same year that Kyōwa Hakkō invented the fermentation method for glutamic acid, Nisshin Shokuhin was drawing up plans to scale instant ramen up to mass production. Nisshin Shokuhin released the Chicken Ramen noodle product in 1958. As Andō Momofuku, the president of Nisshin Shokuhin, told Kinoshita Shukuo many times, the flavoring products made cheap by fermentation greatly aided Nisshin Shokuhin's ramen business.[88] The expansion and diversification of the flavorings industry was a result of how scientific information functioned to give ideas to company researchers, and the resulting high technology concentrated the production into large companies, and the research of companies into similar directions. The novel flavoring products of the era and the growth of the responsive new market were linked to developments in microbiology.

ASK MICROBES, AND THEY WILL NEVER BETRAY YOU

In the new climate of science-based competition between companies, the academy served as an easy information channel for ideas. As one microbiologist remarked, the country was small and one knew where everyone else was.[89] This was significant because, according to researchers, it was the *ideas*, born out of integrating various fields, that were the most important forces guiding research in microbiology, and not the broad targets set by companies or the government (such as finding a cure for cancer, or eliminating human disease with an antibiotic). These included, for example, an idea of how exactly to detect a new microbe, using various techniques appropriately brought together, or an idea like Kyōwa Hakkō Kōgyō's that one might be able to use a microbe to make any amino acid in existence.[90]

The somewhat centralized nature of the training structure helped create the personal links that encouraged information exchange. Professor Sakaguchi Kin'ichirō's laboratory at the Agricultural Chemistry Department at

Tōdai's Faculty of Agriculture, in particular, trained numerous graduates who went on to work for different companies. Some of the graduates also became university faculty, whose students in turn went to work for companies.[91] This was why a dispute between Ajinomoto and Kyōwa Hakkō could divide the academic community, since the vertical chains of commitment persisted long after graduation. Inevitably, these pyramidal links also meant large amounts of horizontal information exchange across competing sites.[92] However, the academy was more than simply an information channel, as it also played a key role in building and promoting a discipline of fermentation science. Because fermentation science was not a wholly applied field, the existence of "basic" areas strengthened the academic coordination of approaches between different and competitive sites of research.

As an institution builder, Sakaguchi Kin'chirō placed the academy at the center of the new developments. In 1953 he set up the Institute of Applied Microbiology (Ōyō biseibutsu kenkyūjo; Ōbiken or IAM), at Tōdai. One sees a certain nostalgia for the wartime and occupation command model in the design of the institution.[93] The IAM was intended to help coordinate, under a single supra-institution, the microbiological research which was spread across experimental institutes in different government ministries, including the Institute for Fermentation, Osaka (Hakkō kenkyūjo), under the Ministry of International Trade and Industry; the Fermented Foods Section of the Food Institute (Shokuryō kenkyūjo hakkō shokuhinbu), under the Ministry of Agriculture and Forestry; and the Brewing Experiment Station (Jōzō shikenjo) under the Tax Department of the Ministry of Finance.[94]

The IAM was an academic institution and the research members, like the institution itself, were university affiliates; but it would also respond to more practical lines of research put forward by the various ministries' institutes. Sakaguchi referred to the triple-pronged importance of universities, public research institutes, and company research laboratories in the historical development of science in Japan. He stressed that the IAM should be a new form of university-style institute that would at the same time be more like a national institute in terms of scale, financial liabilities, and shared responsibilities. It would conduct broad research in line with the more complicated problems of the age, which meant that research organizations had to break down the boundaries between basic and applied science, as well as challenge the mold of the four traditional faculties of science, agriculture, engineering, and medicine that had persisted in the natural sciences since the Meiji period.[95] In the end, in spite of its name, the Institute of Applied Microbiology concentrated much more on "pure" research.[96]

Immediately after Udaka Shigezō's discovery of the glutamic acid-producing microbe, Sakaguchi Kin'chirō also moved to set up the Amino Acid Fermentation Association (Aminosan hakkō shūdankai; later the Amino Acid and Nucleic Acid Association, or Aminosan kakusan shūdankai) in 1957, to promote a new scientific field led by Japan. In a manner reminiscent of the Japan Penicillin Research Association, the association drew research funding from the government, specifically the Ministry of Education, and the association's meetings and journal brought together academic and industrial researchers.[97] In fact, the intense competition between laboratories fueled by constant patent wars made the atmosphere very different from penicillin research (in which the product had not been patented, and which operated under economic conditions where state support had offered the only opportunity for companies' survival).

Nonetheless, it was the precedent of penicillin research that made it possible for the fermentation community to put on a "front" of cooperation in order to reel in public funding.[98] Fermentation scientists' accounts of their own efforts to seek government support, which were motivated by the research itself, contrast with the accounts of economic historians who tend to emphasize the state's imposition of policies from above in order to overcome the barriers to information flow that stemmed from interfirm competition.[99] In this case, although the collaboration initially looked like it might involve "open" sharing of information—including various culture techniques to improve microbial glutamic acid production, or reports on pilot plant runs—the atmosphere rapidly became closed. It became unthinkable to share critical manufacturing information such as the constitution of media for large-scale culture, as researchers had done for penicillin in JPRA meetings, or to share microbial strains. What researchers did exchange, for the most part, were reports of their "basic" research on biochemical mechanisms and microbial taxonomy.[100]

Sakaguchi Kin'ichirō became an icon in the field. Known more widely in Japan as "Dr. Sake" for his popular writings on fermentation, at Tōdai he became what researchers later variously called the "boss" of the fermentation community, or a "shield" to hold up in the face of abundant conflict.[101] He sought to institutionalize fermentation science as a globally cutting-edge, nationally unique field. The sayings that crystallized in the fermentation community were attributed to him; every fermentation scientist in Japan, whether they worked on antibiotics or on genetics, knew to "ask microbes, and they will never betray you" (*biseibutsu ni motomete, uragirareru koto wa nai*; sometimes also translated as "Microbes will give you anything if you only know how to ask"). He was said to have named the new field

FIG. 6.3. *Kinzuka* (microbe mound), Manshuin, Kyoto, November 2010. Photo by author.

of amino acid fermentation "metabolic engineering" (*taisha seigyo hakkō*), though other scientists also claimed to have coined the term.[102] The name highlighted human ingenuity in the novel method of "shifting" the cell's biochemical pathway to demand.[103] When microbiologists built a "microbe mound" (*kinzuka*, fig. 6.3) at the Manshuin temple in Kyoto much later in 1981 in order to commemorate the souls of microbes who had died in the service of humans, Sakaguchi was asked to provide the calligraphic inscription, and he also wrote a poem for the ceremony.[104]

Outside Tokyo, other fermentation science laboratories also existed around the country, especially in the agricultural chemistry departments of the agricultural faculties of Kyoto University, Hokkaidō University, and other former imperial universities. An important center of fermentation science was the city of Osaka, with its cluster of institutions including the Institute for Fermentation, Osaka (IFO), and the Fermentation Engineering Department of the Faculty of Engineering at Osaka University, as well as many of the country's most prominent pharmaceutical businesses. Among other research activities, the IFO maintained the nation's largest collection of microbial strains, holding a greater number of cultures than the Institute of Applied Microbiology in Tokyo. The IFO published its first *List of Cultures* in 1953. The same year, the Ministry of Education published a catalogue of all the microbial strains preserved in Japan. Next to each scientific name was a note on the strain's individual history and the culture collection from which it came.[105] Later in 1963, following Japan's proposal—put forward especially by Hasegawa Takeji, director of the IFO culture collections—UNESCO initiated programs to promote microbiological research, including an international network of culture collections.[106]

The Fermentation Engineering Department in the Faculty of Engineering at Osaka University was the successor to the Brewing Department at the Osaka Higher Technical School. It had changed its name from the Brewing Science Department to the Fermentation Engineering Department during World War II, while research was being carried out there on acetone-butanol fermentation.[107] Whereas the department had once largely trained the sons of brewers, now it had diversified into other topic areas in the chemical industries, and the direct link with small and medium-sized breweries was fading. Traditional breweries had themselves expanded, and the industry further mechanized and underwent concentration in the decades after the war.[108]

Ōshima Yasuji remembers from his student days (he graduated from the department in 1955) how, immediately after the war, sake breweries moved to use the "fast *moto* method" (*sokujōmoto*) for the yeast starter mash. It involved pure-cultured yeasts distributed by the Brewing Society, rather than the customary method of allowing the entry of wild yeasts to make *kimoto*. His teacher, Oda Masao, worried that the original *kimoto* strains would be lost, and sought to isolate and collect yeast strains from the *kimoto* of Nada, Fushimi, and other brewing districts and preserve them in his laboratory. A yeast from Kikumasa would be marked KM followed by a number, and a yeast from Hakutsuru would be marked HT followed by a number.[109]

However, the researchers in the university laboratory did not hold any particular interest in those yeasts from the sake houses. The IFO engaged in some consulting work for breweries, but researchers at Osaka University (Handai) were concerned with their own work in genetics, systematics, or other areas. *Kōji*, too, faded in prominence as a research object, though Ōshima recalls that one of the Handai professors was still interested in *Aspergillus* microbes for investigating organic acid fermentations.[110] For genetic studies, it was far simpler to use yeast than multinucleate *kōji* cells.[111] The graduates themselves were no longer primarily going to breweries to work, and not many of them were brewers' sons anymore. When Ōshima Yasuji was a student at the Handai department, he was the only sake brewer's son among his friends. Now sake brewers' sons tended to enroll elsewhere, such as at the Tokyo University of Agriculture.

Upon graduation, students from the Handai department went to pharmaceutical and chemical companies, and perhaps some of the excellent students would become university researchers. Ōshima himself went to Suntory later, while another peer went to Yakult, and others might go to the chemical companies Kaneka and Orient Kagaku, or to one of the pharmaceuticals in the Osaka region such as Shionogi, Fujisawa, Tanabe, or Takeda. By about 1980, a third of the department's graduates went to pharmaceuticals, another third to chemical companies, perhaps a handful to sake breweries, and others to enzyme-making companies.[112] At the same time, the laboratory held an older history in its collections, and included famous microbes from the past. For example, it held the Taitō Yeasts, grown on molasses from the Taitō sugar company in Taiwan and isolated by Nakazawa Ryōji during the colonial period. These tropical yeasts had the special property of fermenting alcohol at high temperatures, making the fermentation faster and more stable.[113]

Both metropolitan centers, Tokyo and Osaka, held high-profile workshops for training fermentation scientists in making mutant strains. Tokyo's Institute of Applied Microbiology invited Francis J. Ryan to visit from Columbia University in 1955–56 to run a seminar in microbial genetics.[114] A similar workshop took place in Osaka under the auspices of Kikkawa Hideo, a medical professor at Handai who ran a genetics laboratory.[115] It was the age of mutation.[116] Mutants had been used industrially in Japan earlier, when Arima Kei had found a pigmentless mutant of *Penicillium chrysogenum* Q176 that would make penicillin without the yellow color, thus simplifying the refinement process for the drug.[117] Under the spell of the promise of antibiotic research, researchers no longer looked for microbes in the breweries, but sought them in soil and in various raw materials. They aimed to raise the potency (*potenshi*) of microbes using mutations and various treatments and culture methods.[118]

When the focus of fermentation science shifted from alcohols to antibiotics, the new techniques also had to deal with refining and analyzing chemical products that were firmly not in the macroworld but the micro-world, a world now measured in quantities smaller than grams.[119] The change in scale was the reason why the company Kyōwa Hakkō, despite its expertise in fermentation, had encountered such difficulties in produc-ing refined glutamic acid.[120] For microbiologists, on the other hand, once glutamic acid fermentation appeared on the scene, the goal was to look not merely for antibiotics but more broadly for physiologically active sub-stances (*seiri kassei busshitsu*).[121]

The nature of the microbes that were seen as promising was changing, as pharmaceutical and chemical companies became major loci of micro-biological research. Komagata Kazuo's lab in Ajinomoto's Central Labora-tory, where he worked between 1961 and 1968 as chief microbiologist, was tasked with three main lines of work by the company.[122] One was to foresee which types of microbe would become topical for microbial industry and to train the relevant specialists. Another was to maintain a microbial cul-ture collection, and the third was to respond to customer complaints that came in through the sales department, such as getting rid of the mold that appeared in products during shipping. Of these, the most important task was to be able to know which microbes would develop into major lines of research (*mondai ni naru*) in industry in the future.

These company researchers were an integral part of the academic land-scape, enjoying rich facilities and giving away enough information to main-tain their academic profile. Because Ajinomoto's economic outlook was healthy, those were good times for researchers. As Komagata Kazuo re-members it, there were always plans handed down by the company, but the researchers felt relatively free to do as they liked, since the company leaders had respect and sympathy for science. Moreover, as chief microbiologist, Komagata had the prerogative to decide on research topics. There was more or less basic or applied research depending on the section, but Komagata's section was one of the most basic, being involved with microbial systemat-ics. A large culture collection was established to serve the purposes of re-search within the company. At its core were the microbial strains isolated, identified, and studied by the company's scientists themselves, which were labeled with the initials AJ followed by a number. They were preserved by freeze-drying, a great hassle, as the process was not yet automated; but it worked, and the same strains were still alive there decades later.[123]

Naturally, the collection was closed to the public, but because original strains formed the basis of microbial taxonomy, those specific strains that were used in taxonomic research were released. Despite the limitations of company confidentiality, the researchers in Komagata's laboratory published a large number of research reports. Publications were, after all, the sacred rule of academic participation. Microbiologists often moved out of companies' central laboratories to work for universities. Komagata, too, would return to Tōdai's Institute of Applied Microbiology upon leaving Ajinomoto.[124] The significance of scientific expertise for the fermentation industries was by no means limited to a field like microbial systematics, but ranged widely from genetics to biochemistry to the engineering knowledge for mass production. Because all of these fields of expertise were needed, it was said that if somebody stole one useful microbial strain from another company, they would not be able to create a new industry from the strain alone, for they still had no fermentation tank, no experience, and no knowledge—though if they did have all of those things, it would be a different story.[125]

The disciplinary continuity of fermentation experts between the prewar and postwar eras and across industrial, academic, and government research sites is easy to overlook, due to the dramatic swings in consumer attitudes toward processed flavors between the 1950s and 1970s. MSG exemplifies these changes. In 1950s Japan, MSG was a well-regarded staple in households and restaurants, having been since the 1930s a luxury for bourgeois housewives that claimed "rationality and convenience based on scientific research."[126] By the 1970s, the public image of MSG had been through a radical transformation. There was a reversal of the previously widespread belief that MSG was good for one's brain, while headlines ran in Japanese newspapers about American regulatory debates over MSG's role in "Chinese restaurant syndrome." MSG remained on the US Food and Drug Administration's list of GRAS ("generally regarded as safe") substances, but Ajinomoto's MSG sales fell for the first time in 1970.[127] It was a time that marked the height of Japanese public distrust toward large chemical companies in general, in an age of unprecedented industrial growth.[128]

Marketing of processed foods based on the health appeal of beneficial food reconstitution, which was typical in the 1960s and recalled prewar nutritional work in fermentation science, was no longer possible in the 1970s.[129] Nisshin Shokuhin stopped calling attention to how instant ramen's additives, namely vitamin B and the amino acid lysine, supposedly rendered it nutritious. Rather than evoking scientific progress and efficiency, MSG became associated with fears of toxic and carcinogenic properties, along with anxieties surrounding its negative moral effects precisely due to its convenience.[130] The breakthroughs in fermentation technology during this same period enabled Ajinomoto to respond to the new cultural opposition against MSG in two main ways: by diversifying the company's flavor product lines to encompass chemicals other than MSG, particularly the novel nucleotide flavorings IMP and GMP, and also by summoning a more natural image of MSG. People no longer had to wonder whether *ajinomoto* came from somewhere strange, like snakes.[131] Now that MSG was made by fermentation (*hakkō*)—the same word used to describe processes of sake, soy sauce, and miso production—its improved, traditional image helped put people more at ease.[132]

After 2000, research by physiologists in the United States establishing umami as a "fifth basic taste" sparked a partial reversal yet again in the image of MSG, back to that of a celebrated, scientifically prescient gastronomic substance. Ajinomoto's founder, the physical chemist Ikeda Kikunae, had long ago claimed that umami stood alongside the tastes of sweet, sour, salt, and bitter, with its own distinct group of receptors on the tongue; and in the new millennium, findings from physiological research on taste reception vindicated his statements. But at the same time as culinary opinion moved to praise MSG, laboratory research continued to suggest harmful effects of high-MSG diets.[133] Ultimately, the vicissitudes of consumer culture surrounding processed flavors rendered hidden the strong historical continuities of large companies in Japan with older traditions of ecological and environmental thinking, which were maintained via the fermentation expertise of scientists who staffed the company laboratories.[134]

PROBABLY A LIVELY ONE ON THE SHOP FLOOR

Amid these recent developments, in a fermentation science community that was increasingly distant from the world of small and medium-sized breweries and *tanekōji* making, there were nonetheless deep resonances in the approach to microbes. From the perspective of scientists, business thinking tended to envision target-driven research, but in actuality new industries could also be technology-driven, coming out of tools looking for a use, such as the fermentation of amino acids for which the medical uses were still not well understood.[135] Researchers screened with a special sensitivity to uncommon, "powerful" microbes. What was different was the stress on the production of novelty itself: researchers looked for new qualities in known or wild microbes, to bring forth new products with market appeal. The research setup in the companies was effectively oriented to produce novelty, not merely consistency or improvement.

Such changes were paralleled to some degree by shifts in the brewing world. In the 1950s, for example, the *tanekōji* company Akita Konno Shōten picked up a white mutant of a normally "black" *kōji* microbe used for making *shōchū* and other distilled liquors, *Aspergillus usamii*, that was created by gamma irradiation at Iizuka Hiroshi's laboratory at Tōdai, and brought it to market as a new *tanekōji* product (*shirousami B11 kin*).[136] Among microbiologists, an approach combining an older view of microbes as a diverse world harboring stronger and weaker makers with a more recent emphasis on microbes as sources of novelty had taken widespread hold across the Japanese chemical industries by the late 1950s and early 1960s. This may be seen in the account of Imada Akira, who formerly worked in the research laboratories of the pharmaceutical company Takeda Yakuhin.

Takeda Yakuhin did not have a long tradition of fermentation expertise. It was one of the largest pharmaceutical companies in Japan, with its headquarters in the Doshōmachi district in Osaka, alongside other pharmaceutical companies that had been prominent herbal medicine dealers in the Tokugawa period.[137] Takeda's proudest technological achievements lay in synthetic chemistry and especially vitamin synthesis, with products including the vitamin B_1 derivative Alinamin, developed from the research of Fujiwara Motonori at Kyoto University and first marketed in 1954.[138] Like a number of other chemical companies, Takeda had acquired fermentation capability as part of the legacy of the wartime and occupation periods. In 1946 it bought half of the Hikari Naval Arsenal site in Yamaguchi Prefecture, which had been ruined by bombing (Shinnittetsu, or Nippon Steel, bought the other half), and built a large penicillin factory there with a license from the occupation government. The penicillin factory was primarily intended to create work for employees returning to the Japanese home islands from Takeda's factories in Southeast Asia and China.[139] At the Hikari factory, which had the largest fermentation tank in Japan, the company manufactured vaccines, synthetic medicines, and various antibiotics beginning with licenses purchased from overseas, since Takeda had ties to Lederle: neomycin, erythromycin, achromycin, and so on.[140] Thus Takeda's investment in microbial technology was part of the broader postwar expansion of the fermentation community in Japan.

Imada Akira joined the company in 1959 as a researcher at Takeda Yakuhin's central laboratory. While he was in his third or fourth year as an undergraduate in agricultural chemistry at Hokkaidō University, a newspaper had published a series of articles on Japan's new microbiology. The first installment was on Kinoshita Shukuo and glutamic acid fermentation. Enchanted to discover that such a field existed, Imada hoped to seek employment at Kyōwa Hakkō when he graduated. But the company was mainly in the alcohol business. While his mother did not especially mind if Akira worked for something like a beer company, his father, who was a Christian minister, felt that to work for the sake of people's pleasure-seeking was to oppose the principles to which he himself had devoted his whole life. He asked his son to consider some other industry. The second installment in the newspaper series was on Takeda Yakuhin, and as it happened, Imada Akira was recruited by the pharmaceutical company upon graduation.

In the late 1950s and through the 1960s, fermentation research at Takeda was split between the Takeda Central Laboratory and the new Institute for Fermentation, Osaka, located next door. The biochemical section in the Central Laboratory, within which Imada worked, did research similar to that in the microbiological process development section at the IFO. This included work on nucleic acid fermentation. Other sections at Takeda were also engaged in a variety of microbial work related to such areas as antibiotics, ergot alkaloids, and preservatives and bioassays, as well as other research in synthetic, natural product, and analytical chemistry, and in infectious diseases.[141]

Antibiotics remained the focus of fermentation research, since in that age drug development generally meant either vitamins or antibiotics.[142] In Jūsō in Osaka, where the research laboratory was located, a second factory was set up for penicillin manufacture, with an array of different-sized tanks (fig. 6.4) and a refining column, and perhaps a hundred employees going about their work of culturing, isolating, and developing microbes, separating the microbe bodies from the fermentation liquid in the factory, and analyzing and refining the products.[143] The phenomenon of antibiotic resistance meant that companies constantly needed to find new antibiotics as older ones became ineffective.[144] As the Takeda researchers developed new antibiotics, they would test-make them one by one in the large tank at the plant. Underlying these efforts was a problem hanging over the near future: when the next antibiotic lost its commercial value, there would be nothing for the large penicillin factories in Hikari and Osaka to make. In the 1950s, researchers had scrambled to find new antibiotics; but after glutamic acid fermentation, they saw many other possibilities for fermented products that could keep the tanks moving.[145]

The microbes concerned were not for brewing, and yet researchers were driven by a sense they shared with workers in the brewing industry, of how to look for capabilities in a diverse and finely gradated world of microbes, whether those capabilities were old or new, and whether the microbes were known, wild, or engineered mutants. When Imada Akira entered the company his supervisor was Ōgata Kōitsu, a graduate of Kyoto University (Kyō-dai) who worked in the IFO's microbiological process division. Hearing of Sakaguchi Kin'ichirō's work at Tōdai and Kuninaka Akira's at Yamasa

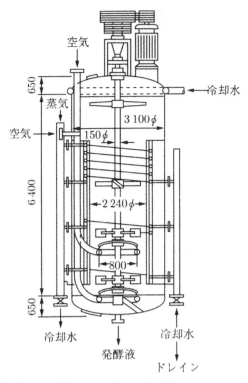

空気

650

蒸気

空気

6 400

冷却水

3 100φ

150φ

2 240φ

800

650

冷却水　　　　　　　冷却水

発酵液

ドレイン

FIG. 6.4. Aerated 50 kL stirred fermentation tank of the type required for penicillin and antibiotic mass production. From Yoshida Toshiomi, "Hakkō sōchi" (2001), fig. 1. Reproduced with permission of Kyōritsu Shuppan and Yoshida Toshiomi.

Shōyu, Ōgata had moved from antibiotics into research on manufacturing ribotide, a mixture of the nucleotides IMP and GMP, by the decomposition of yeast RNA. This flavor product was the aforementioned I no Ichiban, which Takeda Yakuhin began to sell in 1961 in partnership with Kyōwa Hakkō and in competition with Yamasa Shōyu. Afterward, Takeda Yakuhin moved instead to developing nucleotide manufacture by a fermentation process using mutants. This line of research was directed by Ōmura Einosuke (another Kyōdai graduate), while Yoneda Masahiko (a Handai graduate) and Nogami Hoshio (a Kyōdai graduate) also worked on the project.

Imada Akira recalls the contributions that his colleague Nogami Hoshio made to the company's projects. The problem that one often faced was, Which microbe should one use to develop a line of research— gram-positive or gram-negative? What if one used an actinomycete? Or should one proceed with *Bacillus*? The microbial strain collection was at the IFO, immediately next door to the Takeda Central Laboratory; and time and again, the Takeda researchers would find themselves wanting to

FIG. 6.5. Nattō (iStock/LFO62)

use the known microbes with distinguished lineages. But Nogami brought some *nattō* (fermented soybeans, fig. 6.5) from his native Kyoto back to the laboratory and isolated a microbe from it. He reasoned that, since the microbe was used on the shop floor (*genba*), it would probably be a lively one. Somehow the *nattō* microbe that Nogami had isolated always turned out to have a higher production ability than the preserved microbes.

Nattō microbes were *Bacillus subtilis*, and the researchers who worked on nucleotide fermentation at Takeda knew, having investigated various microbes, that *Bacillus* was one of the best for such purposes. Nogami's *nattō* microbe was an extremely good microbe. It was a pity that Takeda Yakuhin did not possess or know much about *Brevibacterium*—which Kyōwa Hakkō would soon use to make inosinic acid (IMP) and guanylic acid (GMP)—because Ajinomoto had already taken out a patent for nucleotide manufacture by *Bacillus subtilis*. That was terribly hard for the Takeda researchers to stomach. For making nucleotide flavorings, they had to turn to using other microbes instead, such as *Bacillus pumilis* and its mutant varieties. However, Nogami's powerful *nattō* microbe was not wasted, as researchers later pursued using it to make ribose as raw material for vitamin B_2, and cytidine as raw material for the drug Nicholin.

Imada's portrayal of the nucleotide fermentation community shares a number of features with the "knowledge communities" of engineering design described by Ann Johnson, including a size small enough for members to know each other; a characterization by appealing problems, or "attractors," which tie the community together and motivate their work; and a disciplinary heterogeneity of members.[146] In highlighting the intellectual aspects of design, which are usually taken for granted in science but not necessarily in engineering, Johnson's approach is apt for understanding reasoning and lived experience in fields of applied and industrial science that do not fall neatly into either category. As Imada Akira recalls, antagonism and mutual understanding knitted together the fermentation community as they worked on the same problems. The Kyōwa Hakkō researchers developing *Brevibacterium glutamicum* for nucleotide fermentation—Kinoshita Shukuo, Nakayama Kiyoshi, Nara Takashi, Tanaka Katsunobu, and so on—were easily visible on the scene, extremely competitive and rather like samurai. The researchers at Ajinomoto who were also working on nucleotide manufacture by *Bacillus subtilis*—including Yamanoi Akio, Shiio Isamu, Momose Haruo, and Yoneki Kōji—were more reserved, but they were an excellent group since many of the top agricultural chemistry graduates went to Ajinomoto; they were sincere and did good, smart work. The Takeda researchers knew the names and faces of the individuals in the group well; and of course Yamanoi, Nogami, Yoneda, and so on would all know each other.

It was through close competition that the researchers understood one another, though at the time the competition was so aggressive that it did not often occur to them to think in such terms. In fact, the sense of community was strong and hopes for the new field were high. Even with antibiotics at the focus of fermentation in the pharmaceutical companies, researchers looked more broadly for possibilities, since it was the activity of discovering new things that maintained their high spirits. As Imada Akira saw it, what motivated researchers like himself was not company-driven competition, but the excitement of being part of that community and the sense of play: the desire to find something interesting and new, and to be remembered for it.[147]

New product development could be pushed along by resources looking for a use, as fermentation scientists considered what to do with the abundance of various kinds of waste product, as they had done in the prewar period.[148] When the Takeda group led by Ōgata Kōitsu was working on manufacturing ribotide from yeast RNA, the researchers wondered whether the 90 percent of the yeast bodies that did not go into the ribotide might harbor something that could become medicine. Imada Akira remembers going to the factory at the pulp company Tōyōbō, on whose waste water the yeast was cultured. The researchers hoped to extract the antioxidant ubiquinone (Coenzyme Q) from the waste yeast. As the product developed, Takeda came to synthesize ubiquinone chemically, and it was for a time Takeda's biggest selling drug, as Idebenone, for use in the treatment of Alzheimer's disease, before it was thought to be an ineffective medicine and discontinued.[149]

Similarly, new lines of research emerged from tools looking for a use. When mutant strain technology rose to prominence in the Takeda laboratories from the 1970s, the researchers sought to employ mutants to create new forms of β-lactam antibiotics, since innovations in semisynthetic cephalosporins had come to the fore.[150] They took mutants that were highly sensitive to β-lactam antibiotics and screened with them, and around 1980 they found a drug the company later marketed as Amasulin.[151] However, the ubiquitous enthusiasm for fermentation belonged to a specific age. Takeda Yakuhin's investment in fermentation narrowed in the 1980s, following the global oil shocks of the 1970s that ended Japan's period of high growth. Eventually, the fermentation tank at Hikari was completely gone.[152]

CONCLUSION

To observe the fermentation community in the 1950s to 1970s is to single out a moment that Japanese microbiologists look back upon as a golden time. The postwar landscape of microbial technology was one of large businesses and large-scale production, chiefly as a result of wartime mobilization. The legacies of the wartime and occupation efforts in nutritional, fuel, and penicillin production under a controlled economy were visible in the postwar market economy, in features including the approaches of fermentation scientists that emphasized making the most of scarce raw materials; mass interdisciplinary coordination among research laboratories; and the presence of large fermentation tanks across the food, fine chemical, and pharmaceutical industries (mainly for antibiotics). The tanks had a low capital cost and were supported by labor-intensive scientific research on— as well as maintenance of collections of—microbes. Growth happened so quickly that systems of commercial mediation and regulation were not yet well defined for the purposes of the new economic structure. Scientists looking for unusual microbes in manure or sewage water sometimes became ill from their subjects.[153] The general goal of applied microbiological research had shifted from the improvement of small- and medium-scale traditional businesses to producing novelty within a fast commercial product cycle in concentrated industries, and the research landscape was dominated by company laboratories rather than government research institutes.

The example of amino and nucleic acid fermentation showcases how academic science and industrial technology interacted in a country with an advanced and rapidly growing industrial economy as well as a large domestic market. In particular, it illustrates how a private sphere used academic information to navigate the consumer market. In the high-technology field of fermentation, resources for the innovation of new technologies lay in scientific information itself. Scientists working in corporations were compelled to share information to maintain their strong ties with government-supported academic networks, and they gained access to information in return. This was possible because of an ideas-based approach to microbial behavior as systems to be both studied *and* designed, and the academy was the information medium that served to coordinate approaches between commercially competing laboratories. As a result, Japanese fermentation experts made key contributions to "pure" scientific discussions in the international fields of biochemistry and molecular biology, often at the same time as they worked within company laboratories and on applied areas distinctive to Japanese industries. Industrial concentration and product development along highly particular lines were other results; to understand the range of products that came to market, one needs to understand the scientific developments.

Led by large firms, this was a resource-saving and labor-intensive structure of scientific research that supported efficient high-growth manufacturing. It contrasts with the landscape of biotechnology in the United States after the 1970s, which predominantly relied on the commercialization of discoveries originating in academic laboratories.[154] In discussions of Japanese biotechnology, the historically significant role of large companies in research is often equated with weakness in basic research.[155] This chapter's examples of microbial research, as well as of many scientists' career mobility between universities, private industry, and government ministries, counter that premise for the high-growth era. The distinctive structure of scientific research in postwar Japan explains why, after the oil shocks of the 1970s, and especially after the burst of Japan's economic bubble in 1989–90, which marked an unambiguous end to the age of manufacturing-based high-speed growth, the same companies could no longer afford to invest in biotechnological research to the same degree as they had done previously, and the Japanese life sciences could find few venture businesses willing to pick up the thread of developments.[156]

Even in the climate of the modern food and pharmaceutical industries, the importance of local skills, approaches, and material traditions stands out, including in advanced technologies that were exported. Japan is often seen as the classic successful case of postwar technology transfer into the local context, guided by the Ministry of International Trade and Industry and exemplified by semiconductor exports. Such a narrative supports the view of "local" science as a mirror image of global trends. Instead, when we look at how fermentation scientists tuned in to hierarchies of power and fame within the microbial world, looking for fine differences between the individual strains in their culture collections precisely because of application, and valuing collections of microbial life as the source of their creativity and reputation, the similarities with traditional *tanekōji* makers are more obvious than the differences—for example, in the precision of techniques that the switch to mutant strains and pharmaceutical products demanded. Instead of looking at how globally widespread technologies were adapted to local circumstances in Japan, this chapter has followed a local approach to the tool kit of the microbial world, which was embodied in a community of skilled researchers with a memory of prior problems in the national context. It demonstrates how local forms were reshaped to fit new globally influenced circumstances, with novel results.

CONCLUSION

The Science of Modern Life

Most histories of human engagement with microbes have focused on pathogens. But microbes are also productive, and microbial collections — or what one scientist called "microbial gardens" — were cultivated by Japanese scientists and technicians in anything from a small brewing shop to a national scientific facility, for the microbes' ability to manufacture a variety of food, pharmaceutical, and chemical goods. My exploration of fermentation science in twentieth-century Japan presents a case for consideration in inquiries on pluralism in science that use historical studies to conduct "complementary science." Complementary science, as originally proposed by Hasok Chang, is a counterfactual endeavor which attempts to recover knowledge that was suppressed when a technical choice was made in the past.[1] Here, I have instead examined fermentation science in twentieth-century Japan as a path of microbiology that *actually happened*. Japanese scientists made an alternative technical choice in which microbes came to be known and manipulated as living workers instead of simply deadly germs.[2]

Japanese microbiology's alternative trajectory illuminates, through comparison, the inherently situated nature of knowledge in modern experimental biology.[3] As I have argued, Japanese fermentation scientists self-consciously created a relatively autonomous tradition in microbiology, and deliberately took advantage of their position on the global periphery, using local or regional materials as objects of experimentation to make international contributions to science. By implication, the path of microbiology that is more familiar to us in the same period, from 1900 to 1980 — running from medical bacteriology to microbial genetics to recombinant DNA biotechnology — is also situated. There is a cultural particularity to the very way we divide the living world into parts. In mid-twentieth-century molecular biology, the rise of genetics, which focused on the cell's nucleus or the "executive" of the cell and its role in reproduction, eclipsed work on metabolism and physiology in the rest of the cell in capturing the center of attention. The latter was mentally relegated to a "housekeeping" role (a gendered metaphor that comes from computer science).[4] This was in spite of the lingering importance of the Pasteur and Delft schools of microbiology, which studied microbes as beneficent natural resources rather than solely origins of disease.[5]

The above series of epistemic choices in cell biology made by the scientific community in the United States and Europe explains why certain kinds of biological knowledge became valued less highly in science. For example, microbial screening processes, which relied on fundamental components of expertise in fermentation science—and for which Ōmura Satoshi later won the 2015 Nobel Prize in Physiology and Medicine—were taken seriously as a research topic in Japan with significant scientific dimensions, while they were generally perceived to be routine and intellectually inconsequential in the United States during in the same period.[6] Such contrasting values also manifested materially, in the strains held in the microbial type culture collections. As detailed in chapter 6, molds and yeasts constituted the vast majority of strains preserved in the Japanese national collections at the Institute for Fermentation, Osaka (IFO), in the decades after World War II. The Dutch Centraalbureau voor Schimmelcultures (CBS) shared an emphasis on molds and yeasts, but bacteria predominated in the American Type Culture Collection (ATCC) and the British National Collection of Type Cultures (NCTC).[7]

Yet wide-ranging discoveries in contemporary biomedical studies have begun to break down such dichotomies of reproduction and metabolism, and of genome and environment.[8] A new consciousness of interdependence and symbiosis with other organisms challenges the biological concepts that have conventionally served as resources for addressing problems in health, biodiversity, and climate change, and brings metabolism back into central consideration alongside genetics.[9] If the aim of complementary science is to revive an overlooked parallel tradition of scientific research, and to examine how it might interact with and contribute to science in the present day, then Japanese fermentation science in the twentieth century offers an ideal historical example with which to explore the potential fruits of complementary science as well as pluralism in science more broadly. In Japanese society, a differing set of scientific values, questions, and contributions to knowledge resulted from an alternative path of microbiology as well as its global interactions with parallel research traditions. Moreover, it shaped a distinctive material culture of everyday life in modern Japan.

Japan's alternate biological universe, where scientists "asked" microbes for what they termed "gifts," highlights how the premodern context continued to matter even in modern science and technology. At the end of the nineteenth century, Meiji government officials as well as other Japanese elites supported the institutionalization of modern science with a particular goal in mind—namely, to promote industrialization. In the process, as I have argued in chapter 1, microbiological concepts and tools imported from Europe demonstrated a striking convergence with earlier commercial developments in the Tokugawa-era brewing industry, which in turn were part of early modern changes that historians agree were "protoindustrial" or "protocapitalist."[10] More than that, economic historians, especially Tanimoto Masayuki, have argued that the temporal significance of Japan's "internal logic" of industrialization, which was driven by small-scale indigenous industries, should be understood to extend beyond the early modern era into the Meiji period.[11] Taking up an examination of the implications for science, I have contended that the persistence of the importance of indigenous industry into the twentieth century produced a subtly—yet significantly—different modern science.

The reason for the strong continuity in Japan of a fermentation-based vision of life throughout the twentieth century, despite the gradual decline of the significance of indigenous industry after the Meiji period, lies in the entanglement of scientific knowledge with institutions that originated in the Meiji era. The history of Japanese fermentation science draws attention to the inseparability of knowledge and institutions. To ask whether this is mainly a scientific story about the production of knowledge, or a technological story about the production of material artifacts, is to privilege historically contingent categories of knowledge over the fact that the same institutions produced both kinds of outcomes, as recent scholarship on applied science has emphasized.[12] Scientific research, technological innovation, business interests, military investment, and government control were all inseparable dimensions of the network of institutions that supported the fermentation community in Japan, even as the primary goals of these institutions shifted through the twentieth century.

When it came to deciding which scientific questions or technological design projects to pursue and how to pursue them, the epistemic choices made by individual members of the fermentation community were consequences not only of the national policy paradigms of the era, nor of business strategies, nor of market or other social forces. They were a result of the collective values of fermentation science, especially in the discipline of agricultural chemistry, which carried an institutional memory of prior problems in the national context.[13] The era of improving traditional craft industries at the turn of the twentieth century left in fermentation science a view of microbes as living workers as much as pathogens. Work on food and alcohol in the interwar and wartime periods further shaped a particular knowledge of microbes as redistributors in a nutritional and industrial ecology, and of how microbes could be used to transform an environment characterized by raw material scarcity to accomplish resource-efficient mass chemical production. Even when the economic and environmental context had utterly changed in the high-growth era, fermentation scientists drew on the disciplinary values accumulated from the past to make distinctive contributions to the present: these were the specific ways in which they knew how to ask microbes to "make things work."[14] Thus, using microbes, they screened for new antibiotics and studied such fundamental cellular components as amino acids and nucleotides, not only for how they carried the genetic secrets of life, but also for how they *tasted*.

In fermentation science, the material culture of everyday life in modern Japan was a product of the tension between the state's ideal visions for political economy and what scientists—as middling statist elites—could achieve. Scientists navigated material constraints, most notably of Japan's local and regional raw materials and industrial technologies, and social constraints, especially the consumer landscape, in their attempt to make things work.[15] In their aim to improve traditional industry in the Meiji period, government experiment stations did not succeed in converting sake brewers to using pure-cultured yeast until after World War II. Yet the private sector rapidly adopted pure culture technologies to make commercial *kōji* mold starters. In response to Malthusian anxieties of resource scarcity in the interwar period, nutrition scientists—whether at Riken or at government research institutes—never managed to make a synthetic food economy, nor to persuade consumers to eat molds and fermented industrial waste. Both they and technicians at brewing companies, however, introduced chemical methods that left a permanent impact on sake and soy sauce manufacture in Japan.

Japanese scientists working in the formal and informal empire up to 1945 could not realize "East Asian" fermentation science as a regionally united field that would compete as such with Western industrial chemistry. Yet in the end they drew on regional knowledge to create what became the institutional and technical foundations for a globally competitive Japanese fermentation industry in the postwar period, beginning with antibiotics. Finally, the corporation-led, information-rich, university-supported nationwide infrastructure of fermentation science and technology, which had been unexpectedly fruitful in the period of high growth, did not sustain its productivity after the oil shocks of the 1970s, nor did it lead naturally into national strength in commercial recombinant DNA biotechnology from the 1980s. But it did create a biology that kept pace with the forefront of biochemical research on gene regulation in the emerging field of molecular biology, at the same time as it generated such globally novel commercial flavor technologies as the fermentation of MSG, IMP, and GMP, which were both widely consumed domestically and exported.

If we see in Japanese fermentation science features that precede the era since the recombinant DNA revolution in the 1970s—for example, the cultivation and genetic alteration of microorganisms for commercial aims, and close relations between academia and industry—then why, commentators ask, did Japanese biotechnology subsequently disappoint as a means of the country's economic revival?[16] Following the early 1970s oil shocks, Japanese science policy promoted the life sciences in industry (as it far more successfully promoted information technology) as a way to help firms diversify into high-tech areas that could cut energy costs. After the advent of recombinant DNA technology in the 1970s, MITI initiatives in the 1980s strategically promoted biotechnology.

However, despite advances in the enzyme industry, the engagement with recombinant DNA technology did not bring the returns hoped for in drug manufacture. Recombinant DNA plant research in Japan was also more heavily regulated than in the United States, and while microbial research remained strong, this meant that genetically modified crop research was weaker.[17] Firms worried that antibiotic research would bring diminishing returns, and there were few venture businesses in the field due to difficult economic times. The nature of the microbial technologies and the resources required to support them had changed; unlike having a good yeast to make alcohol in the past, one could possess genetically modified *Escherichia coli* that could make interferon, and yet the microbe might bring no profit at all.[18] Commentators noted how little Japan contributed to the Human Genome Project completed in 2000.[19]

By the era following the Human Genome Project, agricultural chemistry departments had given themselves more contemporary names with stronger connections to biotechnology. Life sciences disciplines relabeled and rearranged themselves in line with new postgenomic agendas, such as chemical biology.[20] Whatever the disappointment over Japanese biotechnology, microbiologists were occupied with broader questions about life and the environment that were not necessarily related to commercial growth. For example, how might one build a theoretical framework for understanding and managing microbial ecosystems, since without being able to understand what happened when more than one microbe was cultured at a time, one could not understand even how leaves decomposed in soil?[21] What means might there be to grasp or control the nature of microbial mutation, which had led to the diversity of microorganisms in the world?[22] Whatever the view ahead, microbial biosynthesis in Japan is a key part of the historical resources upon which today's researchers draw as they address problems in the life sciences relating to health, industry, and the environment.

While fermentation scientists have long understood the potential of the microbial world to yield gifts for human use, perhaps a time of looking closely at microbes as if they are as important as "macrobes" is only beginning.[23] Within medical microbiology itself, which focused as a matter of necessity on training experts to become "microbe hunters" to eliminate disease, reservations, qualifications, and criticisms of utopian ideas of what scientific medicine could accomplish had begun to appear as early as the 1950s. "By the 1950s the most optimistic dreams of the founders of medical microbiology had been essentially fulfilled in several countries," the French-American microbiologist René Dubos wrote in a critique in 1965. "Most clinicians, public health officers, epidemiologists, and microbiologists felt justified [in proclaiming] that the conquest of infectious diseases had finally been achieved."[24] Against a pervasive belief in the 1950s and 1960s that microbial pathogens could be eradicated by technological control, Dubos was one of a number of scientists who advocated an ecological approach to infectious disease: a view in which microbes were not static but were a rapidly evolving part of the wider environment. This was captured in his motto "Think globally, act locally."[25]

It was in this sense that Dubos emphasized the nineteenth-century observation of Louis Pasteur, made in the course of studies of a French silkworm disease, that the microorganisms present in the silkworm gut were "more an effect than a cause of the disease."[26] This was the same Pasteur whose experiments had first made microbes truly visible to his society as real, living beings.[27] A century and a half later, with HIV/AIDS, SARS, Ebola, Zika, and COVID-19, microbes have returned pressingly to consideration as forces that destabilize what scientists and policymakers thought they knew.[28] Moreover, the same technologies of human control over microbes that had once vividly symbolized modernity and progress now appear to signify the elusiveness of control. Antibiotic and antiviral resistance, the persistent emergence and reemergence of infectious diseases worldwide, and even shifting attitudes toward vaccination in some societies—linked partly to the long-term success of vaccination as a preventive measure—question the logic of the very technologies that have been used against microbes.[29] Yet the urgency to act remains as strong as ever. Dubos's slogan appeared during a time of heightening environmentalism in the 1970s. Today, along with the problems of climate change and industrial pollution, the prominence of emerging and reemerging infectious diseases is part of the collective shift "from localized to globalized risk perceptions."[30]

The aim of complementary science is to enrich present-day science by expanding the range of possible approaches, as well as to inform public understanding by casting the contingency of our present vision in a broader light. The history of Japanese applied and industrial microbiology provides one instantiation of the approach of "microbe smiths," who sought to manage microbial ecologies and their complex interactions with society. It may aid us in considering present-day global strategies for sustainable growth amid the deepening sense of humanity's limits. In the search for technologies of microbial control, antibiotics are often held up as a symbol of undeniable medical progress. The fungus *Penicillium chrysogenum* was itself co-opted by humans since the 1940s to produce the first antibiotic, penicillin, on an industrial scale in order to destroy other microbes. However, we have known for more than half a century that each antibiotic remains effective only for a limited period of time. As we "macrobes" progress, so do microbes. Antibiotic resistance is noted by the World Health Organization as "one of the biggest threats to global health, food security, and development."[31] With all eyes trained on macrobes, the overarching plan for responding to antibiotic resistance through the 1990s and beyond was to look to microbes for more antibiotics.[32] This happened even as the understanding of microbial life had changed radically.

In seeing or not seeing microbes through plans centered around human convenience, what has our vision missed? In the 1950s, research especially in Japan revealed that genes for antibiotic resistance, carried by plasmids, could be transferred easily between bacteria and disseminated through entire bacterial populations.[33] Thus we have known for decades that multiple antibiotic resistance poses a massive clinical problem, even before superbug cases such as MRSA came to the fore in the 1990s. Through the same antibiotic resistance studies, scientists have also long known that there exist mechanisms for widespread gene exchange in microbes, and drug-resistant plasmids became central tools in genetic engineering as the field emerged in the 1970s and 1980s. Nonetheless, it is only in recent years that we have recognized the significant role microbial gene exchange plays in evolution, as well as the fact that as many cells in the human body are microbial as are human. The microbiome has emerged as a site of microbes' unexpectedly complex role in human health, with wide-ranging impacts from metabolism to brain function.[34]

From this perspective, antibiotic resistance is a powerful symbol of a broader set of problems concerning sustainable growth. In 2012, the United Nations Conference on Sustainable Development (Rio+20) highlighted population growth and concomitant increases in consumption as the most significant threat facing the global environment, and emphasized the challenges of achieving a "green economy." In this context, industrialized societies have increasingly turned to the biodiversity of microbes to address sustainability problems in food, energy, and health. Scientific visions of a nutritional future include cultured protein for human diet and animal feed to offset pressures on agricultural resource constraints. In industrial chemistry, *chimie douce* (soft chemistry), in which reactions are performed by microorganisms and other biological materials, is considered for its implications in environmentally friendly production.[35] Antibiotic research itself forms one of the most direct precedents in the search for cellular processes to create materials, especially in areas of medicinal and pharmacological interest. The development of microbial and genetic techniques that look to nature's bounty for innovation in drug and chemical discovery has taken the Nobel Prizes in medicine and chemistry in recent years, and the CRISPR-Cas9 system at the center of the latest gene editing technology originates from an antiviral immune strategy in bacteria.[36] Scientists dream of "borrowing" the power of other life to extend the future of humanity, though the logic behind the ultimate dream of making unlimited food, unlimited health, and unlimited energy is perhaps tempered by Japan's stories of microbial biosynthesis from the twentieth century.

The development of applied and industrial microbiology in twentieth-century Japan offers a comparative perspective on the question: How might states and experts manage microbial ecologies, and with what social implications? History yields a reservoir of possibility in one industrialized society's attempts to engineer microbial control. While work in environmental history has critiqued visions of unlimited capitalist growth by highlighting the costs to life and the environment, I have built on such insights by focusing on how various scientific groups in Japan envisioned alternative ways of managing life and livelihood as means of environmental action through the twentieth century. In an age of ecological degradation, industrial pollution, and irreversible human intervention in the environment, emergent problems compel new ways of approaching the microbial world to address urgent issues in the production of food, energy, and health. In this light, the history of fermentation science and the interrelations between science, environment, and the state in Japan are direct precedents to the more recent recognition of microbial ecologies as an inseparable part of human society in Europe and America.

Acknowledgments

This book is the product of the kindnesses of many people, much as our bodily cells function by the aid of multitudes of microbial cells that are not our own. I sincerely thank all the people and institutions below, as well as many more whose contributions I have not been able to name here, but to whom I remain deeply indebted.

Angela Creager, Benjamin Elman, and Furukawa Yasu all tirelessly shared their wisdom and guided the project from its inception. In Japan, I am especially grateful to the community of fermentation scientists, whose generosity with their time, perspectives, and documents was invaluable to the book: Beppu Teruhiko, Danbara Hirofumi, Imada Akira, Kaneko Yoshinobu, Kitamoto Katsuhiko, Komagata Kazuo, the late Nakase Takashi, Ōnishi Yasuo, Ōshima Yasuji, Tomita Fusao, the late Udaka Shigezō, Yagisawa Morimasa, and Yokota Akira. I thank the microbiologists and *moyashi* makers Konno Hiroshi and Konno Kenji, and the late Konno Eiichi of Akita Konno Shōten. The National Museum of Nature and Science, the National Research Institute of Brewing, RIKEN, and the Society for Biotechnology, Japan, helped with historical materials.

I have been very fortunate to find a sympathetic intellectual home in the department of history at Ohio University, where my research benefited from conversations with all the faculty and especially Michele Clouse, Mariana Dantas, Joshua Hill, Robert Ingram, Jaclyn Maxwell, Paul Milazzo, Assan Sarr, Miriam Shadis, Kevin Uhalde, and Jacqueline Wolf. My department chairs, Katherine Jellison and Brian Schoen, and the administrative staff, Sherry Gillogly and Brenda Nelson, always went beyond their ordinary duties to help. A yearlong residence at the Institut d'études avancées de Paris provided an ideal environment for finishing the book. A two-year postdoctoral fellowship at the Max Planck Institute for the History of Science's "Histories of Planning" group, led by Dagmar Schäfer, was critical to the book's framing.

Financial support from the Chemical Heritage Foundation, the D. Kim Foundation for the History of Science and Technology in East Asia, the Japan Foundation, the Konosuke Matsushita Memorial Foundation, the Ohio University Baker Fund, and the Princeton Institute for International and Regional Studies and Program in East Asian Studies made the research possible, as did language training at the Inter-University Center for Japanese Language Studies, Yokohama. The Department of History and Philosophy of Science and the Institute for Advanced Studies on Asia at the University of Tokyo were welcoming hosts.

At the University of Chicago Press, I am extremely grateful to Karen Darling, Tristan Bates, Deirdre Kennedy, Renaldo Migaldi, and the *Synthesis* board.

For the generous mentorship they offered to an early career scholar, I thank especially Philip C. Brown, Hannah Landecker, Cyrus Mody, and Julia Adeney Thomas. Helen Curry, David Howell, and Federico Marcon provided insightful readings of the entire manuscript. For comments on early drafts of chapters, I thank Jakobina Arch, David Bloor, William Deringer, Sharon Kingsland, Nina Lerman, Jordan Sand, Hallam Stevens, and David Wittner.

I appreciated invitations to present portions of the manuscript at the following forums: the Jeudis de l'histoire et de la philosophie des sciences at École Normale Supérieure, the History of Modern Medicine and Biology Seminar Series at the University of Cambridge, the *Japanese Scientist* workshop at Cardiff University, the Institute for Japanese Studies and Department of History at the Ohio State University, the *Science and Capitalism* workshop at Columbia University, the Institut für Japanologie at Martin-Luther-Universität Halle-Wittenburg, the Séminaire du Centre de recherches sur le Japon and Séminaire du groupe techniques au Centre Alexandre Koyré at École des hautes études en sciences sociales (EHESS), the *Stoffwechsel* workshop at Technische Universität Berlin, the *Unusual Lives* workshop at the Berlin Center for the History of Knowledge, the History and Philosophy of Biology and Medicine Seminar Series at Washington University in St. Louis, the Research Meeting on Economic History at the University of Tokyo, the Biological Unit of the History of Science Society of Japan, and the Kazemi in STS at the Tokyo Institute of Technology.

Exchanges with an extremely supportive community of scholars aided this work in various ways. Particular thanks to Dan Bouk, Emily Brock, Hasok Chang, Howard Chiang, John DiMoia, Kjell Ericson, Yulia Frumer, Michael Gordin, Benjamin Gross, Mathias Grote, Hashimoto Takehiko, Stefan Helmreich, the late Kaji Masanori, Aleksandra Kobiljski, Shigehisa

Kuriyama, Saadi Lahlou, Angela Ki Che Leung, Daniel Liu, Simon Luck, Pierre-Olivier Méthot, Erika Milam, Izumi Nakayama, Takashi Nishiyama, Chris Otter, Heather Paxson, Karen Rader, Carsten Reinhardt, Karen-Beth Scholthof, George Solt, and Tanimoto Masayuki. I am grateful to them all, as well as to many I have not mentioned.

Portions of this material have been reprinted or adapted from "Mold Cultures: Traditional Industry and Microbial Studies in Early Twentieth-Century Japan," in *New Perspectives on the History of Life Sciences and Agriculture*, edited by Denise Phillips and Sharon Kingsland, 231–52 (© 2015 Springer International Publishing Switzerland); "The Microbial Production of Expertise in Meiji Japan," *Osiris* 33:171–90 (© 2018 History of Science Society); and "Microbial Transformations: The Japanese Domestication of Penicillin Production, 1946–1951," *Historical Studies in the Natural Sciences* 48:441–74 (© 2018 Regents of the University of California).

Above all, I am indebted to Joyman Lee, who read drafts, helped with household tasks, and made life beautiful. He was always game for journeys across cultures, languages, and continents, as well as for sampling the assorted fermented foods and drinks that accompanied them. His support and companionship through the years made this book at first conceivable, and eventually possible.

Paris, November 2020

Notes

INTRODUCTION

1 Kitasato Institute, "Splendid Gifts from Microorganisms."

2 This assumption is reflected in a focus in the earlier historiography on problems of knowledge transfer and technology transfer to modern Japan from Europe and the United States. For an overview of Japan's historical tropes of modernization, which have tended to portray Japan as an exception in Asia, see Gluck, "Past in the Present." More recent explorations of the question of continuity across the Meiji transition have stressed the early modern logics that underlay Japanese approaches to nature in ways that were parallel to, and in some cases convergent with, developments in the West. See Thomas, "Reclaiming Ground"; Marcon, *Knowledge of Nature*; Frumer, *Making Time*.

3 Kozai Yoshinao, "Nihonshu jōzō no kairyō ni tsuite."

4 Suzuki Umetarō, "Eiyō kagaku to Manshū no sangyō" (Nutrition science and the industry of Manchuria), in *Kenkyū no kaiko*, 125–34 (keynote lecture at the opening meeting of the Manchuria branch of the Industrial Chemistry Society, 1932).

5 Yamazaki Momoji, *Tōa hakkō kagaku ronkō*. Yamazaki won the Suzuki Prize, one of the Japan Prizes in Agricultural Sciences, for this work in 1945.

6 Tamiya, "The Koji."

7 For recent perspectives on applied science in the twentieth century, see A. Johnson, "Applied Science"; Westwick, *Into the Black*; Lécuyer, *Making Silicon Valley*; Ndiaye, *Nylon and Bombs*; Akera, *Calculating a Natural World*; A. Johnson, *Hitting the Brakes*; Mody, *Instrumental Community*.

8 Exceptions include Holmes, *Hans Krebs*, 2 vols.; Spath, "C. B. van Niel." Genetic experiments on bacteria, viruses, and fungi were central in the rise of molecular biology. These studies used microbial phenomena to represent genetic phenomena. See Falk, "What Is a Gene?"

9 Bud, *Uses of Life*.

10 Epigenetics: broadly, how environment modifies genes. See Landecker, "Gene Regulation." Microbiome: in the case of humans, the aggregate of microbial

cells in the human body. See Rose, "Human Sciences"; and Thomas, "History and Biology." Horizontal gene transfer: the movement of genes outside of reproduction, which is especially frequent in microbes. See O'Malley, *Philosophy of Microbiology*; Helmreich, *Alien Ocean*. One prominent area of green chemistry is *chimie douce* (soft chemistry), in which reactions are performed by microorganisms and other biological materials. See Bensaude-Vincent et al., "School of Nature."

11   Russell, *Evolutionary History*; in contrast to the account in which humans always have the upper hand in Latour, *Pasteurization of France*.

12   Paxson, *Life of Cheese*; Tsing, *Mushroom*; Spackman, "Formulating Citizenship."

13   For a bestselling popular account of the new microbiology, see Yong, *I Contain Multitudes*; and compare the contrasting older account of de Kruif, *Microbe Hunters*. On fermented cuisine, see Katz, *Art of Fermentation*. On microbes in human health and the environment, see *Fantastic Fungi*, directed by Louie Schwartzberg (Moving Art, 2019).

14   Chang, *Inventing Temperature*; Chang, *Is Water H₂O*?

15   On the role of science and technology in constructing modern Japanese nationalism and empire, see Mizuno, *Science for the Empire*; Mimura, *Planning for Empire*; D. Yang, *Technology of Empire*; Moore, *Constructing East Asia*; Nishiyama, *Engineering War and Peace*; Wittner and Brown, *Science, Technology, and Medicine*.

16   Tanimoto, *Role of Tradition* argues that looking at technology transfer from America and Europe is insufficient to fully understand Japanese industrialization, and that scholars need to take small-scale indigenous industries into account.

17   On modern Japanese scientific institutions, see Bartholomew, *Formation of Science*. Work on the history of Japanese chemistry has noted that roughly half of the articles published in Japanese chemical journals concerned traditional industries, and that the foreign consultants brought over by the Meiji state as educators and technical advisers performed a substantial amount of work on problems related to traditional industry. See Furukawa Yasu, "Dentō sangyō kara kindai sangyō e"; Kikuchi, "Analysis, Fieldwork and Engineering." Scholars of Meiji-period science have also underlined the persistence of local problems, knowledge, and skills in scientific work. See Clancey, *Earthquake Nation*; Onaga, "Toyama Kametaro and Vernon Kellogg."

18   Morris-Suzuki, *Technological Transformation*, 84–103; Sugihara, "Development of an Informational Infrastructure."

19   Bartholomew, *Formation of Science* focuses primarily on the science and medical faculties of the imperial universities, as well as the role of government policymakers trained in engineering and the sciences.

20   This remains to be studied in detail, but see Cwiertka, "Soy Sauce Industry in Korea"; Cwiertka, *Cuisine, Colonialism*; C. Huang, "Postwar Taiwan Experience"; D. Yang, "Resurrecting the Empire?" On the pre-1945 impact, see also N. Smith, *Intoxicating Manchuria*. For a perspective from the history of the social sciences, see Joyman Lee, "Economics."

21 Historians of science recognize that the diffusionist view of modern science is problematic. Multiple alternative frameworks for discussing science in a global context are under debate. See Sivasundaram, "Focus: Global Histories"; McCook, "Focus: Global Currents"; Raj, "Beyond Postcolonialism . . . and Postpositivism."

22 The classic critique of the problem of analytical categories in modern non-Western history is Chakrabarty, *Provincializing Europe*. For an important study that demonstrates how names for categories that appear imported and universal can obscure fundamental institutions that function distinctively in the Japanese context, see Howell, *Geographies of Identity*. The entire constructivist literature on science as practice, especially on early modern Europe, has argued that present-day categories of "science" and "technology" restrict our understanding of knowledge practices concerning the natural world. For an overview, see Golinski, *Making Natural Knowledge*. For examples that work to overcome field categories, see Long, *Openness, Secrecy, Authorship*; P. H. Smith, *Body of the Artisan*.

23 This may account for the lack of scholarship on Japanese fermentation as such in both the English- and the Japanese-language literatures.

24 On the science/technology boundary, see especially Kline, "Construing 'Technology'"; and the article collection in Bud, "Focus: Applied Science." On the technology/art boundary, see Oldenziel, *Making Technology Masculine*; Paxson, *Life of Cheese*, 128–57.

25 Schäfer, *Crafting of the 10,000 Things*, 12–13.

26 To do so, I draw on recent scholarship on the material culture of everyday life and especially food in modern Japan, including Sand, *House and Home*; Cwiertka, *Modern Japanese Cuisine*; Rath and Assmann, *Japanese Foodways*; Solt, *Untold History of Ramen*.

27 The two classic accounts representing the top-down and bottom-up approaches are, respectively, C. Johnson, *MITI and the Japanese Miracle*; and George, *Minamata*.

28 Julia Adeney Thomas has argued that differing conceptions of the relationship between nature and society exist in political discourse in the modern era, both across different societies and within Japanese society; Thomas, *Reconfiguring Modernity*. Recent scholarship on modern Japanese ecologies includes Stolz, *Bad Water*; Golley, *When Our Eyes No Longer See*; Driscoll, *Absolute Erotic, Absolute Grotesque*; Walker, *Toxic Archipelago*; Miller, *Nature of the Beasts*. Moreover, instead of dismissing science as outside of culture, this work has also pushed us to see scientific thinking as part of a broader cultural "grammar" in Japan, with all its ethical implications. Gregory Golley argues, for example, that a desire to understand "identity and difference . . . in terms of interconnection and mutual transformation," shared by the science and modernist literature of the late 1910s to the mid-1930s, becomes in the writer Miyazawa Kenji's work an impulse "to imagine and cherish as real a world that may live and die unseen by humans." Golley believes that Miyazawa's ecological formulation had much more to do with his scientific awareness (he taught chemistry and soil science

in an agricultural college in Tōhoku) than his Buddhist influences, though the latter has previously received the bulk of attention from scholars. Golley, *When Our Eyes No Longer See*, 9, 69–70.

29 The literature is too large to summarize adequately here. This brief synopsis draws especially on Bartholomew, *Formation of Science*, and Morris-Suzuki, *Technological Transformation*.

30 Itakura and Yagi, "Japanese Research System."

31 For an example involving the Osaka Higher Technical School, see chapter 1.

32 On prewar consulting work, see Hashimoto, "Hesitant Relationship Reconsidered."

33 See, for example, chapter 4. On technological development in the Japanese empire and its relation to that in the home islands more broadly, see D. Yang, *Technology of Empire*; Mizuno, *Science for the Empire*; Mimura, *Planning for Empire*; Moore, *Constructing East Asia*.

34 See especially Mimura, *Planning for Empire*; Mizuno, *Science for the Empire*.

35 On scientific and technological institutions in Japan since 1945, see Nakayama, Gotō, and Yoshioka, *Social History of Science and Technology*.

36 C. Johnson, *MITI and the Japanese Miracle*.

37 See, for example, chapter 6. On corporate research laboratories, see Hashimoto, "Hesitant Relationship Reconsidered." On research associations, see Hashimoto, "Technological Research Associations."

38 Yoshioka, introduction to *Social History of Science and Technology*, vol. 4.

39 Bud, *Uses of Life*, 106–7, 146–47; M. V. Brock, *Biotechnology in Japan*; Collins, *Race to Commercialize Biotechnology*, 121–24; Fukushima, "Policy Process," 16–17.

40 In doing so, the book contributes to a growing literature on biotechnology in the Asian context; Ong and Chen, *Asian Biotech*.

CHAPTER 1

1 On the development of pure yeast culture and the introduction of the technique to the German beer industry, see Ceccatti, "Science in the Brewery." On the English context, see Teich, "Fermentation Theory and Practice."

2 After the 1880s, the national government shifted the focus of its policies from transplanting Western industries to supporting the "traditional" sectors that constituted the bulk of the economy. See Morris-Suzuki, *Technological Transformation*, 98–103.

3 Recent work on the significance of traditional industry in twentieth-century Japanese science includes Clancey, *Earthquake Nation*; Onaga, "Silkworm, Science and Nation." On their importance to the modern Japanese economy, see Tanimoto, *Role of Tradition*.

4 Fujiwara Takao, *Kindai Nihonshuzōgyōshi*, 185; Tanimoto, "Capital Accumulation," 301.

5 Morris-Suzuki, *Technological Transformation*, 98–103; Morris-Suzuki, "Great Translation."

6 Marcon, *Knowledge of Nature*, 276–77.

7 In his study of early modern Japanese *honzōgaku* (materia medica), Federico Marcon points to the homology between species and commodities as abstractions concealing human labor. He argues that the production of species transformed the natural world into "a collection of objects to analyze, represent, manipulate, control, and produce . . . devoid of any metaphysical or sacred aura." Marcon, *Knowledge of Nature*, 296–97. In the Meiji era, as this chapter shows, the parallel between the species of fermentation microbes produced by scientists and the commercial brands of spores produced by *tanekōji* makers became a literal one; both were microbes isolated, identified, and preserved by methods of pure culture, and characterized not only by their morphology as experienced by the senses or under a microscope, but by their physiological activity in the context of brewing processes.

8 In the history of technology and science, Japan has often been accorded "exceptional" status in East Asia for breaking rapidly with a traditional past or possessing traditions that were closer to European than to Asian countries. On rapid science and technology transfer under state promotion by the Meiji government and then under MITI (Ministry of International Trade and Industry) in the post–World War II period, the authoritative works remain Bartholomew, *Formation of Science*, and C. Johnson, *MITI and the Japanese Miracle*. On Western-style traditions in early modern Japan, there is a large literature on *rangaku* ("Dutch studies," or studies of Western science and medicine based on translations of Dutch texts). For one example, see Jannetta, *Vaccinators*.

9 Takahashi Teizō, *Jōzō bairon*, 3–4. On comparable trends in entomology, see Setoguchi Akihisa, *Gaichū no tanjō*.

10 Takahashi, "Preliminary Note"; Takahashi and Yamamoto, "Physiological Differences."

11 Tanimoto, "Capital Accumulation," 301–2.

12 Howell, "Urbanization, Trade, and Merchants," 356–57.

13 Francks, "Inconspicuous Consumption," 149.

14 Tanimoto, "Capital Accumulation," 301–5.

15 Katō Hyakuichi, "Nihon no sakazukuri no ayumi," 168; Francks, "Inconspicuous Consumption," 153.

16 T. C. Smith, *Native Sources of Japanese Industrialization*; Hayami, *Japan's Industrious Revolution*; Hayami Akira, "Keizai shakai no seiritsu to sono tokushitsu"; Howell, *Capitalism from Within*; Wigen, *Making of a Japanese Periphery*; Pratt, *Japan's Protoindustrial Elite*.

17 Pratt, *Japan's Protoindustrial Elite*, chap. 1.

18 Pratt, *Japan's Protoindustrial Elite*, 71–72; Morris-Suzuki, *Technological Transformation*, 49–50; Gordon, *Modern History of Japan*, 33.

19 Francks, "Inconspicuous Consumption," 155–56.

20 Katō Hyakuichi, "Nihon no sakazukuri no ayumi," 212–15.

21 Tonomura, "Gender Relations in the Age of Violence," 275.

22 Tanimoto, "Capital Accumulation"; Fruin, *Kikkoman*, chap. 1.

23 Katō Hyakuichi, "Nihon no sakazukuri no ayumi," 215–29.

24 Katō Hyakuichi, "Nihon no sakazukuri no ayumi," 215–16.

25 Fujiwara Takao, *Kindai Nihonshuzōgyōshi*, 185. From the late 1920s to the late 1930s, Marxist intellectuals looked back upon Meiji-era capitalism in an intense debate over how it should be situated analytically. The *Kōza-ha* (lectures faction), which followed the line given by the Comintern, emphasized the feudal elements remaining in the special or hybrid case that was Japanese capitalism, and argued that state institutions were upheld by a base of semifeudal production relations in the countryside. The *Rōnō-ha* (worker-farmer faction) took a different view that placed Japanese capitalism among the imperialist finance capitalisms of the world, and which saw the Meiji Restoration of 1868 as a bourgeois revolution with roots that reached back to the Tokugawa period. See Barshay, *Social Sciences in Modern Japan*.

26 Pratt, *Japan's Protoindustrial Elite*, 35.

27 Fujiwara Takao, *Kindai Nihonshuzōgyōshi*, 185–99. On translations of Western technical knowledge in general in the 1870s, see Meade, "Translating Technology."

28 Pratt, *Japan's Protoindustrial Elite*, 37–38.

29 Morris-Suzuki, *Technological Transformation*, 98–103; Sugihara, "Development of an Informational Infrastructure."

30 Tanimoto, *Role of Tradition*; Tanimoto, "Peasant Economy to Urban Agglomeration."

31 Pratt, *Japan's Protoindustrial Elite*, 6. On the class background of scientists in the early twentieth century, see Bartholomew, *Formation of Science*, 52–63.

32 Bartholomew, *Formation of Science*, 93.

33 Murakami Hideya, "Kōjikin," 58, citing Matsubara Shinnosuke's report on Hermann Ahlburg's work in 1878, the translation of Robert. W. Atkinson's work in 1881, the scientifically trained brewer Seki Gorōmatsu's prominent book in 1894, and the brewing science work of Kozai Yoshinao and student Yagi Hisatarō in 1895.

34 Shimoyama Jun'ichirō, "Seishu no jōzō ni tsuite"; Kozai Yoshinao, "Nihonshu jōzō no kairyō ni tsuite."

35 Shimoyama Jun'ichirō, "Seishu no jōzō ni tsuite"; Furukawa Sōichi, "Nyūsankin

to kōbo no kyōson to kyōsei," with thanks to Furukawa Yasu (no relation) for the reference.

36 The name given by Hermann Ahlburg in 1876 was *Eurotium oryzae*, which was then changed to *Aspergillus oryzae* by Ferdinand Cohn in 1883 and Oskar Kellner in 1895. Murakami Hideya, "Kōjikin," 57; Murakami Hideya, "Sake to kōji," 321.

37 Saitō Kendō, *Tōkyō zeimusho kantokukyoku Saitō Kendō chōsa.*

38 Akiyama Hiroichi, "Sake to kōbo," 395.

39 On the Japanese doctoral degree, see Bartholomew, *Formation of Science,* 50–52.

40 The lecture is published as Kozai Yoshinao, "Nihonshu jōzō no kairyō ni tsuite."

41 That Kozai used an English rather than a German term for yeast was probably due to the prominent work of the British chemist Robert W. Atkinson in Japan. Atkinson taught applied chemistry at the Kaisei Gakkō (the predecessor of Tokyo Imperial University) and described at length the processes of sake brewing in microbiological terms. His descriptions were published in a 1881 pamphlet, *The Chemistry of Sake Brewing,* which was translated into Japanese and widely known among brewers. Other Europeans in Meiji Japan who researched on sake but were less well known to brewers included the German bacteriologist Oskar Korschelt, who worked with the Kaitakushi Brewery to help produce Sapporo Lager, and the German botanist Hermann Ahlburg, who like Korschelt taught at the Tokyo Medical School, both in the 1870s. For details, see Hasegawa Takeji, *Zatsuroku: Biseibutsugakushi,* 71–100; Fujiwara Takao, *Kindai Nihonshu-zōgyōshi,* 49–85.

42 On the German context, see Ceccatti, "Science in the Brewery."

43 For a detailed account, see Fujiwara Takao, *Kindai Nihonshuzōgyōshi,* 148–255.

44 The foreign consultants trained Japanese scientists for government ministries and public projects at the new colleges, while helping government factories to transplant beer and wine technology in Japan. They included Robert W. Atkinson, Oskar Korschelt, and Hermann Ahlburg, discussed earlier in this chapter (see note 42). A significant proportion of foreign consultants' work in chemistry related to traditional industry; see Kikuchi, "Analysis, Fieldwork and Engineering." The first generation of Japanese chemists continued to invest heavily in problems related to traditional industry. Of all the articles published in the first ten years (1880–90) of the *Journal of the Tokyo Chemical Society,* roughly half were on traditional products; Furukawa Yasu counts 42 of 88 in Furukawa Yasu, "Dentō sangyō kara kindai sangyō e."

45 The list of industries is from Furukawa Yasu, "Dentō sangyō kara kindai sangyō e."

46 Sugihara, "Development of an Informational Infrastructure"; Morris-Suzuki, *Technological Transformation,* 71–104.

47 Outside the specialist brewing districts, the *tōji* employed in the multitude of

new, small brewing houses were of mixed quality in terms of skills. Fujiwara Takao, *Kindai Nihonshuzōgyōshi*, 165.

48 Fujiwara Takao, *Kindai Nihonshuzōgyōshi*, 194–99, 312–13.

49 Kozai Yoshinao, "Nihonshu jōzōhō no kairyō to kenkyū kikan no seibi ni tsuki."

50 On Hansen's yeasts, see Müller-Wille, "Hybrids, Pure Cultures."

51 Fujiwara Takao, *Kindai Nihonshuzōgyōshi*, 350.

52 Fujiwara Takao, *Kindai Nihonshuzōgyōshi*, 202–5.

53 Sawamura Makoto, "Mazu tōji no zunō o kairyō su beshi."

54 Kimoto Chōtarō, "Nihonshu jōzō kairyō ni tsuite no kibō."

55 Morris-Suzuki, *Technological Transformation*, 100–103.

56 Takayama Jintarō, "Jōzō shikenjo no hitsuyō."

57 Fujiwara Takao, *Kindai Nihonshuzōgyōshi*, 20–48.

58 Ono Ryōzō, "Shuzō jō no daiteki 'bakuteriya' ni tsuki kotau."

59 Yamagata Unokichi, "Jōzōgyō to seiketsu."

60 Kimoto Chōtarō, "Shuzō genryōchū 'bakuteriya' no kensahō."

61 Ōtsuka Ken'ichi, "Kenkyū no rekishi to tenbō," 1.

62 Murai Toyozō, "Tanekōji konjaku monogatari," 42.

63 Figures are for the year 1985. Narahara Hideki, "Moyashi," 36. Compared to its impact on the industrial structure of the *tanekōji* sector, pure culture's impact on the quality of sake or soy sauce production is harder to measure; for an overview of a related discussion, see V. Lee, "Wild Toxicity, Cultivated Safety."

64 *Moyashi* is the traditional term for what is now often called *tanekōji*.

65 Konno Eiichi, Konno Hiroshi, and Konno Kenji, interview.

66 Akita Konno shōten kabushiki gaisha, *Konno Moyashi 101 nen no ayumi*, 26–27.

67 Hyakushūnen kinen jigyōkai, *Hyakunenshi: Ōsaka daigaku kōgakubu jōzō, hakkō, ōyō seibutsu kōgakuka*, 10–13.

68 Tsuboi Sentarō, "Jōzōkai no kakumei jigyō."

69 Advertisements, *Jōkai* 10 (1902).

70 Akita Konno shōten kabushiki gaisha, *Konno Moyashi 101 nen no ayumi*, 27–28.

71 Akita Konno shōten kabushiki gaisha, *Konno Moyashi 101 nen no ayumi*, 28–29. On clocks, industry, and work discipline, see Thompson, "Time, Work-Discipline." On the history of clocks and time measurement in Japan, see Frumer, *Making Time*; Hashimoto, "Japanese Clocks."

72 Kawamata kabushiki gaisha, *Murasaki*, 90.

73 Akita Konno shōten kabushiki gaisha, *Konno Moyashi 101 nen no ayumi*, 26; Kawamata kabushiki gaisha, *Murasaki*, 88–89.

74 Narahara Hideki, "Moyashi," 35; Chikudō Shō, "Tanekōji ni tsuite (1)"; Chikudō Shō, "Tanekōji ni tsuite (2)."

75 Murai Toyozō, "Tanekōji konjaku monogatari," 40; Fukuoka-ken shōyu kumiai, *Fukuoka-ken shōyu kumiai nanajūnenshi*, 158. Kikkōman also claims that it pioneered pure-cultured *tanekōji* in 1904 and that the practice spread out from there. Nakadai Tadanobu, "Kikkōman ni okeru shōyu jōzō gijutsu kaihatsu no rekishi," 4.

76 Yamazaki Yoshikazu, "Higeta shōyu gijutsu enkaku no kaisō," 2.

77 Murai Toyozō, "Tanekōji konjaku monogatari," 40.

78 One such incident occurred in Kyoto in the mid-fifteenth century, when the shogunate attempted to revoke the monopoly law in response to wider discontent, and the struggle with the Kitano Tenmangu shrine, which dominated *kōji* making in and around the capital, resulted in most of the shrine burning down. Koizumi Takeo, *Kōji kabi to kōji no hanashi*, 105.

79 Narahara Hideki, "Moyashi," 35.

80 Narahara Hideki, "Moyashi," 34.

81 Kawamata kabushiki gaisha, *Murasaki*, 86–88.

82 Konno Eiichi, Konno Hiroshi, and Konno Kenji, interview.

83 Konno Eiichi, Konno Hiroshi, and Konno Kenji, interview.

84 Narahara Hideki, "Moyashi," 42–43. *Tane, tomokōji*, and *moyashi* were all terms used prior to the introduction of microbiology. *Moyashi* was the starter product, used for making *kōji*. *Tane* (seeds) were the material that made up *moyashi*. *Tomokōji* were specifically the *tane* deriving from a particular round of *kōji* making. *Tanekōji* was a Meiji-era neologism synonymous with *moyashi*. The term *hōshi* (spores) was in use by some experts as early as 1878, and could be used in place of *tane*. See Matsubara Shinnosuke, "Kōji no setsu," reprinted in Murakami Hideya, "Heruman Aaruburugu to sono shūhen"; and R. W. Atkinson, "Brewing in Japan," as summarized in Shurtleff and Aoyagi, *History of Koji*, 68–69.

85 Narahara Hideki, "Moyashi," 35.

86 Narahara Hideki, "Moyashi," 32–33.

87 Tsuboi Sentarō, "Jōzōkai no kakumei jigyō," 47.

88 On knowledge and technics, see Sigaut, "Technology."

89 Saitō Kendō, *Hakkō biseibukki*, 224.

90 Saitō Kendō, *Hakkō biseibukki*, 224. Saitō recounts the date as 1901, slightly earlier than the date given by accounts from the Brewing Experiment Station.

91 Saitō Kendō, "Shōyu jōzō ni kansuru biseibutsugakuteki kenkyū" (Microbiological research concerning soy sauce brewing), in *Tōkyō zeimusho kantokukyoku Saitō Kendō chōsa*, 1–24.

92 Saitō Kendō discovered and named *Saccharomyces soja*.

93 Saitō Kendō, "Kūki oyobi mizu no biseibutsugakuteki kenkyū" (Microbiological research on air and water), in *Tōkyō zeimusho kantokukyoku Saitō Kendō chōsa*, 25–39.

94 Saitō Kendō, "Taikichū no fuyū kinshi."

95 Saitō Kendō, "Higashi Ajia no yūyō hakkōkin."

96 Saitō Kendō, "Shina kōjikin ni kansuru chōsa" (Investigation of Chinese *kōji* microbes), in *Tōkyō zeimusho kantokukyoku Saitō Kendō chōsa*, 41–47.

97 Saitō Kendō, *Hakkō biseibukki*, 79.

98 Hasegawa Takeji, "Nihon no biseibutsukabu hozon jigyō 1," 5.

99 For example, see Saitō's contributions in Lodder and Kreger-Van Rij, *The Yeasts*.

100 In the 1950s the Institute for Fermentation, Osaka (IFO), was the largest culture collection in Japan. The IFO collection included substantial contributions from the Manchuria collection, as well as from the Government Research Institute of Formosa collection supervised by agricultural chemist Nakazawa Ryōji, formerly Saitō's colleague at the Brewing Experiment Station in the 1910s. See chapter 6; Hasegawa Takeji, "Nihon no biseibutsukabu hozon jigyō 1," 5–6; Higher Education and Science Bureau, Ministry of Education, Japan, *A General Catalogue of the Cultures of Micro-organisms Maintained in the Japanese Collections*.

101 What was formerly called the Brewing Experiment Station (Jōzō shikenjo) is now known as the National Research Institute of Brewing (Shurui sōgō kenkyūjo). Hasegawa Takeji, "Nihon no biseibutsukabu hozon jigyō 1," 5.

102 Hasegawa Takeji, "Nihon no biseibutsukabu hozon jigyō 1," 6–7. When Takahashi's seminar split off in 1924, the microbial chemist Yabuta Teijirō took over the seminar in agricultural products. Sakaguchi Kin'ichirō, "Michi e no gunzō."

103 Hasegawa Takeji, "Nihon no biseibutsukabu hozon jigyō 1," 7.

104 The two departments had been set up separately in the 1870s. When the departments separated again after briefly uniting in 1886, basic soil science with geology at its foundation was taught in agricultural science, while soil science similar to that envisioned by Justus von Liebig was taught in agricultural chemistry. Plant nutrition was subsumed under the Agricultural Chemistry Department's seminars in fertilizer science and plant physiology. As a result of the splitting of the two departments, topics directly related to agricultural production, such as soils, fertilizers, plant nutrition, and animal feed, came to take a marginal place within agricultural chemistry. Agricultural chemistry became a broad, heterogeneous discipline embracing chemistry, biology, and microbiology. Kumazawa Kikuo, "Riibihi to Nihon no nōgaku."

105 Initially, agricultural chemistry at the College of Agriculture as taught by *oyatoi gaikokujin* (foreign consultants) had been based on Justus von Liebig's original vision centered on fertilizers and soil science, expanding into plant and animal physiology. One *oyatoi gaikokujin*, Oscar Leow, introduced the teaching of bacteriology and fermentation science, and subsequently Kozai Yoshinao was

intrigued by the development of bacteriology under Louis Pasteur in France while he studied in Berlin. Through Leow and Kozai's teaching, agricultural chemistry came to include microbiology as well. Kumazawa Kikuo, "Riibihi to Nihon no nōgaku." On agricultural chemistry in France and Germany, see Jas, *Au carrefour de la chimie et de l'agriculture.*

106 Takahashi Teizō, "Kōbo no chozōhō; Takahashi Teizō et al., "Seishu kōbo no henshu ni tsuite."

107 Takahashi and Sakaguchi, *Summaries of Papers*, 8, item 30.

108 The different research materials reflected the fact that sake companies purchased *tanekōji* from specialist houses, whereas the largest soy sauce companies maintained their own *tanekōji*. Takahashi, "Preliminary Note"; Takahashi and Yamamoto, "Physiological Differences."

109 Akiyama Hiroichi, "Sake to kōbo," 396; Takahashi Teizō, "Yo ga iwayuru yodan shikomi shijō kakuchi shijō no kekka gaihyō." Fujiwara Takao, *Kindai Nihon-shuzōgyōshi*, gives an account of the program but overlooks its failure in this period.

110 Nihon jōzō kyōkai, *Nihon jōzō kyōkai nanajūnenshi*, 3, 107-11.

111 Akiyama Hiroichi, "Sake to kōbo," 397; Takahashi Teizō et al., "Seishu kōbo no henshu ni tsuite," 4.

112 Akiyama Hiroichi, "Sake to kōbo," 396-97.

113 Akiyama Hiroichi, "Sake to kōbo," 403.

114 See Takahashi and Sakaguchi, *Summaries of Papers.*

115 Sakaguchi Kin'ichirō, "Michi e no gunzō," 198. On studies of organic acid fermentations in Europe and the United States around the same period, see Bud, *Uses of Life*, chap. 2. On the significance of research on organic acid fermentations to penicillin production in the 1940s, see chapter 5 of this book.

116 Beppu Teruhiko, "Biseibutsugaku no kiso to ōyō."

117 Sakaguchi Kin'ichirō, "Michi e no gunzō," 191-208.

118 Chapter 6; Sakaguchi Kin'ichirō, "Michi e no gunzō," 195.

119 Cryns, "Influence of Hermann Boerhaave's Mechanical Concept." On Europe, see Riskin, *Restless Clock*. For a perspective from the history of recent cell biology, see Grote, *Membranes to Molecular Machines.*

120 He liked to examine slime molds under an old-fashioned single-lens microscope; Nakazawa Shin'ichi, *Mori no barokku*, 227. He began collecting specimens from the age of nineteen during his four-year visit in the United States, ranging from Michigan to Florida with a venture outside to Cuba, and he continued collecting during his eight-year sojourn in London from 1892 to 1900, followed by decades of long walks in the forests of his native Wakayama Prefecture in Japan. He published fifty-one contributions in the scientific journal *Nature*, mostly during his time in London. Together with his followers, he amassed a substantial collection of slime molds and continued to send speci-

mens to the naturalists Arthur and Gulielma Lister at the British Museum over
the years; Tamura, "English Essays of Minakata Kumagusu." Other brief por-
traits of Minakata Kumagusu may be found in Thomas, *Reconfiguring Modernity*, 188–93; Driscoll, *Absolute Erotic, Absolute Grotesque*, 1–21; Godart, *Darwin,
Dharma, and the Divine*, 92–103.

121   Nakazawa Shin'ichi, *Mori no barokku*, 253–55.

122   On the rise of protoplasm theory in the context of cell theory, see Liu, "The Cell
and Protoplasm."

123   I follow the interpretation of Minakata's thought in Nakazawa Shin'ichi, *Mori
no barokku*, 263–69. The literature on comparative anatomy and physiology is
extensive; see, for example, Foucault, *The Order of Things*; Foucault, *Birth of
the Clinic*; Lesch, *Science and Medicine in France*.

124   Coleman, *Biology in the Nineteenth Century*, chap. 2; Mendelsohn, "Lives of the
Cell"; Otis, *Müller's Lab*.

125   Matta, "Science of Small Things." Similarly, Christoph Gradmann argues in "Isolation, Contamination" that concepts of purity and the stability of species were
not prioritized in medical bacteriology until the late 1870s. On species in immunology, see Mazumdar, *Species and Specificity*.

126   Minakata's letters are reproduced in Nakazawa Shin'ichi, *Mori no barokku*,
260–62. Here my paraphrasing combines Minakata's descriptions with Nakazawa's analysis.

127   Amsterdamska, "Medical and Biological Constraints," 667. For a similar
example from brewing, see Müller-Wille, "Hybrids, Pure Cultures."

128   On the contrast with Tokugawa-era intellectual paths, see Bartholomew, *Formation of Science*, chap. 2; E. G. Nakamura, *Practical Pursuits*, 166–70; Sawada,
Practical Pursuits, chap. 4.

CHAPTER 2

1   Sato, "Formation of the Resource Concept"; Satō Jin, *"Motazaru kuni" no
shigenron*.

2   For population figures, see Schumpeter, *Industrialization of Japan and Man-
chukuo*, 72. On agricultural production, see Schumpeter, *Industrialization of
Japan and Manchukuo*, 62, 124–27. On political violence, see Siniawer, *Ruffians,
Yakuza, Nationalists*. On the rice riots, see Lewis, *Rioters and Citizens*.

3   This was like Europe and unlike the United States in the same period. Metzler,
Lever of Empire, 115.

4   Gordon, *Labor and Imperial Democracy*. In the 1920s, military ideologues
including Ishiwara Kanji appealed to the concept of "deadlock" to argue for
imperial expansion; Metzler, *Lever of Empire*, 235. Similarly, resource-scarcity
arguments were used by technical experts directly in support of imperialism;

Sato, "Formation of the Resource Concept," 155n9; Dinmore, "Small Island Nation." On emigration, see Lu, *Making of Japanese Settler Colonialism.*

5   Sand, *House and Home,* chap. 5.

6   Speech by Minister of Finance Inoue Junnosuke at the All-Kansai Women's Federation on August 15, 1929, quoted in Metzler, *Lever of Empire,* 206. According to Metzler, per capita consumption in general rose toward the mid-1920s and remained steady in the late 1920s, but per capita spending specifically on food (which accounted for more than half of total personal spending) was in fact already falling somewhat in the late 1920s before decreasing significantly during Inoue's depression-era deflationary policies. This trend was reflected in decreasing overall consumption levels of sake, tea, and tobacco. Metzler, *Lever of Empire,* 208–9.

7   Apple, *Vitamania.*

8   Harootunian, *Overcome by Modernity,* 27.

9   Spary, *Feeding France,* chap. 8, also 320. The historian of Japanese food Eric Rath suggests that gastronomy as a mass cultural phenomenon was burgeoning rather than firmly established in early twentieth-century Japan, as Spary likewise argues for eighteenth-century France. Rath also notes Susan Hanley's observation that food and taste are barely mentioned in extant diaries from the early modern period (1603–1868). Rath, "Foods of Japan," 319; Hanley, *Everyday Things,* 88.

10 Suzuki Umetarō, "Hakkō kagaku no shinpo," 776–77.

11 Kamminga, "Vitamins."

12 On how the rise of nutrition science was tied to national security concerns of the state, especially in Britain and Germany, see Kamminga and Cunningham, *Science and Culture of Nutrition.* On soy in the Chinese context, see Fu, *The Other Milk.* The discovery of vitamins is generally credited to the Polish biochemist Casimir Funk, rather than to Suzuki Umetarō. This is in spite of the pivotal importance of beriberi, which at the time was especially prevalent in Japan and was widely perceived to be an Asian disease. For more on the international recognition accorded to Japanese scientists in the early twentieth century in comparison to their European counterparts, see Bartholomew, "Japanese Nobel Candidates."

13 Sumner, *Brewing Science, Technology and Print.*

14 The full effects of industrialization upon the Japanese diet would be seen only after World War II. By the industrialization of diet, I am referring to a pattern in industrialized societies known as the "nutrition transition" and highlighted by historians such as Chris Otter. It involved a shift to consumption habits oriented toward foods that were high in calories, high in fat, low in fiber, and rich in animal protein. The shift in consumption habits was powered by an economy heavily reliant on energy from fossil fuels (most prominently to produce synthetic nitrate fertilizer); and the new diet provided the bodily energy for manufacturing-based, high-speed economic growth. Otter, "Industrializing

Diet, Industrializing Ourselves"; Otter, "British Nutrition Transition"; Mintz, *Sweetness and Power*. A detailed picture of the nutrition transition in Japan, which was supplied partly by imports of cheap surplus wheat from the United States, is painted in Solt, *Untold History of Ramen*.

15   Ceccatti, "Science in the Brewery"; Sumner, *Brewing Science, Technology and Print*.

16   On Tsuboi Sentarō, see also chapter 1.

17   Tsuboi Sentarō, "Nikushoku bōkokuron."

18   The high proportions of impurities in Japanese coal became especially problematic in coke manufacture. Kobiljski, "Poor Planning and Good Ideas."

19   Neswald and Smith, introduction to *Setting Nutritional Standards*, 7.

20   Treitel, *Eating Nature*, chap. 3.

21   On metabolism as exchange, see Landecker, "Food as Exposure," 170–74.

22   For example, those displayed in John Harvey Kellogg's *Shall We Slay to Eat?* (1899). See Warren, *Meat Makes People Powerful*, 58–59.

23   On the moral dimension to critiques of meat-eating, see Hong, "Digesting Modernity," 113.

24   Oberländer, "Western 'Scientific Medicine' in Japan," 21.

25   Oberländer, "Western 'Scientific Medicine' in Japan," 29. See also Bay, *Beriberi in Modern Japan*; H. A. Smith, *Forgotten Disease*.

26   In 1901, the government introduced a series of new tax laws in which *konseishu* was so heavily taxed that making it legally was no longer feasible economically, thus ending the brief boom of *konseishu*. Fukinbara Takashi, "Riken no gōseishu," 135.

27   Hyakushūnen kinen jigyōkai, *Hyakunenshi: Ōsaka daigaku kōgakubu jōzō, hakkō, ōyō seibutsu kōgakuka*, 30–32. For more on these concerns in the brewing industry, see chapter 1.

28   Akai Akinori, "Tsuboi Sentarō no 'katsuryokuso.'"

29   On the patent medicine market in modern Japan, see Burns, "Marketing 'Women's Medicines'"; Burns, "Japanese Patent Medicine Trade."

30   Bud, *Uses of Life*, 36.

31   Compare especially 1930s arguments for "vegetarianism" (which did not preclude consumption of fish); Mitsuda, "'Vegetarian' Nationalism."

32   For an account of the vitamin priority debate, see Fujitani, "Child of Many Fathers."

33   For years before World War I, a number of Japanese scientists believed that Japan needed a physical sciences research institution that would serve as an equivalent of the Kaiser Wilhelm Gesellschaft in Germany and similar institutions in Britain, France, and the United States. During the war, the disruption

in imports of science-based commodities, including industrial chemicals such as aniline dyes and drugs as well as precision instruments, especially from Germany, caused a crisis in Japanese industry. It made the promotion of physical sciences research a high priority for businessmen, bureaucrats, and scientists alike. See Bartholomew, *Formation of Science*, 123, 199–201, 212–17.

34 Ikeda Kikunae was simultaneously a professor at Tokyo Imperial University. Miyata Shinpei, *"Kagakusha no rakuen" o tsukutta otoko*, 113–14.

35 According to lore, the researcher Ikeda spoke to, Isobe Hajime, developed the desiccant Adosōru for refrigeration and air conditioning. It became Riken's first commercial product.

36 Compare, for example, the justifications for climate control technology in terms of facilitating imperialism that are directed to a British audience in Markham, *Climate and the Energy of Nations*.

37 Riken's ideals were to promote fundamental research without special regard for the commercialization of discoveries, but the revenue from Riken Bitamin was more than welcome to the institute's director. Suzuki's laboratory later followed up the product with a series of commercial vitamin preparations, which all contributed to Riken's finances. Miyata Shinpei, *"Kagakusha no rakuen" o tsukutta otoko*, 113–15.

38 On the links between the emergence of nutrition science and the growth of the Japanese military, see Cwiertka, *Modern Japanese Cuisine*, 67-68, 120–21.

39 Miyata Shinpei, *"Kagakusha no rakuen" o tsukutta otoko*, 113.

40 Suzuki Umetarō, "Hatsuiku ni kansuru eiyō mondai" (Nutritional problems concerning growth), in *Kenkyū no kaiko*, 99-112 (essay originally written in 1929).

41 Suzuki Umetarō, "Vitamin to horumon," (Vitamins and hormones), in *Kenkyū no kaiko*, 60 (lecture at Manchuria Medical University given in 1937).

42 Suzuki Umetarō, "Hatsuiku ni kansuru eiyō mondai," in *Kenkyū no kaiko*, 100.

43 Tsutsui, *Manufacturing Ideology*.

44 Kamminga, "Vitamins," 81–87.

45 Kamminga, "Vitamins," 88–93.

46 Kamminga, "Vitamins," 89.

47 Tsutsui, *Manufacturing Ideology*, 10–11.

48 Suzuki Umetarō, "Kome no shōhi setsuyaku" (Economizing consumption of rice), in *Kenkyū no kaiko*, 233–34 (lecture at the Food Meeting in the House of Peers, 1922); see also Suzuki Umetarō, "Hakkō kagaku no shinpo."

49 Suzuki Umetarō, "Kome no shōhi setsuyaku," in *Kenkyū no kaiko*, 233–34.

50 Suzuki Umetarō, "Hakkō kagaku no shinpo."

51 Suzuki Umetarō, "Kome no shōhi setsuyaku," in *Kenkyū no kaiko*, 234–35.

52 Suzuki Umetarō, "Kome no shōhi setsuyaku," in *Kenkyū no kaiko*, 236–38.

53 Tsutsui, *Manufacturing Ideology*, 8, 25.

54 Thanks to Nicolas Rasmussen for the suggestion that food supply acted as a "reverse salient" in industrialization.

55 Suzuki Umetarō and Katō Shōji, "Rikenshu ni tsuite."

56 Schumpeter, *Industrialization of Japan and Manchukuo*, 72.

57 For a comparison between the food policies of Britain and Germany during World Wars I and II, see Otter, "British Nutrition Transition."

58 Suzuki Umetarō and Katō Shōji, "Rikenshu ni tsuite," 362; Miyata Shinpei, *"Kagakusha no rakuen" o tsukutta otoko*, 117.

59 This was similar to beer adulteration in the nineteenth-century British brewing industry. Sumner, *Brewing Science, Technology and Print*.

60 Suzuki Umetarō and Katō Shōji, "Rikenshu ni tsuite," 362.

61 Kamminga, "Vitamins"; Suzuki Umetarō and Katō Shōji, "Rikenshu ni tsuite."

62 Suzuki Umetarō, "Eiyō to wa nani zo ya" (Oh, what is nutrition?), in *Kenkyū no kaiko*, 52 (lecture at the Mothers' Meeting hosted by the Ministry of Education, 1933).

63 Suzuki Umetarō, "Vitamin to horumon," in *Kenkyū no kaiko*, 61.

64 Household reformers: Sand, *House and Home*, 199.

65 Suzuki Umetarō, "Vitamin to horumon," in *Kenkyū no kaiko*, 61–75.

66 See also Suzuki Umetarō, "Hatsuiku ni kansuru eiyō mondai," in *Kenkyū no kaiko*, 108–11.

67 Suzuki Umetarō, "Eiyō to wa nani zo ya," in *Kenkyū no kaiko*, 49–59.

68 On eugenics in Japan, see Y. J. Chung, *Struggle for National Survival*; Otsubo, "Between Two Worlds."

69 Suzuki Umetarō, "Vitamin to horumon," in *Kenkyū no kaiko*, 60–82.

70 Suzuki Umetarō, "Eiyō kagaku to Manshū no sangyō" (Nutrition science and the industry of Manchuria), in *Kenkyū no kaiko*, 125–34 (keynote lecture at the Opening Meeting of the Manchuria Branch of the Industrial Chemistry Society, 1932).

71 On intensification, see Christian, *Maps of Time*.

72 Suzuki Umetarō, "Eiyō kagaku to Manshū no sangyō," in *Kenkyū no kaiko*.

73 Hashitani Yoshitaka, "Kōbo seizai."

74 On the strong opposition to nutrition science among professors in the Faculty of Medicine at Tokyo Imperial University, see Bay, *Beriberi in Modern Japan*, chap. 5. On the 1930s vitamin craze in Britain, see Bud, *Penicillin*, 12–14.

75 Osamu Chonan, "Advertisement Feature—Yakult's Research Activities: The

Inheritance and Practice of Shirota-ism," Nature.com, https://www.nature
.com/articles/d42473-019-00016-8 (accessed November 12, 2019).

76 Treitel, *Eating Nature.*

77 Takada Ryōhei, *Senjika shokuryō taisaku,* 85.

78 Mizuno, *Science for the Empire,* 43–68; Mimura, *Planning for Empire.*

79 In 1937 the military budget comprised three-fourths of all central government
spending, compared to approximately one-third in 1930; Gordon, *Modern History of Japan,* 192.

80 Cwiertka, *Modern Japanese Cuisine,* 121; Katarzyna Cwiertka, "Western Food,"
132–33.

81 On Kita Gen'itsu and the department of industrial chemistry at Kyoto, see Furukawa, "Gen-itsu Kita"; Furukawa Yasu, *Kagakushatachi no Kyōtō gakuha.*

82 Takada Ryōhei, "Biseibutsu shokuryōka no gainen to kōjikin shokuryōka ni chakushu shita riyū."

83 "Takada Ryōhei narabi ni kyōdō kenkyūsha ni yoru ronbun mokuroku." Document supplied by the Society for Biotechnology, Japan.

84 Out of this grew the beginnings of the synthesis of vitamin B_1 as the pharmaceutical company Wakamoto Seiyaku set up a pilot plant for B_1 synthesis. Later
in 1948, Takeda Yakuhin picked up on this research and developed a method of
B_1 synthesis that put it at the forefront of the vitamin market in Japan with its
product Alinamin. Takada Ryōhei, "Bitamin shinpen zatsuwa," 164.

85 Suzuki Umetarō, "Gōsei shokumotsu" (Synthetic food), in *Kenkyū no kaiko,* 87.

86 Takada Ryōhei, "Kōgyō kara mita shokuryō mondai."

87 Suzuki Umetarō, "Ikani shite shokuryō o jikyū subekika" (How should we
achieve self-sufficiency in food?), in *Kenkyū no kaiko,* 90–98 (article in *Tōkyō
nichi nichi shimbun,* 1941).

88 Takada Ryōhei, *Senjika shokuryō taisaku,* 16–18.

89 Takada Ryōhei, "Kōgyō kara mita shokuryō mondai."

90 By this figure, Takada was probably referring to food imported from outside the
Japanese formal empire, without including imports from Korea and Taiwan.

91 Takada Ryōhei, "Bitamin shigen yori mita Nihon."

92 Takada Ryōhei, *Senjika shokuryō taisaku,* 1–5.

93 Takada Ryōhei, *Senjika shokuryō taisaku,* 14.

94 Takada Ryōhei, *Senjika shokuryō taisaku,* 19–26, 28.

95 Takada Ryōhei, *Senjika shokuryō taisaku,* 29.

96 Takada Ryōhei, *Senjika shokuryō taisaku,* 96–97.

97 Takada Ryōhei, *Senjika shokuryō taisaku,* 48–52, 74–86.

98 Takada Ryōhei, *Senjika shokuryō taisaku*, 52–55.

99 On overfishing of herring, see Howell, *Capitalism from Within*. For general figures related to the decline in the fishing industry in wartime, see Garon, "Home Front," 46.

100 Takada Ryōhei, *Senjika shokuryō taisaku*, 55–64. On the development of inosinic acid as a flavoring product, see chapter 6.

101 Takada Ryōhei, *Senjika shokuryō taisaku*, 64–73.

102 Takada Ryōhei, "Bitamin shinpen zatsuwa," 167–71.

103 For background on this debate, see Bay, *Beriberi in Modern Japan*, chap. 6.

104 Takada Ryōhei, *Senjika shokuryō taisaku*, 74–86.

105 Takada Ryōhei, *Senjika shokuryō taisaku*, 73.

106 Takada Ryōhei, "Bitamin shinpen zatsuwa," 172–75.

107 Takada Ryōhei, *Senjika shokuryō taisaku*, 85.

108 Takada Ryōhei, "Bitamin," 121–22.

109 Thanks to John S. Ceccatti for this observation.

110 Treitel, *Eating Nature*, 223–24; Kamminga, "Vitamins"; Warner, *Sweet Stuff*; Apple, *Vitamania*.

111 Most famously, one new company, Suzuki Tōzaburō's Nihon Shōyu Jōzō, had at one point threatened to kill the entire soy sauce market in both east and west Japan by devising a method that brewed huge quantities of soy sauce at an unprecedented rate. In 1909, to the relief of other soy sauce manufacturers, a scandal broke out around the company adding illegal amounts of saccharin to the products. And to put the last nail in the coffin, their factory burned down. Kawamata kabushiki gaisha, *Murasaki*, 96–97.

112 Ichikawa Kunisuke, "Aminosan chōmiryō," 111–12.

113 Cwiertka, "Soy Sauce Industry in Korea."

114 Ichikawa Kunisuke, "Aminosan chōmiryō," 111–12.

115 The Amino Acid Manufacturing Association was established in 1939; from 1941 it distributed rationed raw materials. In 1941 soy sauce and miso became rationed products, and the Japan Amino Acid Control Company was established. By the end of 1942, the manufacturing association had merged with the control company, which had thirty-one factories in 1942 and forty-one factories in 1943. Kawamata kabushiki gaisha, *Murasaki*, 141–43.

116 "New Style Soy Sauce No. 1" was soy sauce lees decomposed with dilute hydrochloric acid, neutralized, and then matured with *kōji* grown on soy sauce lees. The "New Style Soy Sauce No. 2" process used dilute hydrochloric acid to decompose soybean meal, pressed soybean, and other soy products, and then neutralized the resulting mixture, and matured it with *kōji* cultured on either copra meal from coconut or soy sauce lees. Shurtleff and Aoyagi, "History of Soy Sauce, Shoyu, and Tamari," 5–6.

117 Fukinbara Takashi, "Riken no gōseishu," 135–41.

118 The producers' association was the Japan New-Style Sake Union, founded in 1932. By this time, Riken's Sake Laboratory had been renamed the Fermentation Laboratory for the purposes of researching other fermentation processes needed in war, such as butanol fermentation. It also began research on producing flavin (part of the vitamin B complex) by fermentation, which was eventually made in a Riken factory.

119 Sakurai Yoshito, "Shokuhin no kyōka"; Kawashima Shirō, "Senjichū no Nihon rikugun ni okeru bitamin hokyū."

CHAPTER 3

1 Yamazaki Momoji, *Tōa hakkō kagaku ronkō*. For brief overviews of the history of Japanese anthropology, see Atkins, *Primitive Selves*, chap. 2; Tierney, *Tropics of Savagery*, 78–85.

2 Chapter 1; Saitō Kendō, "Higashi Ajia no yūyō hakkōkin"; Saitō Kendō, *Hakkō biseibukki*, 79.

3 On traditional food, economy, and regional identity, see also Ceccarelli, Grandi, and Magagnoli, *Typicality in History*.

4 Tierney, *Tropics of Savagery*, 33.

5 Tierney, *Tropics of Savagery*, 21.

6 Mizuno, *Science for the Empire*. See also Mimura, *Planning for Empire*.

7 *Kōji* was designated the national microbe (*kokkin*) of Japan by the Brewing Society of Japan on October 12, 2006; see Scientific Conference of Brewing Society Japan, "Declaration," Brewing Society of Japan, October 12, 2006, rev. November 28, 2013, https://www.jozo.or.jp/gakkai/wp-content/uploads/sites/4/2020/01/gakkai_koujikinnituite2.pdf. On the common knowledge that *kōji* originates from China: Kitamoto Katsuhiko, in discussion.

8 Peattie, introduction to *Japanese Colonial Empire*, 7–8, 22–23. The term "statist elites" is used aptly in Moore, *Constructing East Asia*, 13.

9 Tierney, *Tropics of Savagery*, 148. On the historical background, see Duus, "Japan's Informal Empire in China."

10 Suzuki Sadaharu, "Nihon no arukōru no rekishi," 168.

11 Atkins, *Primitive Selves*, 7, citing Schmid, "Colonialism and the 'Korea Problem.'"

12 Gluck, "Past in the Present," 68.

13 On continuities between postwar technical assistance and the Japanese empire, see Mizuno, Moore, and DiMoia, *Engineering Asia*; D. Yang, "Resurrecting the Empire?"

14 Tierney, *Tropics of Savagery*, 32.

15 Saitō Yoshio et al., "'Utsunomiya kōnō' to nōgei kagaku sono 1."

16 "Kōtō nōrin gakkō jidai 1922–1928" (The Higher Agricultural and Forestry School era 1922–1928), Utsunomiya University Library, http://www.lib.utsu-nomiya-u.ac.jp/tenjikai2012-rekishi2.pdf (accessed November 10, 2013).

17 Saitō Yoshio et al., "'Utsunomiya kōnō' to nōgei kagaku sono 1," 257–58.

18 Saitō Yoshio et al., "'Utsunomiya kōnō' to nōgei kagaku sono 1," 256.

19 Saitō Yoshio et al., "'Utsunomiya kōnō' to nōgei kagaku sono 1," 256.

20 Saitō Yoshio et al., "'Utsunomiya kōnō' to nōgei kagaku sono 1," 256.

21 Kawamura Shunzan, *Dekoboko jinseiki*, 259.

22 Saitō Yoshio et al., "'Utsunomiya kōnō' to nōgei kagaku sono 1," 257.

23 Kawamura Shunzan, *Dekoboko jinseiki*, 259.

24 Yamazaki Momoji and Mukui Masao, "Shaoshinchū ni tsuite (1)"; Yamazaki Momoji and Mukui Masao, "Shaoshinchū ni tsuite (2)"; Yamazaki Momoji, "Shinasan hakkō kinrui oyobi hakkō seihin no kenkyū."

25 Yamazaki Momoji, "Shinasan 'kiku' ni tsuite," 133.

26 On Saitō Kendō, see chapter 1. On fermentation science at the Taiwan Government-General Central Research Institute, see chapter 4.

27 Chapter 6; Hasegawa Takeji, "Nihon no biseibutsukabu hozon jigyō 1," 5–6.

28 Hasegawa Takeji, "Nihon no biseibutsukabu hozon jigyō 1," 7–8.

29 Sakaguchi Kin'ichirō, "Kōjikin kenkyū no omoide," 290.

30 Saitō Yoshio et al., "'Utsunomiya kōnō' to nōgei kagaku sono 1," 256. On the Institute of Applied Microbiology, see chapter 6.

31 Reynolds, "Training Young China Hands," 226.

32 Duus, "Japan's Informal Empire in China," xxvi.

33 Duus, "Japan's Informal Empire in China," xxiii–xxiv.

34 Banno, "Anti-Japanese Boycotts in China."

35 Duus, "Japan's Informal Empire in China," xxiv–xxix. For the general background on Sino-Japanese relations, see Joyman Lee, "Closest Model, Rival, and Fateful Enemy"; Jansen, *Japan and China*.

36 S. Lee, "Foreign Ministry's Cultural Agenda," 276.

37 Reynolds, "Training China Hands," 258; S. Lee, "Foreign Ministry's Cultural Agenda," 276–77.

38 S. Lee, "Foreign Ministry's Cultural Agenda," 301.

39 S. Lee, "Foreign Ministry's Cultural Agenda," 295.

40 S. Lee, "Foreign Ministry's Cultural Agenda," 289.

41 S. Lee, "Foreign Ministry's Cultural Agenda," 281.

42 Yamazaki Momoji, "Taishi bunka jigyō to nōgaku," 10. On Japanese engineers' notions and material implementations of *kōa*, especially via large-scale infrastructure projects in Manchuria, see Moore, *Constructing East Asia.*

43 Yamazaki Momoji, "Taishi bunka jigyō to nōgaku," 10.

44 Yamazaki Momoji, "Taishi bunka jigyō to nōgaku," 13.

45 Yamazaki Momoji, "Taishi bunka jigyō to nōgaku," 13–14.

46 Yamazaki Momoji, "Taishi bunka jigyō to nōgaku," 14.

47 Yamazaki Momoji, "Shina gakuto e no kotoba." On the context of agricultural science in Nanjing, see Stross, *Stubborn Earth.*

48 On the European context, see Somsen, "History of Universalism."

49 For an analysis of this classic in the Chinese context, see Nappi, *Monkey and the Inkpot.*

50 On Takadiastase, see chapter 4.

51 Yamazaki Momoji, "Shina gakuto e no kotoba," 448.

52 There is a priority dispute over whether Suzuki Umetarō should be recognized as the discoverer, since there were problems with his structural characterization of the chemical. But it is nonetheless clear that he played a pioneering role in vitamin research and founded the field of vitamin science in Japan (chapter 2).

53 Suzuki Umetarō, "Vitamin kenkyū no kaiko" (Recollections of vitamin research), in *Kenkyū no kaiko*, 2–3 (published in *Kagaku chishiki* in 1931).

54 Kaji, "Transformation of Organic Chemistry," 14 (emphasis added). See also Kaji Masanori, "Majima Rikō to Nihon no yūki kagaku kenkyū dentō no keisei."

55 Yamazaki Momoji, "Shina gakuto e no kotoba," 449.

56 H. T. Huang, *Fermentations and Food Science.*

57 For a historical analysis of the growth in popularity of Japanese cuisine in Europe and the United States since 1945, see Cwiertka, *Modern Japanese Cuisine*, postscript.

58 I follow H. T. Huang, *Fermentations and Food Science*, in translating 酒 as "wine" for its cultural meaning. The closest process equivalent in Europe would actually be beer (barley wine), in that it is made from cereal grain and not fruits, thereby requiring saccharification performed by outside enzymatic action, which grape wine does not require. Some works that cite Yamazaki on this particular question include Sakaguchi Kin'ichirō, "Hakkō"; Ichikawa Jirō, *Nihon no sake*, 114; Kitahara Kakuo, "Arukōru kōgyō ni okeru amiraaze," 239. Some works that give the same answer but do not cite Yamazaki include Katō Hyakuichi, "Kōji no rekishi," 3; Koizumi Takeo, *Kōji kabi to kōji no hanashi*, 57.

59 Ishige, *History and Culture of Japanese Food*, 34–35.

60 H. T. Huang, *Fermentations and Food Science*, 167.

61 "Yellow robe" from the Yüeh ling (月令) chapter of the *Li chi* (禮記), c. 450 BCE–100 CE, quoted in H. T. Huang, *Fermentations and Food Science*, 165; "coat" from the *Chhi min yao shu* (齊民要術), c. 540 CE, quoted in H. T. Huang, *Fermentations and Food Science*, 173.

62 Sakaguchi Kin'ichirō, "Hakkō," 48.

63 H. T. Huang, *Fermentations and Food Science*, chap. 40 d and e; Sakaguchi Kin'ichirō, "Hakkō"; Kitamoto Katsuhiko, in discussion.

64 Kitahara Kakuo, "Arukōru kōgyō ni okeru amiraaze," 239.

65 Yamazaki Momoji, *Tōa hakkō kagaku ronkō*, 433, 302, 350.

66 Yamazaki Momoji, *Tōa hakkō kagaku ronkō*, 302. For a historical discussion of how "mutation" (*totsuzen hen'i*) was understood by Japanese geneticists in the early twentieth century, see Onaga, "Tracing the Totsuzen."

67 Yamazaki Momoji, *Tōa hakkō kagaku ronkō*, 302.

68 Yamazaki Momoji, *Tōa hakkō kagaku ronkō*, 249.

69 Yamazaki Momoji, *Tōa hakkō kagaku ronkō*, 352.

70 Yamazaki Momoji, *Tōa hakkō kagaku ronkō*, 433.

71 Yamazaki Momoji, *Tōa hakkō kagaku ronkō*, 435.

72 Yamazaki Momoji, *Tōa hakkō kagaku ronkō*, 435.

73 On *chimie douce* (soft chemistry) from the late 1970s, see Bensaude-Vincent et al., "School of Nature." On biomaterials research with relevance to "green chemistry," see Linse, "Directed Evolution," 10; Ramakrishnan, "Potential and Risks," 13. For a brief account of the historical background on biomaterials science, see V. Lee, "Bioorganic Synthesis."

74 Yamazaki Momoji, *Tōa hakkō kagaku ronkō*, 435.

75 An analogously thorny question might be the role of rice in premodern Japan. See von Verschuer, *Rice, Agriculture, and the Food Supply*; Andreeva, review in *East Asian Science, Technology and Society*; Brown, review in *Monumenta Nipponica*.

76 H. T. Huang, *Fermentations and Food Science*; Sakaguchi Kin'ichirō, "Hakkō."

77 H. T. Huang, *Fermentations and Food Science*, 163–67.

78 Robert Gipe, "Weedeater: An Illustrated Novel" (faculty symposium sponsored by the Charles J. Ping Institute for the Teaching of the Humanities, Ohio University, Athens, OH, March 30, 2018).

79 Mizuno, *Science for the Empire*, chap. 5.

80 Two examples of studies that consider the problem of resolving Asia conceptually in modern Japanese scientific identity are Jung Lee, "Between Universalism and Regionalism"; and Trambaiolo, "Vaccination." For a treatment of the early modern period, see Elman, "Sinophiles and Sinophobes."

81 Sakaguchi Kin'ichirō, "Hakkō," 42.

82 On *tanekōji*, see Takeo Koizumi, "Mystery of Fermentation." On *kōji* domestica-
    tion, see V. Lee, "Wild Toxicity, Cultivated Safety"; Murakami, "Classification of
    the Kōji Mold." The point concerning yeast versus *kōji* as a model organism was
    made by Kitamoto Katsuhiko in discussion.

83 Finlay, "China, the West," 267. For a recent reassessment of the legacy of Joseph
    Needham's Science and Civilisation in China series, see Hsia and Schäfer, "Sec-
    ond Look at Joseph Needham." On the reception of the series in Taiwan and
    Korea, see Chu, "Needham in Taiwan"; Lim, "Joseph Needham in Korea";
    Tsukahara and Mei, "Putting Joseph Needham in the East Asian Context."

84 Jung Lee, "Between Universalism and Regionalism," 678; Elman, "Great
    Reversal."

85 Fish and soy: Ishige, "Fermented Fish Products in Asia," 22fig6. Nation-microbe
    chart: Sakaguchi Kin'ichirō, "Hakkō," 50figs. Ethnochemistry: for example, the
    reenactment laboratory work of Ueda Seinosuke and Teramoto Yūji, cited in
    H. T. Huang, *Fermentations and Food Science*, 166.

86 H. T. Huang, *Fermentations and Food Science*, 157n41. Yamazaki is cited in a
    review of the literature by Fang.

87 H. T. Huang, *Fermentations and Food Science*, 608.

88 Pure culture: H. T. Huang, *Fermentations and Food Science*, 282. Quote: H. T.
    Huang, *Fermentations and Food Science*, 377.

89 For a critique of the "Needham question," see Sivin, "Scientific Revolution."

90 H. T. Huang, *Fermentations and Food Science*, 155–56.

91 H. T. Huang, *Fermentations and Food Science*, 275.

92 H. T. Huang, *Fermentations and Food Science*, 282, 189, 278, 605.

93 The point has been made by Sigaut in "Technology," 425–26.

94 Yamazaki Momoji, *Tōa hakkō kagaku ronkō*, 5.

95 Yamazaki Momoji, *Tōa hakkō kagaku ronkō*, 26.

96 Yamazaki Momoji, *Tōa hakkō kagaku ronkō*, 20.

97 Yamazaki Momoji, *Tōa hakkō kagaku ronkō*, 26.

98 Yamazaki Momoji, Tōa hakkō *kagaku ronkō*, 431.

99 Yamazaki Momoji, *Tōa hakkō kagaku ronkō*, 77fig6; H. T. Huang, *Fermentations
    and Food Science*, 600fig136.

100 Yamazaki Momoji, *Tōa hakkō kagaku ronkō*, 67–110.

101 Yamazaki Momoji, *Tōa hakkō kagaku ronkō*, 109–243. He uses the following
    as the main sources for each period: Neolithic: no documents; Northern Wei:
    Seimin yōjutsu 斉民要術 (c. 540 CE); Song: *Hokusan shukei* 北山酒経 (1117 CE).

102 H. T. Huang, *Fermentations and Food Science*, 1–13.

103   H. T. Huang, *Fermentations and Food Science*, 166, 263.

104   Yamazaki Momoji, *Tōa hakkō kagaku ronkō*, 301.

105   Yamazaki Momoji, *Tōa hakkō kagaku ronkō*, 301–8.

106   Yamazaki Momoji, *Tōa hakkō kagaku ronkō*, 320–21.

107   Yamazaki Momoji, *Tōa hakkō kagaku ronkō*, 321–49.

108   H. T. Huang, *Fermentations and Food Science*, 157–68; Sakaguchi Kin'ichirō, "Hakkō," 51–56.

109   Yamazaki Momoji, *Tōa hakkō kagaku ronkō*, 70.

110   Sakaguchi Kin'ichirō, "Hakkō," 51–56.

111   Yamazaki Momoji, *Tōa hakkō kagaku ronkō*, 293–99.

112   Sakaguchi Kin'ichirō, "Hakkō," 52.

113   Yamazaki Momoji, *Tōa hakkō kagaku ronkō*, 326.

114   H. T. Huang, *Fermentations and Food Science*, 166.

115   Katō Hyakuichi, "Nihon no sakazukuri no ayumi," 78–79. On Shinoda Osamu, see Cwiertka and Chen, "Shadow of Shinoda Osamu."

116   Howell, *Capitalism from Within*; Matsumura, *Limits of Okinawa*.

117   Tierney, *Tropics of Savagery*; Atkins, *Primitive Selves*.

118   Tierney, *Tropics of Savagery*; Atkins, *Primitive Selves*.

119   Tierney, *Tropics of Savagery*, 32.

120   Tierney, *Tropics of Savagery*, 12, citing Trouillot, "Anthropology and the Savage Slot"; Tierney, *Tropics of Savagery*, 81.

121   Aboriginal winemaking traditions in Taiwan included a millet-based mold preparation method, as well as the chewing method; Fan, "Liquor Industry of Taiwan," 1–2.

122   Yamazaki Momoji, *Tōa hakkō kagaku ronkō*, 348–51.

123   Similarly, recent popular cultural depictions portray *kuchikamizake* as a primeval Japanese tradition, as in the anime *Kimi no na wa* (Your name), directed by Shinkai Makoto (CoMix Wave Films, 2016).

124   Atkins, *Primitive Selves*.

125   Atkins, *Primitive Selves*, 56.

126   Yamazaki Momoji, *Tōa hakkō kagaku ronkō*, 363–64.

127   Yamazaki Momoji, *Tōa hakkō kagaku ronkō*, 362–63. His main source is the *Sanrin keizai* 山林経済 (end of the 18th century CE).

128   Yamazaki Momoji, *Tōa hakkō kagaku ronkō*, 381–92.

129   Yamazaki Momoji, *Tōa hakkō kagaku ronkō*, 392.

130 Yamazaki Momoji, *Tōa hakkō kagaku ronkō*, 400; Park, "Rise of Soju."

131 Yamazaki Momoji, *Tōa hakkō kagaku ronkō*, 430.

132 For a critique of these dichotomies, see Hart, "Beyond Science and Civilization."

133 "Leaving Asia": as urged in the infamous newspaper editorial, "Datsu A ron," published anonymously in *Jiji shinpō* on March 16, 1885, and later attributed to Fukuzawa Yukichi.

134 Saitō Kendō, *Hakkō biseibukki*, 56.

CHAPTER 4

1 Kitahara Kakuo, "Arukōru kōgyō ni okeru amiraaze," 242; Kinoshita Toshiaki, "Arukōru seizō gijutsu no shinpo," 297.

2 Hasegawa Takeji, "Nihon no biseibutsukabu hozon jigyō 1," 5; Komagata Kazuo, in discussion.

3 Komagata Kazuo, in discussion.

4 The question of the adaptation of Asian knowledge in modern Japan has been better explored in medicine; Otsuka, "Chinese Traditional Medicine in Japan." A few examples of studies on Meiji-period transfer of organizational patterns and technology are Westney, *Imitation and Innovation*; Hashimoto, "Introducing a French Technological System"; Nakaoka, "Transfer of Cotton Manufacturing Technology"; Wittner, *Technology and the Culture of Progress*.

5 Calls in Japanese history to address empire's contribution to metropolitan modernization, rather than simply assuming empire to be the outcome of metropolitan modernity, include Schmid, "Colonialism and the 'Korea Problem'"; Driscoll, *Absolute Erotic, Absolute Grotesque*.

6 Mizuno, *Science for the Empire*; Mimura, *Planning for Empire*; and Moore, *Constructing East Asia* share a focus on Manchuria. Compare the accounts of scientific modernization of agriculture in Taiwan in Amsden, "Taiwan's Economic History"; Ka, "Agrarian Development"; C. Huang, "Postwar Taiwan Experience."

7 Similarly, see the multidirectional scientific translations detailed in T. Yang, "Selling an Imperial Dream." An account of colonial and post-1945 agricultural development (with Japanese technical assistance) that follows the *nanshin* directionality while criticizing the impact on local farmers is Fujihara, "Colonial Seeds, Imperialist Genes." On *nanshin* and the Japanese state vision of agricultural and industrial expansion in Southeast Asia and the Pacific, especially using Taiwan as a base, see Peattie, "Nanshin," 199, 205–6; Peattie, "Nan'yō," 192–94; Schneider, "Business of Empire," 129, 214.

8 Iwai Ki'ichirō, "Honpō inryō shusei narabi ni gōsei seishu no hattenshi."

9 Ka Ushiko, "Shusei no hanashi sono ichi"; Togano Meijirō, "Shusei no hanashi"; Gōdō shusei shashi hensan iinkai, *Gōdō shusei shashi*, 48–49.

10 Ishidō Toyota, "Shusei seizō dan."

11 Sakamoto Tatsunosuke, *Kamiya Denbei*, 82.

12 Yukimatsu Bunta, "Jōryūshu (shusei, shōchū) seizōgyō no genzai oyobi shōrai," 2. On the *ranbiki* distillation apparatus, see Katō Hyakuichi, "Nihon no sakazukuri no ayumi," 307-10.

13 Suzuki Sadaharu, "Nihon no arukōru no rekishi," 59.

14 Gōdō shusei shashi hensan iinkai, *Gōdō shusei shashi*, 7-8.

15 Suzuki Sadaharu, "Nihon no arukōru no rekishi," 59.

16 Iwai Ki'ichirō, "Honpō inryō shusei narabi ni gōsei seishu no hattenshi."

17 Sakamoto Tatsunosuke, *Kamiya Denbei*, 68-69.

18 Ishidō Toyota, "Shusei seizō dan."

19 Shurtleff and Aoyagi, *Jokichi Takamine (1854–1922) and Caroline Hitch Takamine (1866–1954)*, 5-8; Tamiya, "The Koji." For more on Takamine's life, see Iinuma Kazumasa and Kanno Tomio, *Takamine Jōkichi no shōgai*; Yamashima, "Jokichi Takamine."

20 Takamine Jōkichi, "Takajiasutaaze ni tsuite."

21 Iwai Ki'ichirō, "Honpō inryō shusei narabi ni gōsei seishu no hattenshi."

22 Kinoshita Toshiaki, "Arukōru seizō gijutsu no shinpo," 297-98.

23 Yukimatsu Bunta, "Jōryūshu (shusei, shōchū) seizōgyō no genzai oyobi shōrai."

24 Gōdō shusei shashi hensan iinkai, *Gōdō shusei shashi*, 101-3.

25 Iwai Ki'ichirō's use of the technique at the Uji factory was inspired by German beer brewing. Some plants, including the Uji factory, though not all, ordered *Lactobacillus delbrueckii* from Germany for the purpose. Iwai Ki'ichirō, "Honpō inryō shusei narabi ni gōsei seishu no hattenshi"; Oana Fujio, "Kyūshū chihō shusei kōjō shisatsu ki."

26 Walker, *Lost Wolves of Japan*, chap. 4. The Japanese beer industry, which began in Hokkaidō, continued to import malts and hops from Germany for quality and marketing purposes until after World War II. See Alexander, *Brewed in Japan*, 22.

27 Sakamoto Tatsunosuke, *Kamiya Denbei*, 59-139.

28 C. Huang, "Postwar Taiwan Experience," 27.

29 Ho, "Colonialism and Development," 371.

30 Ho, "Colonialism and Development," 371-74; Amsden, "Taiwan's Economic History," 343-48.

31 C. Huang, "Postwar Taiwan Experience," 25.

32 Suzuki Sadaharu, "Nihon no arukōru no rekishi," 88-90.

33 Nakano Masahiro, "Taiwan ni okeru arukōru sangyō," 179.

34 Nakano Masahiro, "Taiwan ni okeru arukōru sangyō," 193–96.

35 Yukimatsu Bunta, "Jōryūshu (shusei, shōchū) seizōgyō no genzai oyobi shōrai"; Ka Ushiko, "Shusei no hanashi sono ni"; Niki Etsutarō, "Kiki ni hinseru shōchū-kai (shuseishiki)"; Kawakita Michitada, "Nenryō toshite shusei no ōyō"; Usami Keiichirō, "Nenryō mondai to nenryō shusei."

36 Kohata Kengorō, "Waga kuni ni okeru shusei seizōgyō no kinkyō (yon)."

37 Maejima Chitoku, "Taiwan ni okeru shusei seizō no gaikan"; Oana Fujio, "Kyūshū chihō shusei kōjō shisatsu ki."

38 Suzuki Umetarō, who sought a source of alcohol for synthetic sake in 1922, believed that eschewing fermentation methods entirely in favor of chemical synthetic methods, if it could be done, would be the best choice because of the need to feed the population. He argued that there was not much room to grow more sugar in Taiwan. Similarly, growing potatoes destined for alcohol took acres of farmland in Hokkaidō, whereas in Germany the industry used a synthetic method to make alcohol from carbide and saved potatoes for food. Suzuki Umetarō, "Kome no shōhi setsuyaku" (Economizing consumption of rice), in *Kenkyū no kaiko*, 236–37 (lecture at the Food Meeting in the House of Peers, 1922). On some of the historical background to the biofuel debates, see Bud, *Uses of Life*, 133, 141–45.

39 Barshay, *Social Sciences in Modern Japan*; Myers and Yamada, "Agricultural Development in the Empire." In the late 1930s, Taiwan and Korea supplied more than 90 percent of the sugar and 98 percent of the rice imported by Japan, but Taiwan and Korea imported large amounts of food despite both being food-surplus countries. Ho, "Colonialism and Development," 370–71. "By the end of the 1920s, the average Japanese consumed almost twice as much rice as the average Korean, who had to supplement his diet with millet, maize, and barley, mostly imported from Manchuria." Han, *History of Korea*, 480, quoted in Peterson, *Brief History of Korea*, 146.

40 Calmette, "Contribution à l'étude des ferments de l'amidon."

41 Hc had initially aimed to create a vaccination center, but the colonial adminis-tration in Cochinchina turned out to be more interested in local revenue gen-eration. Sasges, *Imperial Intoxication*, chap. 2. On Calmette's contributions to medical bacteriology and the study of tuberculosis, see Guénel, "First Over-seas Pasteur Institute"; K. Chung and Biggers, "Albert Leon Charles Calmette"; Bonah, "Experimental Stable"; Hawgood, "Albert Calmette (1863–1933) and Camille Guérin (1872–1961)." On the Pasteur Institutes worldwide, see Velmet, *Pasteur's Empire*; Brès and Chambon, "Pasteur Institutes Worldwide."

42 Sasges, *Imperial Intoxication*; Peters, "Taste, Taxes, and Technologies."

43 Katō Benzaburō, "Amirohō ni kansuru kenkyū"; Zenda Naozō, "Jōryūshu jōzōyō toshite no 'amiro'-hō ni tsuite"; Sugiyama Shinsaku, "Mippei tanku ni yoru shusei hakkō."

44 Sasges, *Imperial Intoxication*, 45–46. For reasons of taste, brewers in Japan also continued to use mixed rather than pure cultures of yeast until after World

War II, and brewers in China used mixed cultures for both mold and yeast until at least the turn of the millennium. Suzuki Umetarō, "Kome no shōhi setsuyaku," 232; H. T. Huang, *Fermentations and Food Science*, 282.

45 Revells, "Filched Fungi?"

46 The monopoly replaced the system of taxation by which the government-general had previously extracted revenue from the alcohol industry. Aboriginal homebrewing was tacitly permitted to continue. Fan, "Liquor Industry of Taiwan."

47 Kinoshita Toshiaki, "Arukōru seizō gijutsu no shinpo," 298.

48 Excerpts from Kamiya Shun'ichi's field records are reproduced in Nakano Masahiro, "Taiwan ni okeru arukōru sangyō," 205–7.

49 Nakano Masahiro, "Taiwan ni okeru arukōru sangyō," 215.

50 Nakano Masahiro, "Taiwan ni okeru arukōru sangyō," 212; Max and Delbrück, *Handbuch der Spiritusfabrikation*.

51 Zenda Naozō, "Jōryūshu jōzōyō toshite no 'amiro'-hō ni tsuite."

52 Katō Benzaburō, "Amirohō ni kansuru kenkyū."

53 Kitahara Kakuo, "Arukōru kōgyō ni okeru amiraaze," 240.

54 Iwate Hayami, "Honpō shusei kōgyō hattenshi"; Doi Shinji, "Saikin ni okonawa-retaru shusei ni kansuru kenkyū ni tsuite."

55 Nakano Masahiro, "Taiwan ni okeru arukōru sangyō," 210–15.

56 Suzuki Sadaharu, "Nihon no arukōru no rekishi," 150.

57 Hanada Kazō, "Nenryōyō shusei taisaku ni tsuite"; Katsume Suguru and Tanabe Osamu, "Saikin Itari ni okeru nenryō shusei kōgyō seisaku"; Nakamura Shizuka, "Shusei nenryō kokusaku jūritsu ni taisuru ikkōsatsu."

58 Gōdō shusei shashi hensan iinkai, *Gōdō shusei shashi*, 276; Iwai Ki'ichirō, "Honpō inryō shusei narabi ni gōsei seishu no hattenshi."

59 Suminoe Kinshi and Minato Chikamomo, "Mippeishiki shusei hakkō ni tsuite"; Sugiyama Shinsaku, "Mippei tanku ni yoru shusei hakkō."

60 Gōdō shusei shashi hensan iinkai, *Gōdō shusei shashi*, 102; Miyazaki Shizu, Shōchū, shusei sono hoka seishu nado no moromi no seizōhō (Manufacturing process of moromi of shōchū, alcohol and also sake), Japan Patent 110127, filed December 17, 1932, and issued March 29, 1935. Kohata Kengorō also worked to implement the amylo method using sweet potato at the alcohol company Dai-nippon Shurui. Gōdō shusei shashi hensan iinkai, *Gōdō shusei shashi*, 289.

61 Kinoshita Toshiaki, "Arukōru seizō gijutsu no shinpo," 302.

62 Kurono Kanroku, Takizawa Sumie, and Iwashita Nobuo, "Amiro hakkōhō o toku."

63 Gōdō shusei shashi hensan iinkai, *Gōdō shusei shashi*, 279–83; Honda Norimoto, "Ōshū ni okeru nenryōyō shusei konyō no kinkyō." For overviews of the debates

surrounding ethanol as a fuel additive or substitute in Japan, see Ōsaka jōzō gakkai, *Nenryō shusei kōenshū*; Nakamura Shizuka and Matsui Tōji, "Musui shusei ni kansuru mondai."

64 Kurono Kanroku, "Shōwa 13 nendo nenryō shusei kōgyō no gaisetsu"; Hirayama Yoichi, *Chōsen shuzō gyōkai yonjūnen no ayumi*, 107–9; Kurono Kanroku, "Honpō nenryō shusei kōgyō no genkyō."

65 Kurono Kanroku, "Honpō nenryō shusei kōgyō no genkyō"; Koike Yoshiyuki, "Senjika ni okeru shusei kōgyō ni tsuite."

66 Iwai Ki'ichirō, "Honpō inryō shusei narabi ni gōsei seishu no hattenshi."

67 Iwai Ki'ichirō, "Honpō inryō shusei narabi ni gōsei seishu no hattenshi."

68 Hirayama Yoichi, *Chōsen shuzō gyōkai yonjūnen no ayumi*, 78–79, 102–13. In 1908 the Tōyō Takushoku (Oriental Development) company was established in Korea with both public and private capital and management, to promote Japanese settlement and economic exploitation. It became one of the largest landholders in Korea and also had business operations in south China and Southeast Asia. Wartime semigovernmental "national policy companies" (*kokusaku kaisha*), which were established in the mid-1930s across the Japanese empire, were modeled on Tōyō Takushoku. See Duus, "Economic Dimensions of Meiji Imperialism," 160–61; Peattie, "Nanshin," 205–6.

69 Suzuki Sadaharu, "Nihon no arukōru no rekishi," 168.

70 Koizumi, "Biofuel Programs in East Asia," 214.

71 Jones, "Strategic Importance of Butanol," 262.

72 Stranges, "Synthetic Fuel Production."

73 Ethanol came to be widely used as a fuel additive and fuel substitute for pilot training in both the Army and Navy early in 1945. Jones, "Strategic Importance of Butanol," 262. On fuel shortages and Japanese strategy late in World War II, see Yergin, *The Prize*, chap. 18; Nolt, *International Political Economy*, 166; Jones, "Strategic Importance of Butanol," 261–63.

74 On Chaim Weizmann and acetone-butanol fermentation, see Bud, *Uses of Life*, 37–45.

75 Gōdō shusei shashi hensan iinkai, *Gōdō shusei shashi*, 340–43. For a more detailed account, see Jones, "Strategic Importance of Butanol."

76 Ōshima Yasuji, interview.

77 See, for example, Kōgyō gijutsuin hakkō kenkyūjo, *Hakkō kenkyūjo 25 nen no ayumi*, 26; Hyakushūnen kinen jigyōkai, *Hyakunenshi: Ōsaka daigaku kōgakubu jōzō, hakkō, ōyō seibutsu kōgakuka*, 69. Outside of fermentation approaches, during the desperate search for fuel in 1944–45 navy chemists initiated a project to extract high-octane fuel from pine root oil. This destroyed a substantial portion of Japan's pine forests, including those pine trees that were seen to be sacred and historic. Such projects contributed to the 15 percent of Japan's forests that were logged between 1941 and 1945. Tsutsui, "Landscapes in the Dark Valley," 299–300.

78   Iwate Hayami, "Honpō shusei kōgyō hattenshi."

79   Nakano Masahiro, "Taiwan ni okeru arukōru sangyō," 223.

80   The linkages between fermentation research and postwar biotechnology were underpinned by continuities in institutions and personnel. For example, Katō Benzaburō, who performed important studies comparing the enzyme systems of the *kōji* and amylo methods in the 1920s, later became the president of Kyōwa Hakkō Kōgyō, the company that critically pioneered the fermentative production of L-glutamic acid as described in chapter 6 of this book. Katō Benzaburō, "Katō Benzaburō." A number of the leading fermentation scientists of the postwar period, such as Sakaguchi Kin'ichirō, were engaged in research on alcoholic fermentations during the 1930s and 1940s. See, e.g., Takahashi and Sakaguchi, *Summaries of Papers*. Many of the institutions that supported research on alcoholic fermentations during the wartime period had a strong tradition in fermentation science and biotechnology generally; see, e.g., Hyakushūnen kinen jigyōkai, *Ōsaka daigaku kōgakubu jōzō, hakkō, ōyō seibutsu kōgakuka*; Kōgyō gijutsuin hakkō kenkyūjo, *Hakkō kenkyūjo 25 nen no ayumi*.

81   On images of China as the "sick man of Asia" and their global circulation, see Heinrich, *Afterlife of Images*. On the development of tropical medicine as an academic field, see Worboys, "Manson, Ross"; Tilley, *Africa as a Living Laboratory*, chap. 4. Calmette's colleague, Alexandre Yersin, as well as the Japanese bacteriologist Kitasato Shibasaburō were involved in the search for the plague microbe in Asia. Cunningham, "Transforming Plague"; Summers, *Great Manchurian Plague*.

82   Mendelsohn, "Cultures of Bacteriology," 35.

CHAPTER 5

1   Screening is a carefully designed search system for a microbial strain with specific properties, especially the ability to produce a chemical of interest in large quantities. In other words, screening is the process of culturing and assaying by which researchers identify and select a desired strain.

2   Nihon penishirin kyōkai, *Penishirin no ayumi*, 3.

3   Neushul, "Mass Production of Penicillin."

4   Umemura, *Japanese Pharmaceutical Industry*, chap. 3; Yongue, "Introduction of American Mass Production Technology"; Cozzoli, "Penicillin and the Reconstruction of Japan."

5   By 2002, Japanese laboratories had developed more than 117 useful antibiotics and other bioactive microbial metabolites. See Kumazawa and Yagisawa, "Antibiotics." On Japanese contributions to production methods and antibiotic resistance research, see, respectively, Bud, *Uses of Life*, and Creager, "Adaptation or Selection?"

6   It is easy to see any kind of mid-twentieth-century knowledge transfer from the United States to Japan as being part of a broader story of international develop-

ment with American visions and ideals at its core, not least because the Japanese example has often been invoked in the context of modernization theory. For a concise summary of the literature on American models of international development in science and technology during the Cold War, see Wolfe, *Competing with the Soviets*, chap. 4. Yet this would be a superficial reading of Japanese penicillin production in the early occupation years. The domestication of penicillin production was mostly achieved before the Cold War began to shape the occupation government's policies, and well before the height of influence of modernization theory. Moreover, it is well known that the occupation state—though centralized and authoritarian—was thin on the ground and relied on indirect rule through the existing bureaucracy. See Dower, *Embracing Defeat*, 212–13. The Ministry of Health and Welfare oversaw developments in the pharmaceutical industry.

7   C. Johnson, *MITI and the Japanese Miracle*.

8   Nishiyama, *Engineering War and Peace*; Setoguchi, "Control of Insect Vectors"; Iijima Wataru, *Mararia to teikoku*.

9   The conception of the problem of "making things work" comes from the Histories of Planning project led by Dagmar Schäfer at the Max Planck Institute for the History of Science.

10 D. Yang, *Technology of Empire*, chap. 4.

11 Mizuno, *Science for the Empire*; Mimura, *Planning for Empire*.

12 D. Yang, *Technology of Empire*, 158.

13 The environmental conditions during wartime have been detailed by William Tsutsui, who suggests that the impact of 1940s material scarcity on postwar practices would make an interesting question for further research: "The profound, often crippling wartime shortages of natural resources—especially fossil fuels—had the effect of driving many Japanese—from housewives to corporate engineers to university scientists—to new extremes of desperation, frugality, and creativity." Tsutsui, "Landscapes in the Dark Valley," 303.

14 Nihon penishirin gakujutsu kyōgikai kiji are published records from the JPRA meetings that are based on minutes kept by Yagisawa Yukimasa, the organization's managing director. The records are printed in the JPRA's *Journal of Penicillin*. Takeda Keiichi, *Penishirin sangyō kotohajime* reproduces large parts of these records with some annotation. Takeda's stated purpose is to draw historians' attention to the records as an important source.

15 On the inseparability of knowledge and institutions, see Lécuyer, *Making Silicon Valley*; Mody, *Instrumental Community*; A. Johnson, *Hitting the Brakes*; Akera, *Calculating a Natural World*.

16 More recently, historians have begun to address the historiographical imbalance—which has favored synthetic organic chemistry—in order to better reflect the importance of biological research to the design, production, and standardization of pharmaceuticals. Studies have considered the significance of "biologics" as a conceptual category in research and regulation. Drugs made

from living organisms also included vaccines and sera. See Gaudillière, "Drug Trajectories," and related articles in the issue; Schwerin, Stoff, and Wahrig, *Biologics*.

17 Bud, *Penicillin*.

18 On how knowledge created at the laboratory bench is scaled up for mass production in industrial fermentation processes, see V. Lee, "Scaling Up from the Bench."

19 Nihon penishirin kyōkai, *Penishirin no ayumi*, 19.

20 Aldous and Suzuki, *Reforming Public Health*, 101–2.

21 Dower, *Embracing Defeat*.

22 Dower, *Embracing Defeat*, 69. The constitution, rewritten in the first years of the occupation, reduced the emperor from an absolute authority to a symbol of the state, and forbade the country from going to war, as well as guaranteeing new civil liberties. The push for democratization and demilitarization lasted briefly until the onset of the Cold War. With the Communist victory in China in 1949 and the outbreak of the Korean War (1950–53), American policy changed to support a conservative order and economic growth in Japan, a trend that would persist beyond the end of the occupation in 1952.

23 Dower, *Embracing Defeat*, 103.

24 On occupation-era public health, see also Sensui Hidekazu, "Beikoku shiseika Ryūkyū no kekkaku seiatsu jigyō."

25 Nihon penishirin gakujutsu kyōgikai kiji I, 57. Here, the Roman numeral refers to the number of the report, while the Arabic numeral refers to the page number within the journal. For further explanation of the records as a source, see note 14 above.

26 Nihon penishirin kyōkai, *Penishirin no ayumi*, 2.

27 Bud, *Penicillin*, 83.

28 General Headquarters, Supreme Commander for the Allied Powers, Public Health and Welfare Section, *Mission and Accomplishments of the Occupation in the Public Health and Welfare Fields*, 23.

29 Umezawa Hamao, *Kōseibusshitsu o motomete*, 13.

30 Tsunoda Fusako, *Hekiso*; Mizoguchi, "Penicillin Production," 551.

31 Nishiyama, *Engineering War and Peace*, 85–104.

32 Kim, "Yoshio Nishina and Two Cyclotrons."

33 Mizoguchi, "Penicillin Production," 548. The prefectural government authorized pharmacies, clinics, and other dealers to receive rations via a purchasing passbook. Umemura, *Japanese Pharmaceutical Industry*, 35n35.

34 Sumiki Yusuke, *Kōseibusshitsu*, 162; Nihon penishirin kyōkai, *Penishirin no ayumi*, 19.

35 Nihon penishirin gakujutsu kyōgikai kiji I, 60; Nihon penishirin gakujutsu kyōgikai kiji II, 125; Nihon penishirin gakujutsu kyōgikai kiji III, 189.

36 Nihon penishirin gakujutsu kyōgikai kiji II, 125–28.

37 General Headquarters, Supreme Commander for the Allied Powers, Public Health and Welfare Section, *Mission and Accomplishments of the Occupation in the Public Health and Welfare Fields*, 22.

38 For example: "The calendar year 1949 may be described in summary as a period of transition from postwar activities emphasizing quantity, to a period with primary emphasis on quality improvement" (118). "As a result of the rehabilitation of the pharmaceutical and medical supply industries, the volume of these imports has rapidly decreased, with commensurate savings in the cost of the occupation of the American taxpayer" (121). "Production of penicillin during the year exceeded all expectations. . . . The volume of production justified the removal of penicillin from ration distribution controls in April 1949, and was the causative factor in reducing prices on all penicillin products by approximately 50% as of 1 October. It was necessary to import 125,000 gallons of corn steep liquor for penicillin production during the year" (123). General Headquarters, Supreme Commander for the Allied Powers, Public Health and Welfare Section, *Public Health and Welfare in Japan, Annual Summary 1949, Volume I*. "During 1950, continued advances were realized which have resulted in a product of proven quality with a decided reduction in price, making the Japanese product a factor in international trade" (83). General Headquarters, Supreme Commander for the Allied Powers, Public Health and Welfare Section, *Public Health and Welfare in Japan, Annual Summary 1950, Volume I*.

39 Neushul, "Mass Production of Penicillin"; Swann, "Search for Synthetic Penicillin"; Rasmussen, "'Small Men,' Big Science."

40 Nihon penishirin kyōkai, *Penishirin no ayumi*, 113.

41 Yongue, "Introduction of American Mass Production Technology," 224–25. There the JPMA is referred to as the JPA, or Japan Penicillin Association.

42 Yongue, "Introduction of American Mass Production Technology."

43 Nihon penishirin gakujutsu kyōgikai kiji II, 123.

44 Jackson W. Foster, preface, *Journal of Penicillin* 1 (1946).

45 Nihon penishirin gakujutsu kyōgikai kiji II, 123.

46 Formerly Tokyo Imperial University, the University of Tokyo was renamed in September 1947. For more on its Department of Agricultural Chemistry as a center of fermentation expertise, see chapter 1 of this book; Kumazawa Kikuo, "Riibihi to Nihon no nōgaku."

47 Nihon penishirin gakujutsu kyōgikai kiji II, 123.

48 Nihon penishirin gakujutsu kyōgikai kiji II, 125.

49 Sumiki Yusuke, *Kōseibusshitsu*, 166.

50 Nihon penishirin gakujutsu kyōgikai kiji II, 127.

51 Supreme Commander for the Allied Powers, *Summation of Non-Military Activities in Japan*, no. 15 (December 1946): 226.

52 Until then, the University of Tokyo's Institute of Infectious Diseases had been assaying its own vaccines. It produced about half of the total Japanese manufacture of vaccines, inoculation materials, and serums. The Kitasato Institute— a private medical laboratory founded by Kitasato Shibasaburō in 1914—also made serums, but since it was a private laboratory, its biological products had to be approved by the Institute of Infectious Diseases. GHQ found it odd that the same institute that made vaccines had the responsibility of approving them. In the end, GHQ decided that the new Japanese NIH would be responsible for assays, and it moved about half of the facilities and members of the Institute of Infectious Diseases there, including the antibiotics section. Umezawa, *Kōseibusshitsu o motomete*, 38–39.

53 Nihon penishirin gakujutsu kyōgikai kiji I, 59. The dilution method involved placing a series of increasingly diluted samples of the antibiotic in media in tubes or plates, inoculating and incubating the samples with an organism, and then deducing the samples' potency from the decrease in the organism's growth across the series. The cup method involved placing a cup filled with the antibiotic into a solid medium seeded with an organism, incubating it, and then measuring the size of the zone where the organism's growth was inhibited around the cup.

54 Nihon penishirin gakujutsu kyōgikai kiji IV, 252.

55 Nihon penishirin gakujutsu kyōgikai kiji II, 128.

56 Nihon penishirin gakujutsu kyōgikai kiji II, 126; Iijima Takashi, *Nihon no kagaku gijutsu*, 133.

57 Nihon penishirin gakujutsu kyōgikai kiji II, 123.

58 Iijima Takashi, *Nihon no kagaku gijutsu*, 132–35; Nihon penishirin gakujutsu kyōgikai kiji III, 191.

59 Nihon penishirin gakujutsu kyōgikai kiji III, 192; Nihon penishirin gakujutsu kyōgikai kiji IV, 253; Nihon penishirin gakujutsu kyōgikai kiji VI, 407.

60 Nihon penishirin gakujutsu kyōgikai kiji III, 191–92; Nihon penishirin gakujutsu kyōgikai kiji IV, 255.

61 Nihon penishirin gakujutsu kyōgikai kiji III, 190.

62 Nihon penishirin gakujutsu kyōgikai kiji VIII, 557; Iijima Takashi, *Nihon no kagaku gijutsu*, 135.

63 Ōyama Yoshitoshi, "Kaken to penishirin puranto," 64; Iijima Takashi, *Nihon no kagaku gijutsu*, 134.

64 Nihon penishirin gakujutsu kyōgikai kiji IX, 623.

65 Nihon penishirin gakujutsu kyōgikai kiji IX, 625–26. Physicians reported that the incidence of side effects had lessened after the product potency increased from about 300 u/mg in September 1947 to 800 u/mg in December 1947.

66 Nihon penishirin gakujutsu kyōgikai kiji X, 75. According to physicians, there were usually no side effects apart from smarting and fever, but occasional side effects included headache and vomiting. Nihon penishirin gakujutsu kyōgikai kiji III, 190. At a February 1948 meeting, physicians reported that about 30 percent of patients experienced side effects from injection of penicillin into the muscle. Nihon penishirin gakujutsu kyōgikai kiji X, 75.

67 See, for example, Mimura, *Planning for Empire*; Mizuno, *Science for the Empire*, 47–49.

68 Capocci, "Chain is Gonna Come."

69 Chapter 1; Morris-Suzuki, *Technological Transformation*, 98–103; Sugihara, "Development of an Informational Infrastructure."

70 Nihon penishirin gakujutsu kyōgikai kiji I, 57.

71 Nihon penishirin gakujutsu kyōgikai kiji I, 59.

72 Gaudillière and Gausemeier, "Molding National Research Systems."

73 Chapter 4.

74 Bud, "Deep Fermentation and Antibiotics."

75 Sakaguchi Kin'ichirō, "Michi e no gunzō"; Takahashi and Sakaguchi, *Summaries of Papers*.

76 Bud, "Deep Fermentation and Antibiotics," 332.

77 On screening, see V. Lee, "Bioorganic Synthesis."

78 Nihon penishirin gakujutsu kyōgikai kiji II, 12; Nihon penishirin gakujutsu kyōgikai kiji II, 127; Nihon penishirin gakujutsu kyōgikai kiji III, 189; Nihon penishirin gakujutsu kyōgikai kiji VII, 485.

79 Nihon penishirin gakujutsu kyōgikai kiji XI, 146.

80 Chapter 1.

81 Nihon penishirin gakujutsu kyōgikai kiji XI, 146.

82 Nihon penishirin gakujutsu kyōgikai kiji XII, 336–38.

83 Nihon penishirin gakujutsu kyōgikai kiji XIII, 413.

84 Chapter 4.

85 Sumiki Yusuke, *Kōseibusshitsu*, 177.

86 Nihon penishirin gakujutsu kyōgikai kiji II, 126.

87 Nihon penishirin gakujutsu kyōgikai kiji II, 125.

88 Nihon penishirin gakujutsu kyōgikai kiji IV, 253.

89 Nihon penishirin gakujutsu kyōgikai kiji VI, 407.

90 Nihon penishirin gakujutsu kyōgikai kiji II, 127.

91 Komagata Kazuo, interview by author. An undergraduate student in agricultural

chemistry at the Morioka College of Forestry and Agriculture during and just after the war, Komagata began his career as a research student at the fermentation laboratory (directed by Sakaguchi Kin'ichirō and Arima Kei) at the Faculty of Agriculture, University of Tokyo, in 1948. He moved to the Institute of Applied Microbiology (working with Iizuka Hiroshi) in 1954, and to the Central Research Laboratory of the Ajinomoto company in 1960. Komagata, "Microbial Systematics."

92   Sumiki Yusuke, *Kōseibusshitsu*, 226.

93   Nihon penishirin gakujutsu kyōgikai kiji II, 125.

94   Nihon penishirin gakujutsu kyōgikai kiji VI, 406.

95   Tanaka Hideo, "Hakkōsō, baiyō sōchi," 27.

96   Nihon penishirin gakujutsu kyōgikai kiji IV, 254.

97   Nihon penishirin gakujutsu kyōgikai kiji II, 127.

98   Nihon yakushi gakkai, *Nihon iyakuhin sangyōshi*, 97.

99   Takeda Keiichi, *Penishirin sangyō kotohajime*, 256–57.

100   Ōyama Yoshitoshi, "Kaken to penishirin puranto," 60–67.

101   Gaudillière and Gausemeier, "Molding National Research Systems."

102   Nihon penishirin gakujutsu kyōgikai kiji XV, 66; Nihon penishirin gakujutsu kyōgikai kiji IX, 623.

103   Nihon penishirin gakujutsu kyōgikai kiji XII, 336.

104   Nihon penishirin gakujutsu kyōgikai kiji XXVI, 747.

105   Condrau and Kirk, "Negotiating Hospital Infections."

106   Nihon penishirin gakujutsu kyōgikai kiji XV, 637; Nihon penishirin gakujutsu kyōgikai kiji XXXIII, 275.

107   Combined with the falling price of penicillin due to competition, the change saw JPMA membership dip from its peak of seventy-two firms in 1946 to nineteen in 1951, and the closure of the JPMA by 1961, even though the JARA continued to exist; Yongue, "Introduction of American Mass Production Technology," 224–25.

108   Umemura, *Japanese Pharmaceutical Industry*, 39.

109   Fujii, "Changes in Antibiotic Consumption."

110   On antibiotic research at the NIH, see Umezawa Hamao, *Kōseibusshitsu o motomete*, 38–39.

111   A few examples include kanamycin, discovered at the Institute of Applied Microbiology at the University of Tokyo and commercialized by Kyōwa Hakkō Kōgyō; bleomycin, discovered at the Institute of Microbial Chemistry and commercialized by Nippon Kayaku; and avermectin, developed at the Kitasato Institute and commercialized by Kyōwa Hakkō Kōgyō.

112 Foster, "View of Microbiological Science," 445; Umemura, *Japanese Pharmaceutical Industry*, 71. Compare Santesmases, "Gender in Research and Industry," 72.

113 Omura, "Philosophy of New Drug Discovery."

114 Santesmases, "Screening Antibiotics"; Santesmases, "Gender in Research."

115 Santesmases, "Screening Antibiotics," 413.

116 Foster, "View of Microbiological Science," 446.

117 Komagata Kazuo, interview; Tōkyō daigaku ōyō biseibutsu kenkyūjo, *Jūnen no ayumi*. Similarly, research associations modeled on the JPRA and supported by funding from state ministries, where scientists from different sites in academia, government, and industry shared results, came to play an important role in the development of other fields beyond antibiotic science. Hashimoto, "Technological Research Associations."

118 Ōyama Yoshitoshi, "Kaken to penishirin puranto," 61.

119 Kitasato Institute, "Splendid Gifts from Microorganisms."

120 Fujii, "Changes in Antibiotic Consumption," 2261.

121 Russell, *War and Nature*, 2.

CHAPTER 6

1 Ōshima's comment regarding 1950s and 1960s fermentation science compares it to much later developments in plant breeding. In 1997 and 2004 the Japanese brewing and distilling company Suntory, in collaboration with the Australian biotech venture Florigene, produced the world's first blue carnations and blue roses by genetic engineering.

2 On the broader context of the transition between the wartime, occupation, and postwar market economies, see Okazaki and Okuno-Fujiwara, *Japanese Economic System*. Policy decisions at the national level to focus on promoting economic growth through international trade, rather than domestic self-sufficiency, were linked to the energy transition of Japanese industries from coal to oil in the 1950s. Hein, *Fueling Growth*; Kobori Satoru, *Nihon no enerugii kakumei*. For a recent account of the postwar Japanese economy, see Metzler, *Capital as Will and Imagination*.

3 For several case studies from the heavy industrial and electronics sectors, see Morris-Suzuki, *Technological Transformation*, 182–202. A historical literature on postwar industrial associations discusses the importance of the associations' role for disseminating information, which was distinctive when compared with their counterparts in Europe and the United States, though such studies focus primarily on industrial policy rather than technological change. Yonekura, "Functions of Industrial Associations." Another body of literature examines government support after 1961 for research associations. Unlike their British counterparts, most Japanese research associations were established to solve specific technological problems and dissolved afterward, rather than serving

as a general industry-wide forum, and they involved large instead of small firms. Odagiri and Goto, *Technology and Industrial Development*, 51–56; Goto, "Co-operative Research."

4   Morris-Suzuki, *Technological Transformation*, 186.

5   Takehiko Hashimoto has pointed to the important function of university engineers and academic institutions in Japan's postwar industrial development, as they mediated information between engineers both industrial and academic. Hashimoto, "Hesitant Relationship Reconsidered," 189. Generally, however, other scholars have tended to center their studies of Japanese technology and innovation on the company as a unit of analysis, and on patents as a measurement of innovation, without analyzing scientific developments. As a result, their studies miss the interactions between companies engaging in scientifically related fields. Further, their studies have generally treated technological change in separately defined market sectors (such as pharmaceuticals or food) as independent of each other, and either neglect the strong connections between them or mention them only in passing. For examples focusing on the biomedical industries, see Kneller, "Intellectual Property Rights"; Kneller, *Bridging Islands*; Odagiri and Goto, *Technology and Industrial Development*, chap. 11; Kodama, "Japan's Unique Capacity to Innovate," 154-56.

6   My analysis thus takes the suggestion of Secord, "Knowledge in Transit," to see science as a form of communication, rather than as something to be placed into a bounded and highly localized cultural context. This allows us to move away from the dual images of "creation" and "circulation" of knowledge, and instead to see how information "acting at a distance" can tie together diverse contexts— in this case, between academic and industrial laboratories, and between Japan and global developments.

7   This was comparable to the role of scanning tunneling microscopy in nanotechnology, which Cyrus Mody presents as linking together a pure and applied research community in *Instrumental Community*. Similarly, this chapter's narrative is centered on how one technology mediated the dynamics of the research community, and why those dynamics encouraged the flourishing of the technology across different institutional contexts. The amino and nucleic acid fermentation community also displays many features of a "knowledge community" as elaborated by Ann Johnson in *Hitting the Brakes*. These include the central importance of ideas and of communication of ideas in the design of artifacts, the community's characterization by "attractors" (appealing problems) that draw together a multidisciplinary group of practitioners, and the community's co-production with their objects of study.

8   On the importance of skill as an industrial resource, see Lécuyer, *Making Silicon Valley*. The literature emphasizing the role of technology importation from overseas in Japan's postwar development is too large to cite here, but for examples from the historiography of pharmaceuticals, see chapter 5.

9   Watson, *Double Helix*; Olby, *Path to the Double Helix*; Judson, *Eighth Day of Creation*; Kay, *Who Wrote the Book of Life?*; Morange, *Black Box of Biology*.

10 On metabolism in the historiography of modern biology, see Grote and Keuck,

"Stoffwechsel." For an ethnographic account of metabolic engineering in the twenty-first-century United States, see Roosth, *Synthetic*.

11   In the 1880s, the Meiji government sold off their capital-intensive enterprises, which were unprofitable, to ownership that was ostensibly private but which continued to benefit from strong ties to the government. Morris-Suzuki, *Technological Transformation*, 77–80; T. Nakamura, *Economic Growth in Prewar Japan*, 62. On wartime *zaibatsu* and technocracy, see Mimura, *Planning for Empire*. Following the occupation authorities' (1945–52) moves to dismantle the *zaibatsu*, former *zaibatsu* regrouped more loosely as *keiretsu*. Typically, *zaibatsu* conglomerates were characterized by a holding company at their apex, whereas *keiretsu* linked businesses together by cross-shareholding, with a bank at the top instead of a holding company. On cross-shareholding, see Milhaupt, "Japanese Corporate Governance." In the postwar Japanese economy, national policy encouraged businesses to compete for market share by improving quality rather than lowering prices; Hein, "Growth versus Success," 108–9.

12   For a similar argument pertaining to a range of engineering fields, see Hashimoto, "Hesitant Relationship Reconsidered."

13   On state encouragement of domestic consumption, in contrast to the savings campaigns of the wartime and prewar periods, see Hein, "Growth versus Success," 112–15. For another study of continuity in corporate research between peacetime and wartime, see Ndiaye, *Nylon and Bombs*.

14   On pre-1950s austerity techniques, especially during wartime, see chapters 2, 4, and 5. The argument for strong continuities between the postwar economic structure and wartime transformations has been made by Tetsuji Okazaki and others, and I make a similar argument in this chapter for scientific expertise. See Okazaki and Okuno-Fujiwara, *Japanese Economic System*. My characterization of postwar high-tech industries in Japan as labor-intensive rather than capital-intensive would not be unique to the fermentation sector. For the semiconductor, steel, and automobile sectors, business historians have emphasized Japanese firms' reliance on "high investments in R&D and in plant and equipment coupled with low cost of capital, close interfirm relationships, and management systems and industrial relations conducive to skill formation." Goto, introduction to *Innovation in Japan*, 9. For an exploration of the concept of labor-intensive industrialization in the prewar period, but one that focuses on small-scale industries rather than the large corporations I address here, see Tanimoto, "Peasant Economy to Urban Agglomeration."

15   Rasmussen, "Mid-Century Biophysics Bubble"; Reinhardt, *Shifting and Rearranging*. For comparisons between the American and European contexts, see Creager, "Building Biology across the Atlantic"; Gaudillière, *Inventer la biomédecine*; de Chadarevian, *Designs for Life*; Strasser, *La fabrique d'une nouvelle science*.

16   Two classic accounts are C. Johnson, *MITI and the Japanese Miracle*; and Vogel, *Japan as Number One*. For a more recent examination, see O'Bryan, *Growth Idea*.

17   Francks, *Japanese Consumer*; George, *Minamata*. Notably, trade unions and consumer groups were excluded from the productive information networks

that linked government ministries with public research institutions and private industry, as observed by Morris-Suzuki in *Technological Transformation*, 187.

18 The role of scientific expertise in Japan's high growth is understudied. For an account that emphasizes the importance of in-house engineers in shaping product development within corporations, see Choi, "Technology Importation, Corporate Strategies." On the significance of academic engineers in industrial R&D due to information networks, see Hashimoto, "Hesitant Relationship Reconsidered"; Hashimoto, "Technological Research Associations"; Hashimoto, "Corporations, Universities, and the Government."

19 Jardine, Kowal, and Bangham, "How Collections End." Specifically, microbial type cultures were living collections, and every strain within needed constant reculturing to stabilize its identity. On the American Type Culture Collection (ATCC), see Strasser, *Collecting Experiments*, 33–40.

20 Hasegawa was the director of the collections of the Institute for Fermentation, Osaka (IFO). Communications at the time made it difficult to obtain strains from overseas collections for comparison. Instead, he resorted to looking in domestic laboratories for preserved strains that were related to the original microbes mentioned in past scientific reports, and he compared these strains with each other. Hasegawa Takeji, "Nihon no biseibutsukabu hozon jigyō 2," 55–56.

21 Similarly, Bruno Strasser argues that "the constitutions of collections accompanied and supported the growth of a research community," in *Collecting Experiments*, 41. The Japanese Federation of Culture Collections of Microorganisms (JFCC) was established in 1951 as a national organization. It included seven institutional collections of industrial microorganisms and four institutional collections of pathogenic microorganisms. Hasegawa, "Japanese Culture Collections," 142.

22 Chapter 1.

23 Nakase Takashi, in discussion.

24 Saitō Kendō in Manchuria: chapter 1; Nakazawa Ryōji in Taiwan: chapters 3 and 4; wartime fuel fermentation: chapter 4.

25 The CLMR strains followed Saitō Kendō when he returned to the Osaka Higher Technical School in 1927, and Naganishi Hirosuke when he moved to the Hiroshima Higher Technical School in 1930. Hasegawa Takeji, "Nihon no biseibutsukabu hozon jigyō 1," 5–6.

26 The Aeronautic Fermentation Institute also held copies of the culture collections of the Faculty of Agriculture, Tokyo Imperial University. Copies of the GRIF strains had followed Nakazawa Ryōji, who had led fermentation research at the Taiwan Government-General Central Laboratory before retiring and returning to Japan in 1939, working first as a consultant at Takeda Yakuhin, and then from 1944 as the inaugural director of the Aeronautic Fermentation Institute. Hasegawa Takeji, "Nihon no biseibutsukabu hozon jigyō 1," 6.

27 Takeda nihyakunenshi hensan iinkai, *Takeda nihyakunenshi*, 1081–82.

28 Chapter 5.

29 On the expansion of research laboratories in the private sector in the 1950s and 1960s, see Nakayama, "Central Laboratory Boom."

30 At the end of 1965, the Institute for Fermentation, Osaka (IFO) held 3,536 molds, 1,571 yeasts, 686 bacteria, and 196 actinomycetes. Hasegawa, "Japanese Culture Collections," 142, table 2. This was in contrast to the American Type Culture Collection (ATCC) and British National Collection of Type Cultures (NCTC), where bacteria predominated.

31 On the origins of protein and nucleic acid sequencing in experimental chemistry, see García-Sancho, *History of Molecular Sequencing*. On microbial screening practices, see V. Lee, "Bioorganic Synthesis."

32 Strasser, *Collecting Experiments*, 258. For two classic accounts of biology based on experiments and collections, respectively, see Rader, *Making Mice*; and Nyhart, *Biology Takes Form*.

33 In Japanese fermentation science, we find little trace of the gendered distinctions setting collections apart from the laboratory that were notable in postwar experimental biology elsewhere. Strasser, *Collecting Experiments*. Moreover, despite their interest in biodiversity, Japanese fermentation scientists paid scant attention to the evolutionary issues that were prominent in the United States and Britain in the same period, since microbiology was marginalized in evolutionary biology until the 1980s. Smocovitis, *Unifying Biology*; Sapp, "Microbial Evolutionary Theory." They were inspired instead by such works as Stanier, Doudoroff, and Adelberg, *The Microbial World*, according to Beppu Teruhiko, in discussion.

34 Udaka Shigezō, "Aminosan seisankin no hakken," 62; Udaka Shigezō, "Aminosan hakkō no reimeiki," 211.

35 Kyōwa Hakkō received permission from the occupation authorities to import molasses from the Philippines as raw material for factories to make solvents, in order to support domestic penicillin manufacture; Kyōwa hakkō sōritsu 50 shūnen shashi hensan iinkai, *Sore kara sore e*, 10–22. The occupation government was opposed in principle to state-directed allocation of raw materials, but faced with a lack of alternatives they nonetheless exerted this control, first via industrial associations and then through government ministries. Okazaki and Okuno-Fujiwara, *Japanese Economic System*, 31–32. Kyōwa Hakkō was permitted to buy a license for streptomycin technology from Merck in 1951. This was significant because, unlike penicillin, streptomycin was protected by patents. Success in streptomycin sales helped the company expand through the early 1950s. Kyōwa hakkō sōritsu 50 shūnen shashi hensan iinkai, *Sore kara sore e*, 33–34, 37–39.

36 On the broader context of Japanese scientists studying abroad in the 1950s, see Nakayama, "Overseas Study Leave." On how changes in communications technologies affected scientific research more generally from the 1950s, see Nakayama, "International Exchange."

37 Udaka Shigezō, interview.

38 Kinoshita Shukuo, "Aminosan hakkō no tanjō," 3–9.

39 Direct glutamic acid production by a microbe was not the only method that Udaka pursued. He also explored fermenting α-ketoglutaric acid (as seen in the latest literature of the time) and converting it to glutamic acid, as well as hydrolyzing the highly viscous, mucoidal polymer of glutamate produced by *nattō* (fermented soybean) microbes, but neither yielded success as possible efficient methods for glutamic acid production. Udaka Shigezō, "Aminosan seisankin no hakken," 62–70; Udaka Shigezō, "Aminosan hakkō no reimeiki," 211–14.

40 Udaka Shigezō, "Aminosan seisankin no hakken," 75–76. Kinoshita presented it differently, emphasizing his own foresight in hiring Udaka "fresh" from Chicago and armed with the latest methods. Kinoshita Shukuo, "Aminosan hakkō no tanjō," 11; Kinoshita Shukuo, "Aminosan hakkō kōgyō," 6. According to Udaka, at Chicago there were many scientists who handled microbes and taught microbiology, but only one professor who researched in that field. The overall focus was rather in biochemistry, especially as it related to medicine. Udaka Shigezō, interview.

41 Udaka Shigezō, interview.

42 Chapter 5.

43 Udaka Shigezō, "Aminosan seisankin no hakken," 73.

44 Udaka Shigezō, interview.

45 Udaka Shigezō, interview. Udaka was in middle school when the war ended. Comparable are Takada Ryōhei's views on flavorings, detailed in chapter 2.

46 Udaka Shigezō, "Aminosan hakkō no reimeiki," 211. The argument that responses to scarcity in the immediate postwar period left significant legacies in present-day Japanese food culture has also been made by Solt, *Untold History of Ramen*; and Arch, "Whale Meat."

47 Ajinomoto kabushiki gaisha, *Aji o tagayasu*, 241.

48 Chapter 2; Udaka Shigezō, "Aminosan seisankin no hakken," 61.

49 Kinoshita Shukuo, "Aminosan hakkō no tanjō," 16.

50 Kinoshita Shukuo, "Aminosan hakkō no tanjō," 17.

51 For details on the method and the patent, see Ajinomoto kabushiki gaisha, *Ajinomoto guruupu no hyakunen*, 274–75.

52 Kinoshita Shukuo, "Aminosan hakkō no tanjō," 19; Udaka Shigezō, interview.

53 Kinoshita Shukuo, "Aminosan hakkō no tanjō," 35.

54 Ajinomoto kabushiki gaisha, *Ajinomoto guruupu no hyakunen*, 276–78.

55 Chapter 5; Imada Akira, interview.

56 Udaka Shigezō, "Aminosan seisankin no hakken," 72. Udaka was not allowed to publish his screening method for years until he obtained permission from Kinoshita in 1959 and published it as Udaka, "Screening Method for Microorganisms."

57 Kinoshita Shukuo, "Aminosan hakkō no tanjō," 22.

58 Tamiya, "The Koji."

59 For the report, see Kinoshita et al., "Glutamic Acid Fermentation."

60 Kinoshita Shukuo, "Aminosan hakkō no tanjō," 22.

61 Kinoshita Shukuo, "Aminosan hakkō no tanjō," 36–37.

62 Udaka Shigezō, "Aminosan seisankin no hakken," 76.

63 Udaka Shigezō, interview.

64 Udaka Shigezō, "Aminosan seisankin no hakken," 77. On MSG fermentation in the United States, see Tracy, "Tasty Waste."

65 Udaka Shigezō, interview.

66 Kyōwa hakkō sōritsu 50 shūnen shashi hensan iinkai, *Sore kara sore e*, 75.

67 Kevles, "Ananda Chakrabarty Wins a Patent." Thanks to Emily Pawley for the observation.

68 Udaka Shigezō, interview; Kinoshita Shukuo, "Aminosan hakkō no tanjō," 14.

69 O'Malley, *Philosophy of Microbiology*, chap. 3.

70 Ōshima Yasuji, interview.

71 Kinoshita Shukuo, "Aminosan hakkō no tanjō," 31.

72 Kinoshita Shukuo, *Zoku-zoku-zoku anohi anogoro*, 73. Compare the similar triumph of large-capital manufacturers in the instant noodle market, though in that case the trend was driven by patent litigation rather than technological change. Solt, *Untold History of Ramen*, 101–3.

73 T. D. Brock, *Emergence of Bacterial Genetics*, 34; Kinoshita Shukuo, "Aminosan hakkō no tanjō," 26.

74 Udaka Shigezō, "Aminosan hakkō no reimeiki," 214–15. See also Udaka and Kinoshita, "Studies on L-Ornithine Fermentation I"; Udaka and Kinoshita, "Studies on L-Ornithine Fermentation II."

75 Kinoshita Shukuo, "Aminosan hakkō no tanjō," 25; Beppu Teruhiko, in discussion; Miwa Kiyoshi, "Hakkō biseibutsu." This could be characterized as a special emphasis on metabolism over genetics in early Japanese biotechnology, in contrast to the context of early genetic engineering in the United States as depicted in Curry, *Evolution Made to Order*. On metabolic engineering and synthetic biology today, see Marlière, "The Farther, the Safer," with thanks to Bernadette Bensaude-Vincent for the reference; Campos, "Synthetic Biology That Was"; Roosth, *Synthetic*.

76 They were cited in Umbarger's presentation there; Umbarger, "Feedback Control by Endproduct Inhibition." In the history of molecular biology, this symposium is especially well known for François Jacob and Jacques Monod's presentation of the operon model. On the operon model, see Morange, *Black Box of Biology*.

77 Kay, *Who Wrote the Book of Life?*, 202.

78 T. D. Brock, *Emergence of Bacterial Genetics*, 312–14; Creager and Gaudillière, "Meanings in Search of Experiments."

79 Kuninaka Akira and Inemori Kazuo, "Yamasa ni okeru kakusankei chōmiryō, iyakuhin no rekishi to shōyu gijutsu no shinpo," 13.

80 In this period the Japanese patent system worked by priority of application, not by priority of discovery as in the United States.

81 Kuninaka Akira and Inemori Kazuo, "Yamasa ni okeru kakusankei chōmiryō, iyakuhin no rekishi to shōyu gijutsu no shinpo," 13–15.

82 Ajinomoto kabushiki gaisha, *Ajinomoto guruupu no hyakunen*, 289–92.

83 Takeda nihyakunenshi hensan iinkai, *Takeda nihyakunenshi*, 625–26; Kinoshita Shukuo, *Zoku-zoku-zoku anohi anogoro*, 71.

84 Takeda nihyakunenshi hensan iinkai, *Takeda nihyakunenshi*, 627.

85 Kinoshita Shukuo, *Zoku-zoku-zoku anohi anogoro*, 71; Kuninaka Akira and Inemori Kazuo, "Yamasa ni okeru kakusankei chōmiryō, iyakuhin no rekishi to shōyu gijutsu no shinpo," 16.

86 Kinoshita Shukuo, "Aminosan hakkō no tanjō," 56.

87 Kinoshita Shukuo, "Aminosan hakkō no tanjō," 56.

88 Kinoshita Shukuo, *Zoku-zoku-zoku anohi anogoro*, 75–76.

89 Komagata Kazuo, interview.

90 Imada Akira, interview; Komagata Kazuo, interview.

91 Sakaguchi Kin'ichirō had succeeded his teacher, Takahashi Teizō, at Tōdai's department of agricultural chemistry, and his career had spanned the war and a wide range of subjects in microbiology. His long career would continue late into the twentieth century.

92 Ōshima Yasuji, interview; Tomita Fusao, in discussion; Yagisawa Morimasa, in discussion.

93 For prewar and wartime examples, see chapters 4 and 5; Mizuno, *Science for the Empire*, 47–49, on the Continental Science Institute; Mimura, *Planning for Empire*.

94 Tōkyō daigaku ōyō biseibutsu kenkyūjo, *Jūnen no ayumi*, 1.

95 Sakaguchi Kin'ichirō, "Shokan" (Impressions), in Tōkyō daigaku ōyō biseibutsu kenkyūjo, *Jūnen no ayumi*.

96 Komagata Kazuo, interview. See also the descriptions of the IAM's areas of research in Tōkyō daigaku ōyō biseibutsu kenkyūjo, *Jūnen no ayumi*.

97 Chapter 5.

98 Imada Akira, interview.

99 For example, see Morris-Suzuki, *Technological Transformation*, 180, 186; Yonekura, "Functions of Industrial Associations."

100 Komagata Kazuo, interview. See also the contents of the Amino Acid Fermentation Association's journal, *Hakkō to taisha*, 1–5 (1959–62), and note the contrast between articles over the first two years and subsequent articles.

101 Imada Akira, interview; Udaka Shigezō, interview.

102 Kinoshita Shukuo, "Aminosan hakkō no tanjō," 26; Udaka Shigezō, interview.

103 Komagata Kazuo, interview. Compare American plant breeders' attempts to make new varieties to demand around the same period, detailed in Curry, *Evolution Made to Order*.

104 *Kinzuka ni yosete*.

105 Hasegawa Takeji, "Nihon no biseibutsukabu hozon jigyō," 55–56; Higher Education and Science Bureau, Ministry of Education, Japan, *A General Catalogue of the Cultures of Micro-organisms Maintained in the Japanese Collections*. The IFO culture collections were transferred to the National Biological Resource Center (NBRC) of the National Institute of Technology and Evaluation (NITE) in 2002.

106 This picked up a thread for promoting international preservation of strains that had begun in 1946, in the wake of worldwide damage to microorganisms during the war, but which had been dropped in the 1950s. Hasegawa, "Japanese Culture Collections," 141–43.

107 Hyakushūnen kinen jigyōkai, *Hyakunenshi: Ōsaka daigaku kōgakubu jōzō, hakkō, ōyō seibutsu kōgakuka*, 66. The department changed its name in 1943. Acetone was a key industrial solvent, while butanol was used in airplane fuel. See chapter 4.

108 Yamamura Tazaemon, "Shuzōgyō hattatsu no katei to kongo no kadai"; Komagata Kazuo, interview.

109 Ōshima Yasuji, interview. On the Brewing Society yeasts ("Kyōkai yeasts"), see chapter 1.

110 Ōshima Yasuji, interview.

111 Kitamoto Katsuhiko, in discussion.

112 Ōshima Yasuji, interview.

113 Ōshima Yasuji, interview. See also chapter 4.

114 On Francis Ryan's contributions to microbial genetics, see Kohler, "Systems of Production"; Sapp, *Where the Truth Lies*.

115 Imada Akira, interview.

116 Komagata Kazuo, interview; Campos and Schwerin, *Making Mutations*.

117 Komagata, "Microbial Systematics," 33.

118 Ōshima Yasuji, interview.

119 Komagata Kazuo, interview.

120 Udaka Shigezō, interview.

121 Komagata Kazuo, interview.

122 Komagata Kazuo, interview.

123 Komagata Kazuo, interview. Similarly, on preservation methods at the American Type Culture Collection (ATCC), see Strasser, *Collecting Experiments*, 38–39.

124 Komagata Kazuo, interview; Komagata, "Microbial Systematics."

125 Komagata Kazuo, interview.

126 Sand, "Short History of MSG," 40.

127 MSG and the brain: Jordan Sand traces this notion to an influential 1958 book authored by a medical professor at Keio University, Hayashi Takashi's *Zunō* (The Brain). Jordan Sand, personal communication, March 27, 2014. Regulatory debates in the United States: Sand, "Short History of MSG," 43–44.

128 George, *Minamata*; Upham, "Movements for Place"; Ui, *Industrial Pollution in Japan*; Walker, *Toxic Archipelago*.

129 Chapter 2; Solt, *Untold History of Ramen*, 97, 108, 114–15.

130 Sand, "Short History of MSG," 45. On moral critiques of "instant" products, see Solt, *Untold History of Ramen*, 113–18. Kyōwa Hakkō Kōgyō's attempt to expand the market for fermented lysine, by having it added as a nutritional supplement to the bread in the national school lunch program, ended in failure in 1975 due to protest from a consumer movement. Nihon shōhisha renmei, "Katsudōreki." On contemporaneous concerns in the United States, Europe, and elsewhere surrounding the carcinogenicity of products derived from the petrochemical industries, see Sellers, "From Poison to Carcinogen."

131 Komagata Kazuo, interview. The rumor had originated in a Tokyo magazine in the 1920s. Sand, "Short History of MSG," 44.

132 Ōshima Yasuji, interview.

133 Ajinomoto played a role in sponsoring research on flavor physiology. Sand, "Short History of MSG," 45–47. On the establishment of umami as a fifth basic taste in the United States, see Tracy, "Delicious Molecules."

134 One telling comparison is the example of lysine manufacturing at Dutch company DSM in the late 1950s and 1960s, using a synthetic process. The process built on advanced expertise in organic chemistry within DSM, just as lysine fermentation at Kyōwa Hakkō Kōgyō and other Japanese manufacturers developed from in-house research in fermentation science. DSM's lysine enterprise was uncompetitive because the task of separating the D-isomers from the desired L-lysine product created a bottleneck in lowering production costs, a problem that did not occur in a fermentation process. Rooij, *Company That Changed Itself*, 114–25, with thanks to Ernst Homburg for the reference.

135 Komagata Kazuo, interview.

136 Akita Konno shōten kabushiki gaisha, *Konno moyashi 101 nen no ayumi*, 38.

137 On the history of the pharmaceutical industry in Japan, see Nihon yakushi gakkai, *Nihon iyakuhin sangyōshi*; Umemura, *Japanese Pharmaceutical Industry*; T. Yang, *Medicated Empire*.

138 Editorial Board of "History of Innovative Research and Drug Discovery at Takeda," *History of Innovative Research and Drug Discovery at Takeda*, 7, 10–11, 33.

139 Editorial Board of "History of Innovative Research and Drug Discovery at Takeda," *History of Innovative Research and Drug Discovery at Takeda*, 7; Imada Akira, interview.

140 Imada Akira, interview.

141 Imada Akira, interview; Imada Akira, "Takeda Yakuhin ni okeru 1950 nendai — 1960 nendai no hakkō kenkyū to tasha to no kyōgō."

142 Imada Akira, interview.

143 Imada Akira, interview. I see the work that Imada describes here, as well as microbiological activities within companies in this period more generally (prior to the introduction of recombinant DNA technology), as being labor-intensive rather than capital-intensive; see n14.

144 Imada Akira, interview; chapter 5; Davies and Davies, "Antibiotic Resistance"; Russell, *Evolutionary History*; Landecker, "Antibiotic Resistance."

145 Imada Akira, interview. On fermentation tanks, see Tanaka Hideo, "Hakkōsō, baiyō sōchi"; Yoshida Toshiomi, "Hakkō sōchi"; Sakuma Hideo, "Hakkōsō"; V. Lee, "Scaling Up from the Bench."

146 A. Johnson, *Hitting the Brakes*. See also n7.

147 Imada Akira, interview.

148 Chapter 2.

149 Imada Akira, interview. The best-known form of ubiquinone is Coenzyme Q_{10}, which is found commercially as a dietary supplement and an ingredient in cosmetics. In the European Union, Idebenone has been designated an orphan drug. See "Raxone," European Medicines Agency, December 9, 2019, https://www.ema.europa.eu/en/medicines/human/EPAR/raxone.

150 On antibiotic development in Japan, see Kumazawa and Yagisawa, "Antibiotics."

151 Imada Akira, interview.

152 Takeda closed the fermentation plant at Hikari in the new millennium; Imada Akira, interview. On the context of the global oil shocks as it pertained to decreasing Japanese investments in technology from the 1980s, see Goto and Odagiri, *Innovation in Japan*. The worsening economic environment for scientific research was most evident in falling R&D investment in private companies. Yoshioka, introduction to *Social History of Science and Technology*, vol. 4.

153 Imada Akira, interview.

154 Hughes, *Genentech*; Yi, *Recombinant University*.

155 For just two examples, see Odagiri and Goto, *Technology and Industrial Development*, chap. 11; Pollack, "Japan's Biotechnology Effort."

156 On the relatively low participation of new or small companies and university laboratories compared to old or large companies in biotechnological innovation in Japan in recent decades, in contrast to the United States, see Kneller, *Bridging Islands*; Kneller, "University-Industry Technology Transfer." The latest government policies have attempted to ameliorate these trends. See, for example, Mallapaty, "Japan's Start-Up Gulf"; Monbukagakushō, "Sangakukan renkei no saikin no dōkō oyobi kongo no ronten ni tsuite" (On recent trends in industry-university-government cooperation and issues hereon), Monbukagakushō kaigi shiryō, May 24, 2019, http://www.mext.go.jp/kaigisiryo/2019/05/__icsFiles/afieldfile/2019/05/24/1416712_004.pdf. Trends in the Japanese life sciences may help indicate patterns in Japanese scientific research as a whole, since Japan has invested as much or more in the life sciences as in other areas of the sciences (this is atypical for the region, where most countries have focused on the physical sciences and chemistry). Nature Index, "East & Southeast Asia."

CONCLUSION

1 Chang, *Inventing Temperature*; Chang, *Is Water H$_2$O?*

2 Chang's examples are drawn from the physical sciences, and the epistemic choices to which he refers tend to be theory choices. Historians of life sciences, on the other hand, have emphasized the leading importance of tools and materials relative to theory in generating new knowledge in experimental biology. Accordingly, my account has emphasized the material rather than theoretical aspects of epistemic choices. For a few examples of studies on materials in biology, see Clarke and Fujimura, *Right Tools*; Rheinberger, *History of Epistemic Things*; Creager, *Life of a Virus*.

3 On the epistemic benefits of marginality owing to the situated nature of knowledge, see Wylie, "Feminist Philosophy of Science"; Haraway, "Situated Knowledges."

4 Landecker, "Metabolism, Reproduction." On molecular biology, see Morange, *Black Box of Biology*.

5 Experiments on bacteria, which do not have nuclei, played a central role in the rise of molecular biology. On the Pasteurian school, see Mendelsohn, "Cultures of Bacteriology"; Ackert, "Thermodynamics of Life to Ecological Microbiology." On the Delft school, see Spath, "C. B. van Niel." Keith Vernon has depicted British microbiology in the pre–World War II period as a loose practical discipline covering not only medicine but also sewage treatment, breweries, dairies, and agriculture. Nonetheless, he agrees that its orientation was predominantly medical. Vernon, "Pus, Sewage, Beer and Milk." On medical microbiology and its contributions to microbial genetics in the United States,

see especially Amsterdamska, "Medical and Biological Constraints"; Amsterdamska, "Stabilizing Instability"; Amsterdamska, "Between Medicine and Science."

6   Foster, "View of Microbiological Science," 446; Omura, "Philosophy of New Drug Discovery."

7   Chapter 6; Samson, van der Aa, and de Hoog, "Centraalbureau voor Schimmelcultures."

8   Landecker, "Metabolism, Reproduction"; Squier, *Epigenetic Landscapes.*

9   Rose, "Human Sciences"; O'Malley and Dupré, "Size Doesn't Matter"; Grote and Keuck, "Stoffwechsel."

10   T. C. Smith, *Native Sources of Japanese Industrialization*; Hayami, *Japan's Industrious Revolution*; Hayami Akira, "Keizai shakai no seiritsu to sono tokushitsu"; Howell, *Capitalism from Within*; Wigen, *Making of a Japanese Periphery*; Pratt, *Japan's Protoindustrial Elite.*

11   Tanimoto argues: "It is not enough to just look at the heritage of proto-industrialization as the foundation of the factory system. It is necessary to place the non-factory production system on the extended line of economic development linking developments in the late Tokugawa and Meiji era"; in "Role of Tradition," 9–10. Tanimoto builds on Nakamura Takafusa's concept of "balanced growth," which refers to how Meiji-period economic growth relied a dual structure consisting of both "modern industry" and "indigenous industry." T. Nakamura, *Economic Growth in Prewar Japan.* Nakaoka Tetsurō has made an argument similar to Nakamura Takafusa's for the history of technology, whereby "hybrid" technologies that combined indigenous and imported features were significant in driving Meiji economic change. Nakaoka Tetsurō, *Nihon kindai gijutsu no keisei.* Tanimoto's approach explicitly carves off indigenous industries from modern industries as an alternative path of economic growth, which he argues persisted into the Meiji period.

12   On the inseparability of knowledge and institutions, see A. Johnson, *Hitting the Brakes*; Mody, *Instrumental Community*; Akera, *Calculating a Natural World*; Lécuyer, *Making Silicon Valley.*

13   Douglas, *How Institutions Think.*

14   I draw insights from the Histories of Planning group (2013–23) organized by Dagmar Schäfer at the Max Planck Institute for the History of Science.

15   For another study of planning and material culture, see J. L. Smith, *Works in Progress.*

16   For two extended studies, see M. V. Brock, *Biotechnology in Japan*; Collins, *Race to Commercialize Biotechnology.*

17   Tomita Fusao, in discussion; "Japan: Genetech's Late Bloomer."

18   Ōshima Yasuji, interview; Slingsby, "Beyond Borders."

19   Japan contributed 6 percent of sequences, in comparison with 59 percent from

the United States and 31 percent from Britain. Ito, "Science Lessons"; Fukushima, "Constructing Failure."

20 Fukushima, "Policy Process."

21 Komagata Kazuo, interview.

22 Udaka Shigezō, interview.

23 O'Malley and Dupré, "Size Doesn't Matter."

24 Dubos, *Man Adapting*, 163; cited in Méthot and Fantini, "Medicine and
Ecology," 217. An earlier version of this short reflection on medical micro-
biology was published in Victoria Lee, "Unraveling the Search for Microbial
Control in Twentieth-Century Pandemics," *Studies in History and Philosophy
of Biological and Biomedical Sciences* 53: 122–25 (© 2015 Published by Elsevier
Ltd.).

25 Méthot and Fantini, "Medicine and Ecology," 219. An eradication-based
approach was prominent in the science of chronic diseases as well, especially
cancer. Scheffler, *Contagious Cause*. On infectious disease and ecology, see
also Honigsbaum and Méthot, "Microbes, Networks, Knowledge," and related
articles in the collection; Tilley, *Africa as a Living Laboratory*; Mitman, "In
Search of Health"; Anderson, "Natural Histories." On infectious disease's inter-
section with environmental health in relation to the COVID-19 crisis, see vom
Saal and Cohen, "Toxic Chemicals." The COVID-19 pandemic has further
encouraged an already existent turn to evolutionary approaches in scientific
work and popular writing on infectious disease microbiology. See, for example,
Grubaugh et al., "Tracking Virus Outbreaks"; Xue, "Viral Evolution."

26 Quoted in Dubos, *Mirage of Health*, 79.

27 Otis, *Membranes*, 27.

28 These diseases are caused not by bacteria but by viruses, the nature of which
was debated and defined by negative properties until the 1930s. Sankaran,
"Viruses Were Not in Style"; Creager, *Life of a Virus*, 17–46.

29 Landecker, "Antibiotic Resistance"; Biss, *On Immunity*.

30 Méthot and Fantini, "Medicine and Ecology," 219.

31 "Antibiotic Resistance," World Health Organization, July 31, 2020, https://
www.who.int/news-room/fact-sheets/detail/antibiotic-resistance.

32 Podolsky, *Antibiotic Era*; Kirchhelle, *Pyrrhic Progress*.

33 Plasmids are small circular strands of DNA that exist separately from the chro-
mosome. Creager, "Adaptation or Selection?"; Davies and Davies, "Antibiotic
Resistance"; Grote, "Hybridizing Bacteria."

34 On evolution, see Sapp, "Bacterium's Place in Nature." The microbiome, for
humans, is the aggregate of microbial cells in the human body. An imbalance
in gut microbiota may contribute to a variety of diseases including autoimmune
disorders, diabetes, and allergies. For a scholarly historical account, see Sangodeyi, "Making of the Microbial Body." For a popular presentation of the micro-

biome, see *The Secret World inside You*, American Museum of Natural History, New York, November 7, 2015–August 14, 2016.

35 Since these reactions take place at room temperature and ambient pressure without requiring toxic metals or large quantities of organic solvent, scientists' expectation is that replacing harsh industrial processes with milder biotechnology will lessen waste and reduce the consumption of energy and fossil fuels. Bensaude-Vincent et al., "School of Nature."

36 V. Lee, "Bioorganic Synthesis"; Nobel Media AB, "The Nobel Prize in Physiology or Medicine 2015," NobelPrize.org, 2015, https://www.nobelprize.org /prizes/medicine/2015/summary/; Nobel Media AB, "The Nobel Prize in Chemistry 2018," NobelPrize.org, 2018, https://www.nobelprize.org/prizes /chemistry/2018/summary/; Nobel Media AB, "The Nobel Prize in Chemistry 2020," NobelPrize.org, 2020, https://www.nobelprize.org/prizes/chem istry/2020/summary/; Jinek et al., "Programmable Dual-RNA-Guided DNA Endonuclease"; Doudna and Sternberg, *Crack in Creation*.

Bibliography

INTERVIEWS

Imada Akira (former coordinator of academia-industry relations at the Ministry of Education, Culture, Sports, Science and Technology, including at the Kyoto Environmental Nanotechnology Cluster; former professor, Nara Institute of Science and Technology; former microbiologist at Takeda Yakuhin), interview by author, Suita, Osaka, Japan, August 17, 2012.

Komagata Kazuo (professor emeritus, Faculty of Agriculture, University of Tokyo; former microbiologist at Ajinomoto), interview by author, Tokyo, July 8, 2012.

Konno Eiichi (nephew of Konno Seiji), Konno Hiroshi (president of Akita Konno Shōten and grandnephew of Konno Seiji), and Konno Kenji (former president of Akita Konno Shōten and nephew of Konno Seiji), interview by author, Kariwano, Daisen, Akita, Japan, February 20, 2012.

Ōshima Yasuji (professor emeritus, Faculty of Engineering, Osaka University; former microbiologist at Kotobukiya, now Suntory), interview by author, Takatsuki, Osaka, Japan, June 6, 2012.

Udaka Shigezō (professor emeritus, Faculty of Agriculture, Nagoya University; former microbiologist at Kyōwa Hakkō Kōgyō), interview by author, Tokyo, November 12, 2012.

INFORMAL DISCUSSIONS

Beppu Teruhiko (professor emeritus, Faculty of Agriculture, University of Tokyo), in discussion with the author, Tokyo, September 7, 2012.

Kitamoto Katsuhiko (professor, Faculty of Agriculture, University of Tokyo), in discussion with the author, Tokyo, June 4, 2012.

Nakase Takashi (former director, Japan Collection of Microorganisms

[JCM], Riken), in discussion with the author, Suita, Osaka, Japan, June 7, 2012.

Tomita Fusao (professor emeritus, Faculty of Agriculture, Hokkaidō University; former microbiologist at Kyōwa Hakkō Kōgyō), in discussion with the author, Tokyo, June 26, 2012.

Yagisawa Morimasa (professor, Faculty of Pharmacy, Keio University), in discussion with the author, Tokyo, November 2, 2012.

JAPANESE-LANGUAGE SOURCES

Unpublished Documents, Privately Printed Sources, and Other Records

Imada Akira. "Takeda Yakuhin ni okeru 1950 nendai–1960 nendai no hakkō kenkyū to tasha to no kyōgō" (Fermentation research and competition with other companies in the 1950s and 1960s at Takeda Yakuhin). Unpublished document supplied by Imada Akira on August 17, 2012.

Jōzōkai 15 (autumn 1924). Osaka, Japan: Konnō shōten eigyōbu. Document supplied by Akita Konno Shōten.

Kinoshita Shukuo. *Zoku-zoku-zoku anohi anogoro* (Those days, those times; 3rd sequel). Sakura, Chiba, Japan: Genkōbō, 2004. Privately printed.

Kinzuka ni yosete (To the microbe mound). Takarazuka, Japan: Kasabō Takeo, 1981. Privately printed.

Nihon penishirin gakujutsu kyōgikai kiji (Records of the Japan Penicillin Research Association). Printed in *Journal of Penicillin*.

Shudō Nagatoshi, ed. *Jōzō daijiten* (Great dictionary of brewing). Nishigō-chō, Japan: Shudō Nagatoshi, 1931. Document supplied by the Society for Biotechnology, Japan.

"Takada Ryōhei narabi ni kyōdō kenkyūsha ni yoru ronbun mokuroku" (Catalog of works by Takada Ryōhei and collaborative researchers). Document supplied by the Society for Biotechnology, Japan.

Institutional Histories

Ajinomoto kabushiki gaisha, ed. *Aji o tagayasu: Ajinomoto hachijūnenshi* (Cultivating taste: Eighty-year history of Ajinomoto). Tokyo: Ajinomoto, 1990.

———. *Ajinomoto guruupu no hyakunen: Shinkachi sōzō to kaitakusha seishin, 1909–2009* (100 years of the Ajinomoto Group: Creation of new value and pioneering spirit, 1909–2009). Tokyo: Ajinomoto, 2009.

Akita Konno shōten kabushiki gaisha, ed. *Konno moyashi 101 nen no ayumi*

(History of 101 years of Konno Moyashi). Daisen, Akita, Japan: Akita Konno shōten, 2011.

Fukuoka-ken shōyu kumiai, ed. *Fukuoka-ken shōyu kumiai nanajūnen-shi* (Seventy-year history of the Fukuoka Prefecture Soy Sauce Association). Fukuoka, Japan: Fukuoka-ken shōyu kōgyō kyōdō kumiai, Fukuoka-ken shōyu jōzō kyōdō kumiai, 1979.

Gōdō shusei shashi hensan iinkai, ed. *Gōdō shusei shashi* (History of Gōdō Shusei). Tokyo: Gōdō shusei, 1970.

Hyakushūnen kinen jigyōkai, ed. *Hyakunenshi: Ōsaka daigaku kōgakubu jōzō, hakkō, ōyō seibutsu kōgakuka* (Hundred-year magazine: Department of Brewing Science–Fermentation Science–Applied Bioengineering, Faculty of Engineering, Osaka University). Suita, Japan: Ōsaka daigaku kōgakubu jōzō, hakkō, ōyō seibutsu kōgakuka, 1996.

Kawamata kabushiki gaisha, ed. *Murasaki: Sakai no shōyuya Kawamata, Daishō 200 nen no ayumi* (Murasaki: 200-year history of Kawamata-Daishō, soy sauce brewer of Sakai). Sakai, Japan: Kawamori Mikio, 2000.

Kōgyō gijutsuin hakkō kenkyūjo, ed. *Hakkō kenkyūjo 25 nen no ayumi* (Fermentation Institute 25-year history). Chiba, Japan: Kōgyō gijutsuin hakkō kenkyūjo, 1965.

Kyōwa hakkō sōritsu 50 shūnen shashi hensan iinkai, ed. *Sore kara sore e: Kyōwa hakkō 50 nen no kiseki to shinseiki e no ishizue* (Since then, toward that: 50-year path of Kyōwa Hakkō and foundations for the new century). Tokyo: Kyōwa hakkō kōgyō, 2000.

Nihon jōzō kyōkai, ed. *Nihon jōzō kyōkai nanajūnenshi* (Seventy-year history of the Brewing Society of Japan). Tokyo: Nihon jōzō kyōkai, 1975.

Nihon penishirin kyōkai, ed. *Penishirin no ayumi* (History of penicillin). Tokyo: Nihon penishirin kyōkai, 1961.

Takeda nihyakunenshi hensan iinkai, ed. *Takeda nihyakunenshi* (Two-hundred-year history of Takeda). Osaka, Japan: Takeda yakuhin kōgyō, 1983.

Tōkyō daigaku ōyō biseibutsu kenkyūjo, ed. *Jūnen no ayumi: 1953–1963* (Ten years' history: 1953–1963). Tokyo: Tōkyō daigaku ōyō biseibutsu kenkyūjo, 1964.

Published Sources

Advertisements. *Jōkai* 10 (1902).

Akai Akinori. "Tsuboi Sentarō no 'katsuryokuso'" (Tsuboi Sentarō's "Katsuryokuso"). *Ōsaka daigaku toshokanpō* 45, no. 2 (2001): 12.

Akiyama Hiroichi. "Sake to kōbo" (Sake and yeast). In *Nihon no sake no rekishi: Sakazukuri no ayumi to kenkyū* (History of sake in Japan: Sake brewing history and research), edited by Katō Benzaburō, 393–440. Tokyo: Kyōwa hakkō kōgyō, 1977.

Beppu Teruhiko. "Biseibutsugaku no kiso to ōyō" (Basic and applied in microbiology). In *Nōgei kagaku no 100 nen* (100 years of agricultural chemistry), edited by Nihon nōgei kagakkai, 136–42. Tokyo: Nihon nōgei kagakkai, 1987.

Chikudō Shō. "Tanekōji ni tsuite (1)" (On *tanekōji* [1]). *Jōzō kyōkai zasshi* 6, no. 7 (1911): 47–52.

———. "Tanekōji ni tsuite (2)" (On *tanekōji* [2]). *Jōzō kyōkai zasshi* 6, no. 8 (1911): 32–41.

Doi Shinji. "Saikin ni okonawaretaru shusei ni kansuru kenkyū ni tsuite" (On recent research on alcohol). *Nihon jōzō kyōkai zasshi* 33, no. 12 (1938): 1308–20.

Fujiwara Takao. *Kindai Nihonshuzōgyōshi* (History of the modern sake industry). Kyoto, Japan: Mineruva shobō, 1999.

Fukinbara Takashi. "Riken no gōseishu" (Riken's synthetic sake). *Tokushū: Rikagaku kenkyūjo 60 nen no ayumi* (Special issue: 60-year history of the Institute of Physical and Chemical Research), *Shizen* 394, appendix (1978): 135–41.

Furukawa Sōichi. "Nyūsankin to kōbo no kyōson to kyōsei" (The coexistence and symbiosis of lactic acid bacteria and yeast). *Seibutsu kōgaku kaishi* 90 (2012): 188–91.

Furukawa Yasu. "Dentō sangyō kara kindai sangyō e: Meiji no kagaku to kagaku kōgyō" (From traditional industry to modern industry: Meiji-period chemistry and chemical industry). Unpublished lecture at Hokkaidō University, July 27, 2000.

———. *Kagakushatachi no Kyōto gakuha: Kita Gen'itsu to Nihon no kagaku* (The chemists' Kyoto School: Kita Gen'itsu and chemistry in Japan). Kyoto, Japan: Kyōto daigaku gakujutsu shuppankai, 2017.

Hakkō to taisha 1–5 (1959–1962).

Hanada Kazō. "Nenryōyō shusei taisaku ni tsuite" (On fuel alcohol measures). *Jōzōgaku zasshi* 13, no. 9 (1935): 918–26.

Hasegawa Takeji. "Nihon no biseibutsukabu hozon jigyō 1: Yōran jidai" (Microbial strain preservation projects in Japan 1: Early period). *Microbiology and Culture Collections* 12, no. 1 (1996): 1–10.

———. "Nihon no biseibutsukabu hozon jigyō 2: Kokunai renmei to sono katsudō" (Microbial strain preservation projects in Japan 2: Domestic leagues and their activities). *Microbiology and Culture Collections* 12, no. 2 (1996): 55–66.

———. *Zatsuroku: Biseibutsugakushi* (Miscellaneous records: History of microbiology). Osaka, Japan: Hakkō kenkyūjo, 2001.

Hashitani Yoshitaka. "Kōbo seizai" (Yeast preparations). In *Bitamin kenkyū gojūnen* (Fifty years of vitamin research), edited by Takada Ryōhei and Katsura Eisuke, 176–81. Tokyo: Daiichi shuppan, 1961.

Hayami Akira. "Keizai shakai no seiritsu to sono tokushitsu" (The emergence of the economic society and its characteristics). In *Atarashii Edo jidai shizō o motomete* (Searching for a new historical view of the Edo era), edited by Shakai keizaishi gakkai, 3–18. Tokyo: Tōyō keizai shinpōsha, 1977.

Hirayama Yoichi. *Chōsen shuzō gyōkai yonjūnen no ayumi* (Forty-year history of the brewing industry in Korea). Tokyo: Yūhō kyōkai, 1969.

Honda Norimoto. "Ōshū ni okeru nenryōyō shusei konyō no kinkyō" (Recent conditions of fuel alcohol mixing in Europe). *Nihon jōzō kyōkai zasshi* 32, no. 7 (1937): 724–27.

Ichikawa Jirō. *Nihon no sake: Sono kigen to rekishi* (Sake in Japan: Its origin and history). Yamato, Kanagawa, Japan: Tōa bunbutsu konwakai, 2002.

Ichikawa Kunisuke. "Aminosan chōmiryō" (Amino-acid seasoning). In *Chōmiryō: Sono kagaku to seizō* (Seasoning: Its science and manufacturing), edited by Takada Ryōhei, 112–20. Tokyo: Kōseikan, 1966.

Iijima Takashi. *Nihon no kagaku gijutsu: Kigyōshi ni miru sono kōzō* (Chemical technology in Japan: Its structure as seen in business history). Tokyo: Kōgyō chōsakai, 1981.

Iijima Wataru. *Mararia to teikoku: Shokuminchi igaku to Higashi Ajia no kōiki chitsujo* (Malaria and empire: Colonial medicine and the East Asian regional order). Tokyo: Tōkyō daigaku shuppankai, 2005.

Iinuma Kazumasa and Kanno Tomio. *Takamine Jōkichi no shōgai: Adorenarin hakken no shinjitsu* (The life of Takamine Jōkichi: The truth about the discovery of adrenalin). Tokyo: Asahi shimbunsha, 2000.

Ishidō Toyota. "Shusei seizō dan" (Talk on alcohol manufacture). *Kōgyō kagaku zasshi* 7, no. 2 (1904): 125–31.

Iwai Ki'ichirō. "Honpō inryō shusei narabi ni gōsei seishu no hattenshi" (History of development of drinking alcohol and synthetic sake in Japan). *Ōsaka jōzō gakkai 30 shūnen kinen tokushū: Hakkō kōgyō no tenbō* (Commemoration of the 30th anniversary of the Osaka Society of Brewing Science, special issue: Survey of the fermentation industries), *Hakkō kōgaku zasshi* 30, no. 10 (1952): 148–51.

Iwate Hayami. "Honpō shusei kōgyō hattenshi" (History of development of the alcohol industry in this country). *Ōsaka jōzō gakkai 30 shūnen kinen tokushū: Hakkō kōgyō no tenbō* (Commemoration of the 30th anniversary of the Osaka Society of Brewing Science, special issue: Survey of the fermentation industries), *Hakkō kōgaku zasshi* 30, no. 10 (1952): 144–47.

Ka Ushiko. "Shusei no hanashi sono ichi" (On alcohol 1). *Jōzō kyōkai zasshi* 3, no. 4 (1908): 56–62.

———. "Shusei no hanashi sono ni" (On alcohol 2), *Jōzō kyōkai zasshi* 3, no. 5 (1908): 52–56.

Kaji Masanori. "Majima Rikō to Nihon no yūki kagaku kenkyū dentō no keisei" (Majima Rikō and the formation of the organic chemistry research tradition in Japan). In *Shōwa zenki no kagaku shisōshi* (Intellectual history of science in the early Shōwa period), edited by Kanamori Osamu, 185–241. Tokyo: Keisō shobō, 2011.

Katō Benzaburō. "Amirohō ni kansuru kenkyū" (Research on the amylo method). *Jōzōgaku zasshi* 2, no. 12 (1925): 1135–51.

———. "Katō Benzaburō." In *Watashi no rirekisho: Shōwa no keieisha gunzō* (My career path: Managers of the Shōwa period), edited by Nihon keizai shimbunsha, 7–77. Tokyo: Nihon keizai shimbunsha, 1970.

Katō Benzaburō, ed. *Nihon no arukōru no rekishi: Sono jigyō to kenkyū* (History of alcohol in Japan: Its enterprise and research). Tokyo: Kyōwa hakkō kōgyō, 1974.

Katō Hyakuichi. "Kōji no rekishi" (History of *kōji*). In *Kōjigaku* (*Kōji* science), edited by Murakami Hideya, 1–31. Tokyo: Nihon jōzō kyōkai, 1986.

———. "Nihon no sakazukuri no ayumi" (History of sake brewing in Japan). In *Nihon no sake no rekishi: Sakazukuri no ayumi to kenkyū* (History of sake in Japan: Sake brewing history and research), edited by Katō Benzaburō, 41–315. Tokyo: Kyōwa hakkō kōgyō, 1977.

Katsume Suguru and Tanabe Osamu. "Saikin Itari ni okeru nenryō shusei kōgyō seisaku" (Recent fuel alcohol industrial policies in Italy). *Nihon jōzō kyōkai zasshi* 31, no. 6 (1936): 461–63.

Kawakita Michitada. "Nenryō toshite shusei no ōyō" (Application of alcohol as fuel). *Nihon jōzō kyōkai zasshi* 17, no. 4 (1922): 2–3.

Kawamura Shunzan. *Dekoboko jinseiki* (Memoir of an uneven life). Tokyo: Nikkan rōdō tsūshinsha, 1960.

Kawashima Shirō. "Senjichū no Nihon rikugun ni okeru bitamin hokyū" (Vitamin supply in the Japanese army during the war). In *Bitamin ken-*

kyū gojūnen (Fifty years of vitamin research), edited by Takada Ryōhei and Katsura Eisuke, 189-96. Tokyo: Daiichi shuppan, 1961.

Kimoto Chōtarō. "Nihonshu jōzō kairyō ni tsuite no kibō" (Hopes regarding sake brewing improvement). *Jōkai* 3 (1902): 6-8.

———. "Shuzō genryōchū 'bakuteriya' no kensahō" (Inspection methods for "bacteria" in the raw materials of sake manufacture). *Jōkai* 11 (1903): 17-23.

Kinoshita Shukuo. "Aminosan hakkō kōgyō: Sono oitachi to genjō" (Amino acid fermentation industry: Its growth and current status). In *Nōgei kagaku no 100 nen* (100 years of agricultural chemistry), edited by Nihon nōgei kagakkai, 6-8. Tokyo: Nihon nōgei kagakkai, 1987.

———. "Aminosan hakkō no tanjō" (Birth of amino acid fermentation). In *Kaisō: Aminosan hakkō o megutte—sono yōranki o kataru* (Reflections: Concerning amino acid fermentation—narrating the early period), edited by Kinoshita Shukuo, 3-58. Tokyo: Kyōwa hakkō kōgyō, 1990.

Kinoshita Toshiaki. "Arukōru seizō gijutsu no shinpo" (Progress of alcohol manufacturing technology). In *Nihon no arukōru no rekishi: Sono jigyō to kenkyū* (History of alcohol in Japan: Its enterprise and research), edited by Katō Benzaburō, 293-324. Tokyo: Kyōwa hakkō kōgyō, 1974.

Kitahara Kakuo. "Arukōru kōgyō ni okeru amiraaze" (Amylases in the alcohol industry). In *Nihon no arukōru no rekishi: Sono jigyō to kenkyū* (History of alcohol in Japan: Its enterprise and research), edited by Katō Benzaburō, 225-74. Tokyo: Kyōwa hakkō kōgyō, 1974.

Kobori Satoru. *Nihon no enerugii kakumei: Shigen shōkoku no kingendai* (The energy revolution in Japan: The modern and contemporary era of a small energy power). Nagoya: Nagoya daigaku shuppankai, 2010.

Kohata Kengorō. "Waga kuni ni okeru shusei seizōgyō no kinkyō (4)" (Recent conditions of the alcohol industry in our country [4]). *Nihon jōzō kyōkai zasshi* 17, no. 2 (1922): 37-38.

Koike Yoshiyuki. "Senjika ni okeru shusei kōgyō ni tsuite" (On the alcohol industry in wartime). *Nihon jōzō kyōkai zasshi* 33, no. 4 (1938): 398-401.

Koizumi Takeo. *Kōji kabi to kōji no hanashi* (*Kōji* mold and the story of *kōji*). Tokyo: Kōrin, 1984.

Kozai Yoshinao. "Nihonshu jōzō no kairyō ni tsuite" (On the improvement of sake brewing). *Jōzō zasshi* 305 (1901): 23-27.

———. "Nihonshu jōzōhō no kairyō to kenkyū kikan no seibi ni tsuki" (On the improvement of sake brewing and the construction of a research facility). *Jōzō zasshi* 308 (1901): 4-11.

Kumazawa Kikuo. "Riibihi to Nihon no nōgaku: Riibihi tanjō 200 nen ni saishite" (Liebig and agricultural science in Japan: On the occasion of the 200th anniversary of Liebig's birth). *Hiryō kagaku* 25 (2003): 1–60.

Kuninaka Akira and Inemori Kazuo. "Yamasa ni okeru kakusankei chō-miryō, iyakuhin no rekishi to shōyu gijutsu no shinpo" (History of nucleic acid-related flavorings and drugs and the progress of soy sauce technology at Yamasa). *Chiba-ken kōgyō rekishi shiryō chōsa hōkokusho* 3 (1995): 12–26.

Kurono Kanroku. "Honpō nenryō shusei kōgyō no genkyō" (Present condition of the fuel alcohol industry in this country). *Nihon kikaigaku zasshi* 42, no. 262 (1939): 31–36.

———. "Shōwa 13 nendo nenryō shusei kōgyō no gaisetsu" (Outline of the fuel alcohol industry in the year Shōwa 13). *Nihon jōzō kyōkai zasshi* 34, no. 2 (1939): 122–23.

Kurono Kanroku, Takizawa Sumie, and Iwashita Nobuo. "Amiro hakkōhō o toku" (Explaining the amylo fermentation method). *Nihon jōzō kyōkai zasshi* 30, no. 2 (1935): 112–23.

Maejima Chitoku. "Taiwan ni okeru shusei seizō no gaikan" (Survey of alcohol manufacture in Taiwan). *Nihon jōzō kyōkai zasshi* 15, no. 8 (1920): 29–34, 42.

Matsubara Shinnosuke. "Kōji no setsu" (Theory of *kōji*). *Tōkyō iji shinshi* 24 (1878): 12–16.

Miwa Kiyoshi. "Hakkō biseibutsu" (Fermentation microorganisms). In *Hakkō handobukku* (Fermentation handbook), edited by Zaidan hōjin baioindasutorii kyōkai hakkō to taisha kenkyūkai, 456–57. Tokyo: Kyōritsu shuppan, 2001.

Miyata Shinpei. *"Kagakusha no rakuen" o tsukutta otoko: Ōkōchi Masatoshi to rikagaku kenkyūjo* (The man who made a "paradise for scientists": Ōkōchi Masatoshi and the Institute of Physical and Chemical Research). Tokyo: Nikkei bijinesujin bunko, 2001.

Miyazaki Shizu. Shōchū, shusei sono hoka seishu nado no moromi no sei-zōhō (Manufacturing process of moromi of *shōchū*, alcohol and also sake). Japan Patent 110127, filed December 17, 1932, and issued March 29, 1935.

Murai Toyozō. "Tanekōji konjaku monogatari" (Story of *tanekōji* past and present). *Shushi kenkyū* 7 (1989): 39–44.

Murakami Hideya. "Heruman Aaruburugu to sono shūhen: Kōjikin no hakkensha" (Hermann Ahlburg and other persons concerned: The discoverer of *kōji*). *Nihon jōzō kyōkaishi* 89, no. 11 (1994): 889–94.

———. "Kōjikin" (*Kōji* microbes). In *Kōjigaku* (*Kōji* science), edited by Murakami Hideya, 48–216. Tokyo: Nihon jōzō kyōkai, 1986.

————. "Sake to kōji" (Sake and kōji). In *Nihon no sake no rekishi: Sakazukuri no ayumi to kenkyū* (History of sake in Japan: Sake brewing history and research), edited by Katō Benzaburō, 319–50. Tokyo: Kyōwa hakkō kōgyō, 1977.

Murakami Hideya, ed. *Kōjigaku* (*Kōji* science). Tokyo: Nihon jōzō kyōkai, 1986.

Nakadai Tadanobu. "Kikkōman ni okeru shōyu jōzō gijutsu kaihatsu no rekishi" (History of the development of soy sauce brewing technology at Kikkōman). *Chiba-ken kōgyō rekishi shiryō chōsa hōkokusho* 4 (1995): 1–11.

Nakamura Shizuka. "Shusei nenryō kokusaku jūritsu ni taisuru ikkōsatsu" (Investigation on the establishment of alcohol fuel national policy). *Jōzōgaku zasshi* 14, no. 11 (1936): 897–907.

Nakamura Shizuka and Ichino Kazuma. *Saishin arukōru kōgyō* (The latest alcohol industry). Tokyo: Sangyō tosho, 1949.

Nakamura Shizuka and Matsui Tōji. "Musui shusei ni kansuru mondai" (Problems concerning absolute alcohol). In *Ōyō yūki kagaku saikin no shomondai* (Recent various problems in applied organic chemistry), edited by Funakubo Eiichi et al., 1–36. Tokyo: Yōkendō, 1942.

Nakano Masahiro. "Taiwan ni okeru arukōru sangyō" (The alcohol industry in Taiwan). In *Nihon no arukōru no rekishi: Sono jigyō to kenkyū* (History of alcohol in Japan: Its enterprise and research), edited by Katō Benzaburō, 177–223. Tokyo: Kyōwa hakkō kōgyō, 1974.

Nakaoka Tetsurō. *Nihon kindai gijutsu no keisei: "Dentō" to "kindai" no dainamikusu* (The formation of modern Japanese technology: The dynamics of "traditional" and "modern"). Tokyo: Asahi shimbunsha, 2006.

Nakazawa Ryōji and Takeda Yoshito. *Kōso kagaku kōgyō zenshū*, vol. 12, *Taiwan hakkō kōgyō* (Enzyme chemical industries complete series, vol. 12, Taiwan fermentation industries). Tokyo: Kōseikaku, 1940.

Nakazawa Shin'ichi. *Mori no barokku* (Baroque of the woods). Tokyo: Seirika shobō, 1992.

Narahara Hideki. "Moyashi." In *Kōjigaku* (*Kōji* science), edited by Murakami Hideya, 32–47. Tokyo: Nihon jōzō kyōkai, 1986.

Niki Etsutarō. "Kiki ni hinseru shōchūkai (shuseishiki)" (The *shōchū* industry [alcohol-style] facing a crisis). *Nihon jōzō kyōkai zasshi* 16, no. 11 (1921): 12–14.

Nihon shōhisha renmei. "Katsudōreki" (History of activities of the Consumers Union of Japan). Nishoren.net. http://nishoren.net/about_us /our_history. Accessed April 22, 2014.

Nihon yakushi gakkai, ed. *Nihon iyakuhin sangyōshi* (The drug industry in Japan). Tokyo: Yakuji nippōsha, 1995.

Oana Fujio. "Kyūshū chihō shusei kōjō shisatsu ki" (Report on the inspection of alcohol factories in the Kyūshū district). *Nihon jōzō kyōkai zasshi* 16, no. 6 (1921): 18-25.

Ono Ryōzō. "Shuzō jō no daiteki 'bakuteriya' ni tsuki kotau" (Answer regarding "bacteria," the great enemy in sake brewing). *Jōkai* 10 (1902): 34-36.

Ōsaka jōzō gakkai, ed. *Nenryō shusei kōenshū* (Collected lectures on fuel alcohol). Osaka, Japan: Ōsaka jōzō gakkai, 1936.

Ōtsuka Ken'ichi. "Kenkyū no rekishi to tenbō" (Overview and history of research). In *Jōzō shikenjo sōritsu 75 shūnen kinen kōen* (Memorial lectures for the 75th anniversary of the establishment of the Brewing Experiment Station), *Jōzō shikenjo hōkoku* 151 (1979): 1-6.

Ōyama Yoshitoshi. "Kagaku kōgaku no riteihyō 3. Kaken to penishirin puranto" (Milestones in chemical engineering 3. Kaken and the penicillin plant). *Shizen* 24, no. 6 (1969): 60-67.

Saitō Kendō. *Hakkō biseibukki* (Memoir of fermentation microorganisms). Osaka, Japan: Fuminsha, 1949.

———. *Hakkō kinrui kensa benran* (Handbook for examination of fermentation fungi). Tokyo: Maruzen, 1929.

———. "Higashi Ajia no yūyō hakkōkin" (Useful fermentation microbes of East Asia). *Tōyō gakugei zasshi* 23, no. 303 (1906): 507-20.

———. "Taikichū no fuyū kinshi" (Microbes and spores floating in the air). *Tōyō gakugei zasshi* 26, no. 331 (1909): 154-63.

———. *Tōkyō zeimusho kantokukyoku Saitō Kendō chōsa* (Surveys by Saitō Kendō for the Tokyo Tax Office and Inspectorate). Tokyo: Tōkyō zeimusho kantokukyoku, 1905.

Saitō Yoshio, Nishi Hiroshi, Takematsu Tetsuo, Sōtome Shin'ichi, and Komagata Kazuo. " 'Utsunomiya kōnō' to nōgei kagaku sono 1: Utsunomiya kōnō o tazunete" ("Utsunomiya Higher Agricultural" and agricultural chemistry 1: A visit to Utsunomiya Higher Agricultural). *Nippon nōgei kagaku kaishi* 60 (1986): 253-58.

Sakaguchi Kin'ichirō. "Hakkō: Higashi Ajia no chie" (Fermentation: Wisdom of East Asia). In *Shoku no bunka shinpojiumu '81: Higashi Ajia no shoku no bunka* (Food culture symposium '81: Food culture in East Asia), edited by Ishige Naomichi, 41-71. Tokyo: Heibonsha, 1981.

———. "Kōjikin kenkyū no omoide" (Memories of *kōji* microbe research). In *Hakkō to sakegaku* (Fermentation and wine science), 287-339. Tokyo: Iwanami shoten, 1998.

————. "Michi e no gunzō" (Portrayal of a group toward the unknown). In *Hakkō to sakegaku* (Fermentation and wine science), 191–208. Tokyo: Iwanami shoten, 1998.

Sakamoto Tatsunosuke. *Kamiya Denbei* (Kamiya Denbei). Tokyo: Sakamoto Tatsunosuke, 1921.

Sakuma Hideo. "Hakkōsō" (Fermenter). In *Hakkō handobukku* (Fermentation handbook), edited by Baioindasutorii kyōkai hakkō to taisha kenkyūkai, 485–90. Tokyo: Kyoritsu shuppan, 2001.

Sakurai Yoshito. "Shokuhin no kyōka" (Food fortification). In *Bitamin kenkyū gojūnen* (Fifty years of vitamin research), edited by Takada Ryōhei and Katsura Eisuke, 197–99. Tokyo: Daiichi shuppan, 1961.

Satō Jin. *"Motazaru kuni" no shigenron: Jizoku kanō na kokudo o meguru mō hitotsu no chi* (Resource thinking of the "have-nots": Sustainable land and alternative vision in Japan). Tokyo: Tōkyō daigaku shuppankai, 2011.

Sawamura Makoto. "Mazu tōji no zunō o kairyō su beshi" (We should first improve the heads of *tōji*). *Jōkai* 12 (1903): 12–14.

Sensui Hidekazu. "Beikoku shiseika Ryūkyū no kekkaku seiatsu jigyō: BCG o meguru 'dōka to ika no hazama de'" (Tuberculosis control in the Ryūkyūs under American administration: BCG "between assimilation and dissimilation"). In *Teikoku Nihon no kagaku shisōshi* (Intellectual history of science in imperial Japan), edited by Sakano Tōru, 295–349. Tokyo: Keisō shobō, 2018.

Setoguchi Akihisa. *Gaichū no tanjō: Mushi kara mita Nihonshi* (Birth of the harmful insect: Japanese history through the insect). Tokyo: Chikuma shobō, 2009.

Shimoyama Jun'ichirō. "Seishu no jōzō ni tsuite" (On the brewing of sake). In *Jōzō taikashū* (Anthology of great brewing experts), vol. 1, edited by Hirayama Kōnosuke, 1–10. Tokyo: Masuike shōten, 1902.

Shinkai Makoto, dir. *Kimi no na wa* (Your name). CoMix Wave Films, 2016.

Sugiyama Shinsaku. "Mippei 'tanku' ni yoru shusei hakkō" (Alcohol fermentation by closed "tank"). *Nihon jōzō kyōkai zasshi* 20, no. 6 (1925): 59–64.

Sumiki Yusuke. *Kōseibusshitsu* (Antibiotics). Tokyo: Tōkyō daigaku shuppansha, 1961.

Suminoe Kinshi and Minato Chikamomo. "Mippeishiki shusei hakkō ni tsuite" (On closed-style alcohol fermentation). *Nihon jōzō kyōkai zasshi* 18, no. 8 (1923): 5–10.

Suzuki Sadaharu. "Nihon no arukōru kōgyōshi (sono 2)" (History of the alcohol industry in Japan [2]). *Hakkō kyōkaishi* 30, no. 7 (1972): 319–24.

————. "Nihon no arukōru no rekishi" (History of alcohol in Japan). In

Nihon no arukōru no rekishi: Sono jigyō to kenkyū (History of alcohol in Japan: Its enterprise and research), edited by Katō Benzaburō, 31–175. Tokyo: Kyōwa hakkō kōgyō, 1974.

Suzuki Umetarō. "Hakkō kagaku no shinpo" (Progress in fermentation chemistry). *Jōzōgaku zasshi* 3, no. 8 (1926): 776–89.

———. *Kenkyū no kaiko* (Research recollections). Tokyo: Kibundō shobō, 1943.

Suzuki Umetarō and Katō Shōji. "Rikenshu ni tsuite" (On Rikenshu). *Jōzōgaku zasshi* 5, no. 5 (1927): 359–67.

Taiwan sōtokufu senbaikyoku. *Taiwan sake senbaishi* (History of the Taiwan alcohol monopoly). 2 vols. Taipei: Taiwan sōtokufu senbaikyoku, 1941.

Takada Ryōhei. "Biseibutsu shokuryōka no gainen to kōjikin shokuryōka ni chakushu shita riyū" (The concept of turning microbes into food and reasons for beginning to turn *kōji* microbes into food). *Kōgyō kagaku zasshi* 32 (1929): 497–99.

———. "Bitamin" (Vitamins). In *Kabi no riyō kōgyō* (Mold-utilizing industries), edited by Tomoda Yoshinori et al., 121–54. Tokyo: Kyōritsu shuppan, 1956.

———. "Bitamin shigen yori mita Nihon" (Japan from the perspective of vitamin resources). *Kōgyō kagaku zasshi* 46, no. 9 (1943): 989–91.

———. "Bitamin shinpen zatsuwa" (Personal miscellaneous words on vitamins). In *Bitamin kenkyū gojūnen* (Fifty years of vitamin research), edited by Takada Ryōhei and Katsura Eisuke, 164–75. Tokyo: Daiichi shuppan, 1961.

———. "Kōgyō kara mita shokuryō mondai" (Food problem from the industrial perspective). *Kōgyō kagaku zasshi* 44, no. 9 (1941): 756–57.

———. *Senjika shokuryō taisaku* (Wartime food policy). Kyoto, Japan: Taigadō, 1944.

Takahashi Teizō. *Jōzō bairon* (Theory of brewing molds). Tokyo: Maruzen, 1903.

———. "Kōbo no chozōhō" (Method of preservation of yeast). *Jōkai* 10 (1902): 16–26.

———. "Yo ga iwayuru yodan shikomi shijō kakuchi shijō no kekka gaihyō" (Short criticism on the results of test brewing of four-stage mash preparation in several districts). *Jōkai* 30 (1904): 49–61.

Takahashi Teizō, Eda Kamajirō, Okumura Junshirō, Yamamoto Takeshi, Nakazawa Ryōji, Shibukawa Kōzō, and Yukawa Matao. "Seishu kōbo no henshu ni tsuite" (On varieties of sake yeast). *Jōzō shikenjo hōkoku* 54 (1914): 1–66.

Takamine Jōkichi. "Takajiasutaaze ni tsuite" (On Takadiastase). *Kōgyō kagaku zasshi* 5, no. 52 (1902): 405–30.

Takayama Jintarō. "Jōzō shikenjo no hitsuyō" (The necessity of a brewing experiment station). *Jōkai* 1 (1902): 4–7.

Takeda Keiichi. *Penishirin sangyō kotohajime* (Dawn of the penicillin industry). Tokyo: Maruzen puranetto, 2007.

Tanaka Hideo. "Hakkōsō, baiyō sōchi" (Fermenters and bioreactors). *Mireniamu tokubetsu gō: Hakkō kōgaku 20 seiki no ayumi; Baiotekunorojii no genryū o tadoru* (Millennium special issue: 20th-century history of fermentation engineering; Retracing the origins of biotechnology). *Seibutsu kōgakushi* (December 2000): 24–32.

Togano Meijirō. "Shusei no hanashi" (On alcohol). *Jōzō kyōkai zasshi* 6, no. 9 (1911): 40–46.

Tsuboi Sentarō. "Jōzōkai no kakumei jigyō" (Revolutionary projects in the brewing world). *Jōkai* 23 (1903): 45–48.

———. "Nikushoku bōkokuron" (Theory that meat-eating will ruin the country). In *Nijūikka kōwa* (Lectures by twenty-one experts), edited by Miyazaki Saburō, 175–82. Tokyo: Hakubunron, 1908.

Tsunoda Fusako. *Hekiso: Nihon penishirin monogatari* (Hekiso: Japan's penicillin story). Tokyo: Shinchōsha, 1978.

Udaka Shigezō. "Aminosan hakkō no reimeiki" (Dawn of amino acid fermentation). *Kagaku to seibutsu* 47, no. 3 (2009): 211–16.

———. "Aminosan seisankin no hakken" (Discovery of the glutamic acid-producing microbe). In *Kaisō: Aminosan hakkō o megutte—sono yōranki o kataru* (Reflections: Concerning amino acid fermentation—narrating the early period), edited by Kinoshita Shukuo, 61–84. Tokyo: Kyōwa hakkō kōgyō, 1990.

Umezawa Hamao. *Kōseibusshitsu o motomete* (Searching for antibiotics). Tokyo: Bungei shunju, 1987.

Usami Keiichirō. "Nenryō mondai to nenryō shusei" (Fuel problem and fuel alcohol). *Nihon jōzō kyōkai zasshi* 17, no. 5 (1922): 2–8.

Yamagata Unokichi. "Jōzōgyō to seiketsu" (The brewing industry and cleanliness). *Jokai* 11 (1902): 16–21.

Yamamura Tazaemon. "Shuzōgyō hattatsu no katei to kongo no kadai" (Development of the sake industry and problems hereon). *Ōsaka jōzō gakkai 30 shūnen kinen tokushū: Hakkō kōgyō no tenbō* (Commemoration of the 30th anniversary of the Osaka Society of Brewing Science, special issue: Survey of the fermentation industries), *Hakkō kōgaku zasshi* 30, no. 10 (1952): 4–5.

Yamazaki Momoji. "Shina gakuto e no kotoba" (Words to Chinese students). *Warera no kagaku* 2 (1929): 447–49.

———. "Shinasan hakkō kinrui oyobi hakkō seihin no kenkyū" (Research on Chinese-produced fermentation molds and fermented goods). PhD diss., Tokyo Imperial University, 1925.

———. "Shinasan 'kiku' ni tsuite" (On Chinese-produced *kōji*). *Shushi kenkyū* 13 (1995): 133–83.

———. "Taishi bunka jigyō to nōgaku" (Cultural projects in China and agricultural science). *Dainihon nōkaihō* 521 (1924): 10–14.

———. *Tōa hakkō kagaku ronkō* (A study of East Asian fermentation chemistry). Tokyo: Daiichi shuppan, 1945.

Yamazaki Momoji and Mukui Masao. "Shaoshinchū ni tsuite (1)" (On Shaoxing wine [1]). *Nihon jōzō kyōkai zasshi* 12, no. 5 (1917): 14–25.

———. "Shaoshinchū ni tsuite (2)" (On Shaoxing wine [2]). *Nihon jōzō kyōkai zasshi* 12, no. 6 (1917): 30–39.

Yamazaki Yoshikazu. "Higeta Shōyu gijutsu enkaku no kaisō" (Reflections on the history of Higeta Shōyu technology). *Chōmi kagaku* 21, no. 5 (1974): 2–6.

Yoshida Toshiomi. "Hakkō sōchi" (Fermentation reactors). In *Hakkō handobukku* (Fermentation handbook), edited by Baioindasutorii kyōkai hakkō to taisha kenkyūkai, 491–95. Tokyo: Kyōritsu shuppan, 2001.

Yukimatsu Bunta. "Jōryūshu (shusei, shōchū) seizōgyō no genzai oyobi shōrai" (Present and future of the distillation [alcohol and *shōchū*] industries). *Nihon jōzō kyōkai zasshi* 16, no. 8 (1921): 1–5.

Zenda Naozō. "Jōryūshu jōzōyō toshite no 'amiro'-hō ni tsuite" (On the "amylo" method for use in distilled liquor brewing). *Jōzō kyōkai zasshi* 9, no. 9 (1914): 1–12.

WESTERN-LANGUAGE SOURCES

Institutional Histories

Editorial Board of "History of Innovative Research and Drug Discovery at Takeda," ed. *History of Innovative Research and Drug Discovery at Takeda*. Osaka, Japan: Pharmaceutical Research Division, Takeda Pharmaceutical Company, 2011.

General Headquarters, Supreme Commander for the Allied Powers, Public Health and Welfare Section. *Mission and Accomplishments of the Occupation in the Public Health and Welfare Fields*. Tokyo: Supreme Commander for the Allied Powers, 1949.

———. *Public Health and Welfare in Japan, Annual Summary 1949, Volume I.* Tokyo: Supreme Commander for the Allied Powers, 1949.

———. *Public Health and Welfare in Japan, Annual Summary 1950, Volume I.* Tokyo: Supreme Commander for the Allied Powers, 1950.

Higher Education and Science Bureau, Ministry of Education, Japan. *A General Catalogue of the Cultures of Micro-organisms Maintained in the Japanese Collections.* Tokyo: Monbushō Daigaku Gakujutsu Kyoku, 1953.

Supreme Commander for the Allied Powers. *Summation of Non-Military Activities in Japan*, no. 15 (December 1946).

Published Sources

Ackert, Lloyd. "From the Thermodynamics of Life to Ecological Microbiology: Sergei Vinogradskii and the 'Cycle of Life,' 1850–1950." PhD diss., Johns Hopkins University, 2004.

Akera, Atsushi. *Calculating a Natural World: Scientists, Engineers, and Computers during the Rise of U.S. Cold War Research.* Cambridge, MA: MIT Press, 2007.

Aldous, Christopher, and Akihito Suzuki. *Reforming Public Health in Occupied Japan, 1945–52: Alien Prescriptions?* Abingdon, Oxon, UK: Routledge, 2012.

Alexander, Jeffrey W. *Brewed in Japan: The Evolution of the Japanese Beer Industry.* Honolulu: University of Hawai'i Press, 2014 [2013].

Amsden, Alice H. "Taiwan's Economic History: A Case of Etatisme and a Challenge to Dependency Theory." *Modern China* 5 (1979): 341–79.

Amsterdamska, Olga. "Between Medicine and Science: The Research Career of Oswald T. Avery." In *Medicine and Change: Historical and Sociological Studies of Medical Innovation*, edited by Ilana Löwy, 182–212. Paris: Inserm, 1993.

———. "Medical and Biological Constraints: Early Research on Variation in Bacteriology." *Social Studies of Science* 17 (1987): 657–87.

———. "Stabilizing Instability: The Controversy over Cyclogenic Theories of Bacterial Variation During the Interwar Period." *Journal of the History of Biology* 24 (1991): 191–222.

Anderson, Warwick. "Natural Histories of Infectious Disease: Ecological Vision in Twentieth-Century Biomedical Science." *Osiris* 19 (2004): 39–61.

Andreeva, Anna. Review of *Rice, Agriculture, and the Food Supply in Premodern Japan: Gokoku bunka*, by Charlotte von Verschuer. *East Asian Science, Technology and Society* 13 (2019): 155–57.

Apple, Rima D. *Vitamania: Vitamins in American Culture*. New Brunswick, NJ: Rutgers University Press, 1996.

Arch, Jakobina. "Whale Meat in Early Postwar Japan: Natural Resources and Food Culture." *Environmental History* 21 (2016): 467–87.

Atkins, E. Taylor. *Primitive Selves: Koreana in the Japanese Colonial Gaze*. Berkeley: University of California Press, 2010.

Atkinson, R. W. "Brewing in Japan." *Nature* 18 (1878): 521–23.

———. *The Chemistry of Saké-Brewing*. Tokyo: Tōkiō Daigaku, 1881.

Banno Junji. "Japanese Industrialists and Merchants and the Anti-Japanese Boycotts in China, 1919–1928." In *The Japanese Informal Empire in China, 1895–1937*, edited by Peter Duus, Ramon H. Myers, and Mark R. Peattie, 314–29. Princeton, NJ: Princeton University Press, 1989.

Barshay, Andrew E. *The Social Sciences in Modern Japan: The Marxian and Modernist Traditions*. Berkeley: University of California Press, 2004.

Bartholomew, James. *The Formation of Science in Japan: Building a Research Tradition*. New Haven, CT: Yale University Press, 1989.

———. "Japanese Nobel Candidates in the First Half of the Twentieth Century." *Osiris* 13 (1998): 238–84.

Bay, Alexander R. *Beriberi in Modern Japan: The Making of a National Disease*. Rochester, NY: University of Rochester Press, 2012.

Bensaude-Vincent, Bernadette, Hervé Arribart, Yves Bouligand, and Clément Sanchez. "Chemists and the School of Nature." *New Journal of Chemistry* 26 (2002): 1–5.

Biss, Eula. *On Immunity: An Inoculation*. Minneapolis: Graywolf Press, 2014.

Bonah, Christian. "The 'Experimental Stable' of the BCG Vaccine: Safety, Efficacy, Proof, and Standards, 1921–1933." *Studies in History and Philosophy of Biological and Biomedical Sciences* 36, no. 4 (2005): 696–721.

Brès, P., and L. Chambon. "The Pasteur Institutes Worldwide." *Cell* 7 (1982): 83–84.

Brock, Malcolm V. *Biotechnology in Japan*. London: Routledge, 1989.

Brock, Thomas D. *The Emergence of Bacterial Genetics*. Cold Spring Harbor, NY: Cold Spring Harbor Laboratory Press, 1990.

Brown, Philip C. Review of *Rice, Agriculture, and the Food Supply in Premodern Japan: Gokoku bunka*, by Charlotte von Verschuer. *Monumenta Nipponica* 72 (2017): 71–73.

Bud, Robert. "Innovators, Deep Fermentation and Antibiotics: Promoting Applied Science before and after the Second World War." *Dynamis* 31 (2011): 323–42.

———. *Penicillin: Triumph and Tragedy*. Oxford: Oxford University Press, 2007.

———. *The Uses of Life: A History of Biotechnology*. Cambridge: Cambridge University Press, 1993.

Bud, Robert, ed. "Focus: Applied Science." *Isis* 103 (2012): 515–63.

Burns, Susan L. "The Japanese Patent Medicine Trade in East Asia." In *Gender, Health, and History in Modern East Asia*, edited by Angela Ki Che Leung and Izumi Nakayama, 139–65. Hong Kong: Hong Kong University Press, 2017.

———. "Marketing 'Women's Medicines:' Gender, OTC Herbal Medicines and Medical Culture in Modern Japan." *Asian Medicine* 5 (2009): 136–72.

Calmette, Albert. "Contribution à l'étude des ferments de l'amidon: La levure chinoise." *Annales de l'Institut Pasteur* 6 (1892): 604–20.

Campos, Luis. "That Was the Synthetic Biology That Was." In *The Technoscience and Its Societal Consequences*, edited by Markus Schmidt, Alexander Kelle, Agomoni Ganguli-Mitra, and Huib Vriend, 5–21. Dordrecht, Netherlands: Springer, 2009.

Campos, Luis, and Alexander von Schwerin, eds. *Max Planck Institute for the History of Science Preprint 393. Making Mutations: Objects, Practices, Contexts*. Berlin: MPIWG, 2010.

Capocci, Mauro. "'A Chain Is Gonna Come.' Building a Penicillin Production Plant in Post-War Italy." *Dynamis* 31 (2011): 343–62.

Ceccarelli, Giovanni, Alberto Grandi, and Stefano Magagnoli, eds. *Typicality in History: Tradition, Innovation, and Terroir / La typicité dans l'histoire: Tradition, innovation et terroir*. Brussels: P.I.E. Peter Lang, 2013.

Ceccatti, John Simmons. "Science in the Brewery: Pure Yeast Culture and the Transformation of Brewing Practices in Germany at the End of the 19th Century." PhD diss., University of Chicago, 2001.

Chakrabarty, Dipesh. *Provincializing Europe: Postcolonial Thought and Historical Difference*. Princeton, NJ: Princeton University Press, 2000.

Chang, Hasok. *Inventing Temperature: Measurement and Scientific Progress*. Oxford: Oxford University Press, 2004.

———. *Is Water H$_2$O? Evidence, Realism and Pluralism*. Dordrecht, Netherlands: Springer, 2012.

Choi, Hyungsub. "Technology Importation, Corporate Strategies, and the Rise of the Japanese Semiconductor Industry in the 1950s." *Comparative Technology Transfer and Society* 6, no. 2 (2008): 103–26.

Christian, David. *Maps of Time: An Introduction to Big History*. Berkeley: University of California Press, 2004.

Chu, Pingyi. "Needham in Taiwan: Translating *Science and Civilisation in China* as Politics of Modernity and Identity." *East Asian Science, Technology and Society* 14 (2020): 379–92.

Chung, King-Thom, and Charles Biggers. "Albert Leon Charles Calmette (1863–1933) and the Antituberculous BCG Vaccination." *Perspectives in Biology and Medicine* 44 (2001): 379–89.

Chung, Yuehtsen Juliette. *Struggle for National Survival: Eugenics in Sino-Japanese Contexts, 1896–1945.* New York: Routledge, 2002.

Clancey, Gregory. *Earthquake Nation: The Cultural Politics of Japanese Seismicity, 1868–1930.* Berkeley: University of California Press, 2006.

Clarke, Adele E., and Joan H. Fujimura, eds. *The Right Tools for the Job: At Work in Twentieth-Century Life Sciences.* Princeton, NJ: Princeton University Press, 1992.

Coleman, William. *Biology in the Nineteenth Century: Problems of Form, Function, and Transformation.* Cambridge: Cambridge University Press, 1977 [1971].

Collins, Steven W. *The Race to Commercialize Biotechnology: Molecules, Markets and the State in the United States and Japan.* London: Routledge, 2004.

Condrau, Flurin, and Robert G. W. Kirk. "Negotiating Hospital Infections: The Debate between Ecological Balance and Eradication Strategies in British Hospitals, 1947–1969." *Dynamis* 31 (2011): 385–406.

Cozzoli, Daniele. "Penicillin and the Reconstruction of Japan." *Medicina nei secoli arte e scienza* 26, no. 2 (2014): 469–84.

Creager, Angela N. H. "Adaptation or Selection? Old Issues and New Stakes in the Postwar Debates over Bacterial Drug Resistance." *Studies in History and Philosophy of Biological and Biomedical Sciences* 38 (2007): 159–90.

———. "Building Biology across the Atlantic." *Journal of the History of Biology* 36 (2003): 579–89.

———. *The Life of a Virus: Tobacco Mosaic Virus as an Experimental Model, 1930–1965.* Chicago: University of Chicago Press, 2001.

Creager, Angela N. H., and Jean-Paul Gaudillière. "Meanings in Search of Experiments and Vice-Versa: The Invention of Allosteric Regulation in Paris and Berkeley, 1959–1968." *Historical Studies in the Physical and Biological Sciences* 27 (1996): 1–89.

Cryns, Frederik. "The Influence of Hermann Boerhaave's Mechanical Concept of the Human Body in Nineteenth-Century Japan." In *Dodonæus in Japan: Translation and the Scientific Mind in the Tokugawa Period,* edited by W. F. Vande Walle and Kazuhiko Kasaya, 343–63. Leuven, Belgium: Leuven University Press, 2001.

Cunningham, Andrew. "Transforming Plague: The Laboratory and the Identity of Infectious Diseases." In *The Laboratory Revolution in Medicine*, edited by Andrew Cunningham and Perry Williams, 209–47. Cambridge: Cambridge University Press, 1992.

Curry, Helen Anne. *Evolution Made to Order: Plant Breeding and Technological Innovation in Twentieth-Century America*. Chicago: University of Chicago Press, 2016.

Cwiertka, Katarzyna. *Cuisine, Colonialism and Cold War: Food in Twentieth-Century Korea*. London: Reaktion Books, 2012.

———. *Modern Japanese Cuisine: Food, Power and National Identity*. London: Reaktion Books, 2006.

———. "The Soy Sauce Industry in Korea: Scrutinising the Legacy of Japanese Colonialism." *Asian Studies Review* 30 (2006): 389–410.

———. "Western Food and the Making of the Japanese Nation-State." In *The Politics of Food*, edited by Marianne Lien and Brigitte Nerlich, 121–39. Oxford: Berg, 2004.

Cwiertka, Katarzyna, ed. *Food and War in Mid-Twentieth Century East Asia*. Farnham, UK: Ashgate, 2013.

Cwiertka, Katarzyna, and Yujen Chen. "The Shadow of Shinoda Osamu: Food Research in East Asia." In *Writing Food History: A Global Perspective*, edited by Kyri W. Claflin and Peter Scholliers, 181–96. London: Berg, 2013.

Davies, Julian, and Dorothy Davies, "Origins and Evolution of Antibiotic Resistance." *Microbiology and Molecular Biology Reviews* 74 (2010): 417–33.

De Chadarevian, Soraya. *Designs for Life: Molecular Biology after World War II*. Cambridge: Cambridge University Press, 2002.

De Kruif, Paul. *Microbe Hunters: The Classic Book on the Major Discoveries of the Microscopic World*. New York: Mariner Books, 2002 [1926].

Dinmore, Eric Gordon. "A Small Island Nation Poor in Resources: Natural and Human Resource Anxieties in Trans-World War II Japan." PhD diss., Princeton University, 2006.

Doudna, Jennifer A., and Samuel H. Sternberg. *A Crack in Creation: Gene Editing and the Unthinkable Power to Control Evolution*. New York: Mariner Books, 2017.

Douglas, Mary. *How Institutions Think*. Frank W. Abrams Lectures, presented at Syracuse University, March 1985. Syracuse, NY: Syracuse University Press, 1986.

Dower, John W. *Embracing Defeat: Japan in the Wake of World War II*. New York: W. W. Norton, 1999.

Driscoll, Mark. *Absolute Erotic, Absolute Grotesque: The Living, Dead, and Undead in Japan's Imperialism, 1895–1945*. Durham, NC: Duke University Press, 2010.

Dubos, René. *Man Adapting*. New Haven, CT: Yale University Press, 1980 [1965].

———. *Mirage of Health: Utopias, Progress, and Biological Change*. New York: Harper & Bros., 1959.

Duus, Peter. "Economic Dimensions of Meiji Imperialism: The Case of Korea, 1895–1910." In *The Japanese Colonial Empire, 1895–1945*, edited by Ramon H. Myers and Mark R. Peattie, 128–71. Princeton, NJ: Princeton University Press, 1984.

———. "Introduction: Japan's Informal Empire in China, 1895–1937: An Overview." In *The Japanese Informal Empire in China, 1895–1937*, edited by Peter Duus, Ramon H. Myers, and Mark R. Peattie, xi–xxix. Princeton, NJ: Princeton University Press, 1989.

Elman, Benjamin A. "The Great Reversal: The 'Rise of Japan' and the 'Fall of China' after 1895 as Historical Fables." Lecture sponsored by the Fairbank Center for Chinese Studies, Harvard University, Cambridge, MA, April 13, 2011.

———. "Sinophiles and Sinophobes in Tokugawa Japan: Politics, Classicism, and Medicine during the Eighteenth Century." *East Asian Science, Technology and Society* 2 (2008): 93–121.

Falk, Raphael. "What Is a Gene?" *Studies in History and Philosophy of Science Part A* 17 (1986): 133–73.

Fan, Yajiun. "The Locality versus the Extraneous Nature of the Liquor Industry of Taiwan." Paper presented at "Making East Asian Foods: Technologies and Values, 19th-21st Centuries," Hong Kong Institute of Humanities and Social Sciences, University of Hong Kong, May 30–31, 2019.

Finlay, Robert. "China, the West, and World History in Joseph Needham's *Science and Civilisation in China*." *Journal of World History* 11 (2000): 265–303.

Foster, J. W. "A View of Microbiological Science in Japan." *Applied Microbiology* 9 (1961): 434–51.

Foucault, Michel. *The Birth of the Clinic: An Archaeology of Medical Perception*. Translated by A. M. Sheridan Smith. New York: Vintage Books, 1975.

————. *The Order of Things: An Archaeology of the Human Sciences.* New York: Pantheon Books, 1970.

Francks, Penelope. "Inconspicuous Consumption: Sake, Beer, and the Birth of the Consumer in Japan." *Journal of Asian Studies* 68 (2009): 135–64.

————. *The Japanese Consumer: An Alternative Economic History of Modern Japan.* Cambridge: Cambridge University Press, 2009.

Fruin, Mark. *Kikkoman: Company, Clan, and Community.* Cambridge, MA, and London: Harvard University Press, 1983.

Frumer, Yulia. *Making Time: Astronomical Time Measurement in Tokugawa Japan.* Chicago: University of Chicago Press, 2018.

Fu, Jia-Chen. *The Other Milk: Reinventing Soy in Republican China.* Seattle: University of Washington Press, 2018.

Fujihara, Tatsuji. "Colonial Seeds, Imperialist Genes: Hōrai Rice and Agricultural Development." Translated by Hiromi Mizuno and Aaron S. Moore. In *Engineering Asia: Technology, Colonial Development, and the Cold War Order*, edited by Hiromi Mizuno, Aaron S. Moore, and John DiMoia, 137–61. London: Bloomsbury, 2018.

Fujii, Ryochi. "Changes in Antibiotic Consumption in Japan during the Past 40 Years." *Japanese Journal of Antibiotics* 37 (1984): 2261–70.

Fujitani, Kevin. "A Child of Many Fathers: The Question of Credit for the Discovery of Thiamine, 1884–1936." In *Proceedings of the International Workshop on the History of Chemistry 2015, Tokyo (IWHC 2015 Tokyo): Transformation of Chemistry from the 1920s to the 1960s*, edited by Masanori Kaji et al., 90–98. Tokyo: Japanese Society for the History of Chemistry, 2016.

Fukushima, Masato. "Between the Laboratory and the Policy Process: Research, Scientific Community, and Administration in Japan's Chemical Biology." *East Asian Science, Technology and Society* 7 (2013): 7–33.

————. "Constructing Failure in Big Biology: The Socio-Technical Anatomy of Japan's Protein 3000 Project." *Social Studies of Science* 46, no. 1 (2016): 7–33.

Furukawa, Yasu. "Gen-itsu Kita and the Kyoto School's Formation." In *Igniting the Chemical Ring of Fire: Historical Evolution of the Chemical Communities of the Pacific Rim*, edited by Seth G. Rasmussen, 157–68. Singapore: World Scientific, 2018.

García-Sancho, Miguel. *Biology, Computing, and the History of Molecular Sequencing: From Proteins to DNA, 1945–2000.* New York: Palgrave Macmillan, 2012.

Garon, Sheldon. "The Home Front and Food Insecurity in Wartime Japan: A Transnational Perspective." In *Consumer on the Home Front: Second World War Civilian Consumption in Comparative Perspective*, edited by Hartmut Berghoff, Jan Logemann, and Felix Römer, 29–53. Oxford: Oxford University Press, 2017.

Gaudillière, Jean-Paul. "Introduction: Drug Trajectories." *Studies in History and Philosophy of Biological and Biomedical Sciences* 36 (2005): 603–11.

———. *Inventer la biomédecine: La France, l'Amérique et la production des savoirs du vivant (1945–1965)*. Paris: La Découverte, 2002.

Gaudillière, Jean-Paul, and Bernd Gausemeier. "Molding National Research Systems: The Introduction of Penicillin to France and Germany." *Osiris* 20 (2005): 180–202.

George, Timothy S. *Minamata: Pollution and the Struggle for Democracy in Postwar Japan*. Cambridge, MA: Harvard University Press, 2001.

Gluck, Carol. "The Past in the Present." In *Postwar Japan as History*, edited by Andrew Gordon, 64–95. Berkeley: University of California Press, 1993.

Godart, G. Clinton. *Darwin, Dharma, and the Divine: Evolutionary Theory and Religion in Modern Japan*. Honolulu: University of Hawai'i Press, 2017.

Golinski, Jan. *Making Natural Knowledge: Constructivism and the History of Science*. Chicago: University of Chicago Press, 2008.

Golley, Gregory. *When Our Eyes No Longer See: Realism, Science, and Ecology in Japanese Literary Modernism*. Cambridge, MA: Harvard University Press, 2008.

Gordon, Andrew. *Labor and Imperial Democracy in Prewar Japan*. Berkeley: University of California Press, 1991.

———. *A Modern History of Japan: From Tokugawa Times to the Present*, 3rd ed. New York: Oxford University Press, 2014.

Goto, Akira. "Co-operative Research in Japanese Manufacturing Industries." In *Innovation in Japan*, eds. Akira Goto and Hiroyuki Odagiri, 256–74. Oxford: Clarendon Press, 1997.

———. Introduction to *Innovation in Japan*, edited by Akira Goto and Hiroyuki Odagiri, 1–19. Oxford: Clarendon Press, 1997.

Gradmann, Christoph. "Isolation, Contamination, and Pure Culture: Monomorphism and Polymorphism of Pathogenic Micro-Organisms as Research Problem 1860–1880." *Perspectives on Science* 9 (2001): 147–72.

Grote, Mathias. "Hybridizing Bacteria, Crossing Methods, Cross-Checking Arguments: The Transition from Episomes to Plasmids (1961–1969)." *History and Philosophy of the Life Sciences* 30 (2008): 407–30.

————. *Membranes to Molecular Machines: Active Matter and the Remaking of Life*. Chicago: University of Chicago Press, 2019.

Grote, Mathias, and Lara Keuck. "Conference report '*Stoffwechsel*. Histories of Metabolism,' Workshop Organized by Mathias Grote at Technische Universität Berlin, November 28-29th, 2014." *History and Philosophy of the Life Sciences* 37 (2015): 210-18.

Grubaugh, Nathan D., Jason T. Ladner, Philippe Lemey, Oliver G. Pybus, Andrew Rambaut, Edward C. Holmes, and Kristian G. Andersen. "Tracking Virus Outbreaks in the Twenty-First Century." *Nature Microbiology* 4 (2019): 10-19.

Guénel, Annick. "The Creation of the First Overseas Pasteur Institute; or, The Beginning of Albert Calmette's Pastorian Career." *Medical History* 43 (1999): 1-25.

Han, Woo-keun. *The History of Korea*. Seoul: Eul-Yoo Publishing, 1970.

Hanley, Susan B. *Everyday Things in Premodern Japan: The Hidden Legacy of Material Culture*. Berkeley: University of California Press, 1997.

Haraway, Donna. "Situated Knowledges: The Science Question in Feminism and the Privilege of Partial Perspective." *Feminist Studies* 14 (1988): 575-99.

Harootunian, Harry. *Overcome by Modernity: History, Culture, and Community in Interwar Japan*. Princeton, NJ: Princeton University Press, 2000.

Hart, Roger. "Beyond Science and Civilization: A Post-Needham Critique." *East Asian Science, Technology, and Medicine* 16 (1999): 88-114.

Hasegawa, Takezi. "Japanese Culture Collections of Micro-organisms in the Field of Industry: Their Histories and Actual State." *Annual Reports of the Institute for Fermentation, Osaka* 3 (1967): 139-43.

Hashimoto, Takehiko. "A Hesitant Relationship Reconsidered: University-Industry Cooperation in Postwar Japan." In *Historical Essays on Japanese Technology*, 173-92. Tokyo: University of Tokyo Center for Philosophy, 2009. Reprint of "The Hesitant Relationship Reconsidered: University-Industry Cooperation in Postwar Japan," in *Industrializing Knowledge: University-Industry Linkages in Japan and the United States*, edited by Lewis M. Branscomb, Fumio Kodama, and Richard Florida, 234-51. Cambridge, MA: MIT Press, 1999.

————. "Introducing a French Technological System: The Origin and Early History of the Yokosuka Dockyard." *East Asian Science, Technology, and Medicine* 16 (1999): 53-72.

————. "Japanese Clocks and the History of Punctuality in Modern Japan." *East Asian Science, Technology and Society* 2 (2008): 123-33.

————. "The Roles of Corporations, Universities, and the Government before and after 1990." In *Historical Essays on Japanese Technology*, 201–12. Tokyo: University of Tokyo Center for Philosophy, 2009.

———. "Technological Research Associations and University-Industry Cooperation." In *Historical Essays on Japanese Technology*, 193–99. Tokyo: University of Tokyo Center for Philosophy, 2009.

Hawgood, Barbara. "Albert Calmette (1863–1933) and Camille Guérin (1872–1961): The C and G of BCG Vaccine." *Journal of Medical Biography* 15, no. 3 (2007): 139–46.

Hayami, Akira. *Japan's Industrious Revolution: Economic and Social Transformations in the Early Modern Period*. London: Springer, 2015.

Hein, Laura E. *Fueling Growth: The Energy Revolution and Economic Policy in Postwar Japan*. Cambridge, MA: Harvard University Press, 1990.

———. "Growth versus Success: Japan's Economic Policy in Historical Perspective." In *Postwar Japan as History*, edited by Andrew Gordon, 99–122. Berkeley: University of California Press, 1993.

Heinrich, Ari Larissa. *The Afterlife of Images: Translating the Pathological Body between China and the West*. Durham, NC: Duke University Press, 2008.

Helmreich, Stefan. *Alien Ocean: Anthropological Voyages in Microbial Seas*. Berkeley: University of California Press, 2009.

Ho, Samuel Pao-San. "Colonialism and Development: Korea, Taiwan, and Kwantung." In *The Japanese Colonial Empire, 1895–1945*, edited by Ramon H. Myers and Mark R. Peattie, 348–98. Princeton, NJ: Princeton University Press, 1984.

Holmes, Frederic Lawrence. *Hans Krebs*. 2 vols. New York: Oxford University Press, 1991–1993.

Hong, Sookyeong. "Digesting Modernity, Healing with Nature: The Birth of a 'Natural' Food Movement in Meiji Japan, 1905–1910." *Global Environment* 11 (2018): 105–29.

Honigsbaum, Mark, and Pierre-Olivier Méthot. "Introduction: Microbes, Networks, Knowledge—Disease Ecology and Emerging Infectious Diseases in Time of COVID-19." *History and Philosophy of the Life Sciences* 42, no. 28 (2020).

Howell, David L. *Capitalism from Within: Economy, Society, and the State in a Japanese Fishery*. Berkeley: University of California Press, 1995.

———. *Geographies of Identity in Nineteenth-Century Japan*. Berkeley: University of California Press, 2005.

———. "Urbanization, Trade, and Merchants." In *Japan Emerging: Premodern History to 1850*, edited by Karl F. Friday, 356–65. Boulder, CO: Westview Press, 2012.

Hsia, Florence, and Dagmar Schäfer, eds. "History of Science, Technology, and Medicine: A Second Look at Joseph Needham." *Isis* 110 (2019): 91–136.

Huang, Chun-chieh. "Historical Reflections on the Postwar Taiwan Experience from an Agrarian Perspective." In *Postwar Taiwan in Historical Perspective*, edited by Chun-chieh Huang and Feng-fu Tsao, 17–35. Potomac, MD: University Press of Maryland, 1998.

Huang, H. T. *Fermentations and Food Science*. Part 5 of *Biology and Biological Technology*, vol. 6 in the series Science and Civilisation in China. Cambridge: Cambridge University Press, 2000.

Hughes, Sally Smith. *Genentech: The Beginnings of Biotech*. Chicago: University of Chicago Press, 2011.

Ishige, Naomichi. "Cultural Aspects of Fermented Fish Products in Asia." In *Fish Fermentation Technology*, edited by Cherl-Ho Lee, Keith Steinkraus, and P. J. Alan Reilly, 13–32. Tokyo: United Nations University Press, 1993.

———. *The History and Culture of Japanese Food*. London: Kegan Paul, 2001.

Itakura, Kiyonobu, and Eri Yagi. "The Japanese Research System and the Establishment of the Institute of Physical and Chemical Research." In *Science and Society in Modern Japan: Selected Historical Sources*, edited by Shigeru Nakayama, David L. Swain, and Eri Yagi, 158–201. Cambridge, MA: MIT Press, 1974.

Ito, Yoshiaki. "Science Lessons." Review of *Genomu haiboku* (A Defeat in the Genome Project), by Nobuhito Kishi. *Nature* 433 (2005): 107–8.

Jannetta, Ann. *The Vaccinators: Smallpox, Medical Knowledge, and the "Opening" of Japan*. Stanford, CA: Stanford University Press, 2007.

Jansen, Marius B. *Japan and China: From War to Peace, 1894–1972*. Chicago: Rand McNally, 1975.

"Japan: Genetech's Late Bloomer." *Seedling*, March 25, 1998. https://www.grain.org/en/category/62-seedling-march-1998.

Jardine, Boris, Emma Kowal, and Jenny Bangham, eds. "How Collections End." *BJHS Themes* 4 (2019).

Jas, Nathalie. *Au carrefour de la chimie et de l'agriculture: Les sciences agronomiques en France et en Allemagne, 1840–1914*. Paris: Éditions des archives contemporaines, 2001.

Jinek, Martin, Krzysztof Chylinski, Ines Fonfara, Michael Hauer, Jennifer A. Doudna, and Emmanuelle Charpentier. "A Programmable Dual-RNA-Guided DNA Endonuclease in Adaptive Bacterial Immunity." *Science* 337 (2012): 816–21.

Johnson, Ann. *Hitting the Brakes: Engineering Design and the Production of Knowledge.* Durham, NC: Duke University Press, 2009.

———. "What If We Wrote the History of Science from the Perspective of Applied Science?" *Historical Studies in the Natural Sciences* 38 (2008): 610–20.

Johnson, Chalmers. *MITI and the Japanese Miracle: The Growth of Industrial Policy.* Stanford, CA: Stanford University Press, 1982.

Johnston, B. F. *Japanese Food Management in World War II.* Stanford, CA: Stanford University Press, 1953.

Jones, D. T. "The Strategic Importance of Butanol for Japan during WWII: A Case Study of the Butanol Fermentation Process in Taiwan and Japan." In *Systems Biology of Clostridium,* edited by Peter Dürre, 220–72. London: Imperial College Press, 2014.

Judson, Horace Freeland. *The Eighth Day of Creation: The Makers of the Revolution in Biology.* New York: Simon and Schuster, 1979.

Ka, Chih-ming. "Agrarian Development, Family Farms and Sugar Capital in Colonial Taiwan, 1895–1945." *Journal of Peasant Studies* 18 (1991): 206–40.

Kaji, Masanori. "The Transformation of Organic Chemistry in Japan: From Majima Riko to the Third International Symposium on the Chemistry of Natural Products." In *Proceedings of the International Workshop on the History of Chemistry 2015, Tokyo (IWHC 2015 Tokyo): Transformation of Chemistry from the 1920s to the 1960s,* edited by Masanori Kaji et al., 14–19. Tokyo: Japanese Society for the History of Chemistry, 2016.

Kamminga, Harmke. "Vitamins and the Dynamics of Molecularization: Biochemistry, Policy and Industry in Britain, 1914–1939." In *Molecularizing Biology and Medicine: New Practices and Alliances, 1910s–1970s,* edited by Soraya de Chadarevian and Harmke Kamminga, 78–98. London: Taylor & Francis, 2005 [1998].

Kamminga, Harmke, and Andrew Cunningham, eds. *The Science and Culture of Nutrition, 1840–1940.* Amsterdam: Rodopi, 1995.

Katz, Sandor Ellix. *The Art of Fermentation: An In-Depth Exploration of Essential Concepts and Processes from Around the World.* White River Junction, VT: Chelsea Green Publishing, 2012.

Kay, Lily E. *Who Wrote the Book of Life? A History of the Genetic Code.* Stanford, CA: Stanford University Press, 2000.

Kevles, Daniel. "Ananda Chakrabarty Wins a Patent: Biotechnology, Law, and Society, 1972–1980." *Historical Studies in the Physical and Biological Sciences* 25 (1994): 111–35.

Kikuchi, Yoshiyuki. "Analysis, Fieldwork and Engineering: Accumulated

Practices and the Formation of Applied Chemistry Teaching at Tokyo University, 1874–1900." *Historia Scientiarum* 18 (2008): 100–20.

Kim, Dong-Won. "Yoshio Nishina and Two Cyclotrons." *Historical Studies in the Physical and Biological Sciences* 36 (2006): 243–73.

Kinoshita, Shukuo, Katsunobu Tanaka, Shigezo Udaka, and Sadao Akita. "Glutamic Acid Fermentation." In *Proceedings of the International Symposium on Enzyme Chemistry, Tokyo and Kyoto, 1957,* edited by the Organizing Committee, International Symposium on Enzyme Chemistry, Science Council of Japan, 464–68. Tokyo: Maruzen, 1958.

Kirchhelle, Claas. *Pyrrhic Progress: The History of Antibiotics in Anglo-American Food Production.* New Brunswick, NJ: Rutgers University Press, 2020.

Kitasato Institute, Research Center for Biological Function, ed. *"Splendid Gifts from Microorganisms": The Achievements of Satoshi Ōmura and His Collaborators,* 2nd ed. Tokyo: Kitasato Institute, 1998.

Kline, Ronald. "Construing 'Technology' as 'Applied Science': Public Rhetoric of Scientists and Engineers in the United States, 1880–1945." *Isis* 86 (1995): 194–221.

Kneller, Robert. *Bridging Islands: Venture Companies and the Future of Japanese and American Industry.* Oxford: Oxford University Press, 2007.

———. "Intellectual Property Rights and University-Industry Technology Transfer in Japan." In *Industrializing Knowledge: University-Industry Linkages in Japan and the United States,* edited by Lewis Branscomb, Fumio Kodama, and Richard Florida, 307–47. Cambridge, MA: MIT Press, 1999.

Kobiljski, Aleksandra. "Poor Planning and Good Ideas: Technology Transfer and Innovation in the Meiji Industrialization." Paper presented at the annual meeting for the Association for Asian Studies, Philadelphia, March 27–30, 2014.

Kodama, Fumio. "Japan's Unique Capacity to Innovate: Technology Fusion and Its International Implications." In *Japan's Growing Technological Capability: Implications for the U.S. Economy,* edited by. Thomas S. Arrison, C. Fred Bergsten, Edward M. Graham, and Martha Caldwell Harris, 147–64. Washington: National Academy Press, 1992.

Kohler, Robert E. "Systems of Production: Drosophila, Neurospora, and Biochemical Genetics." *Historical Studies in the Physical and Biological Sciences* 22 (1991): 87–130.

Koizumi, Takeo. "Traditional Japanese Foods and the Mystery of Fermentation." *Food Culture* 1 (2000): 20–23. https://www.kikkoman.co.jp/kiifc/foodculture/pdf_01/e_020_023.pdf.

Koizumi, Tatsuji. "Biofuel Programs in East Asia: Developments, Perspectives, and Sustainability." In *Environmental Impact of Biofuels*, edited by Marco Aurelio Dos Santos Bernardes, 207–26. Rijeka, Croatia, and Shanghai: InTech, 2011.

Komagata, Kazuo. "Microbial Systematics, 'Weaving Threads into Cloth.'" *Bulletin of BISMiS* 2, no. 1 (2011): 33–60.

Kumazawa, Joichi, and Morimasa Yagisawa. "The History of Antibiotics: The Japanese Story." *Journal of Infection and Chemotherapy* 8, no. 2 (2002): 125–33.

Landecker, Hannah. "Antibiotic Resistance and the Biology of History." *Body & Society* 22 (2016): 19–52.

———. "Food as Exposure: Nutritional Epigenetics and the New Metabolism." *BioSocieties* 6, no. 2 (2011): 167–94.

———. "From Social Structure to Gene Regulation, and Back: A Critical Introduction to Environmental Epigenetics for Sociology." *Annual Review of Sociology* 39 (2013): 333–57.

———. "Metabolism, Reproduction, and the Aftermath of Categories." *Feminist & Scholar Online* 11 (2013).

Latour, Bruno. *The Pasteurization of France*. Translated by Alan Sheridan and John Law. Cambridge, MA: Harvard University Press, 1988 [1984].

Lécuyer, Christophe. *Making Silicon Valley: Innovation and the Growth of High Tech, 1930–1970*. Cambridge, MA: MIT Press, 2006.

Lee, Joyman. "Closest Model, Rival, and Fateful Enemy: China's Political Economy, Law, and Japan." In *Routledge Handbook of Revolutionary China*, edited by Alan Baumler, 243–57. Abingdon, Oxon, UK: Routledge, 2019.

———. "Economics." In *The Making of the Human Sciences in China: Historical and Conceptual Foundations*, edited by Howard Chiang, 267–82. Leiden: Brill, 2019.

Lee, Jung. "Between Universalism and Regionalism: Universal Systematics from Imperial Japan." *British Journal for the History of Science* 48 (2015): 661–84.

Lee, Sophia. "The Foreign Ministry's Cultural Agenda for China: The Boxer Indemnity." In *The Japanese Informal Empire in China, 1895–1937*, edited by Peter Duus, Ramon H. Myers, and Mark R. Peattie, 272–306. Princeton, NJ: Princeton University Press, 1989.

Lee, Victoria. "Bioorganic Synthesis." In *Between Making and Knowing: Tools in the History of Materials Research*, edited by Joseph D. Martin and Cyrus C. M. Mody, 299–314. Singapore: World Scientific, 2020.

—————. "Scaling Up from the Bench: Fermentation Tank." In *Boxes: A Field Guide*, edited by Susanne Bauer, Maria Rentetzi, and Martina Schlünder, 288–304. Manchester, UK: Mattering Press, 2020.

—————. "Wild Toxicity, Cultivated Safety: Aflatoxin and Kōji Classification as Knowledge Infrastructure." *History and Technology* 35 (2019): 405–24.

Lesch, John E. *Science and Medicine in France: The Emergence of Experimental Physiology, 1790–1855.* Cambridge, MA: Harvard University Press, 1984.

Lewis, Michael. *Rioters and Citizens: Mass Protest in Imperial Japan.* Berkeley: University of California Press, 1990.

Lim, Jongtae. "Joseph Needham in Korea, and Korea's Position in the History of East Asian Science." *East Asian Science, Technology and Society* 14 (2020): 393–401.

Linse, Sara Snogerup. "Scientific Background on the Nobel Prize in Chemistry 2018: Directed Evolution of Enzymes and Binding Proteins." October 3, 2018. https://www.nobelprize.org/uploads/2018/10/advanced-chemistryprize-2018.pdf.

Liu, Daniel. "The Cell and Protoplasm as Container, Object, and Substance, 1835–1861." *Journal of the History of Biology* 50 (2017): 889–925.

Lodder, J., and N. J. W. Kreger-Van Rij. *The Yeasts: A Taxonomic Study.* Amsterdam: North-Holland, 1952.

Long, Pamela O. *Openness, Secrecy, Authorship: Technical Arts and the Culture of Knowledge from Antiquity to the Renaissance.* Baltimore: Johns Hopkins University Press, 2001.

Lu, Sidney Xu. *The Making of Japanese Settler Colonialism: Malthusianism and Trans-Pacific Migration, 1868–1961.* Cambridge: Cambridge University Press, 2019.

Maercker, Max, and Max Delbrück. *Handbuch der Spiritusfabrikation.* Berlin: P. Parey, 1908.

Mallapaty, Smriti. "Japan's Start-Up Gulf." *Nature* 567 (2019): S24–S25.

Marcon, Federico. *The Knowledge of Nature and the Nature of Knowledge in Early Modern Japan.* Chicago: University of Chicago Press, 2015.

Markham, S. F. *Climate and the Energy of Nations.* London: Milford, 1942.

Marlière, Philippe. "The Farther, the Safer: A Manifesto for Securely Navigating Synthetic Species Away from the Old Living World." *Systems and Synthetic Biology* 3 (2009): 77–84.

Matsumura, Wendy. *The Limits of Okinawa: Japanese Capitalism, Living Labor, and Theorizations of Community.* Durham, NC: Duke University Press, 2015.

Matta, Christina. "The Science of Small Things: The Botanical Context of German Bacteriology, 1830–1910." PhD diss., University of Wisconsin–Madison, 2009.

Mazumdar, Pauline M. H. *Species and Specificity: An Interpretation of the History of Immunology*. Cambridge: Cambridge University Press, 1995.

McCook, Stuart, ed. "Focus: Global Currents in National Histories of Science: The 'Global Turn' and the History of Science in Latin America." *Isis* 104 (2013): 773–817.

Meade, Ruselle. "Translating Technology in Japan's Meiji Enlightenment, 1870–1879." *East Asian Science, Technology and Society* 9 (2015): 253–74.

Mendelsohn, J. Andrew. "Cultures of Bacteriology: Formation and Transformation of a Science in France and Germany, 1870–1914." PhD diss., Princeton University, 1996.

————. "Lives of the Cell." *Journal of the History of Biology* 36 (2003): 1–37.

Méthot, Pierre-Olivier, and Bernardino Fantini. "Medicine and Ecology: Historical and Critical Perspectives on the Concept of 'Emerging Disease.'" *Archives internationales d'histoire des sciences* 64 (2014): 213–30.

Metzler, Mark. *Capital as Will and Imagination: Schumpeter's Guide to the Postwar Economic Miracle*. Ithaca, NY: Cornell University Press, 2013.

————. *Lever of Empire: The International Gold Standard and the Crisis of Liberalism in Prewar Japan*. Berkeley: University of California Press, 2006.

Milhaupt, Curtis J. "A Relational Theory of Japanese Corporate Governance: Contract, Culture, and the Rule of Law." *Harvard International Law Journal* 37, no. 3 (1996): 3–64.

Miller, Ian Jared. *The Nature of the Beasts: Empire and Exhibition at the Tokyo Imperial Zoo*. Cambridge, MA: Harvard University Press, 2013.

Mimura, Janis. *Planning for Empire: Reform Bureaucrats and the Japanese Wartime State*. Ithaca, NY: Cornell University Press, 2011.

Mintz, Sidney W. *Sweetness and Power: The Place of Sugar in Modern History*. New York: Viking Penguin, 1985.

Mitman, Gregg. "In Search of Health: Landscape and Disease in American Environmental History." *Environmental History* 10 (2005): 184–210.

Mitsuda, Tatsuya. "'Vegetarian' Nationalism: Critiques of Meat Eating for Japanese Bodies, 1880–1938." In *Culinary Nationalism in Asia*, edited by Michelle T. King, 23–40. London: Bloomsbury Academic, 2019.

Mizoguchi, Hazime. "Penicillin Production and the Reconstruction of the Pharmaceutical Industry." In *A Social History of Science and Technology in Contemporary Japan*, vol. 2, *Road to Self-Reliance 1952–1959*, edited by Shigeru Nakayama, Kunio Gotō, and Hitoshi Yoshioka, 541–54. Melbourne: Trans Pacific Press, 2005.

Mizuno, Hiromi. *Science for the Empire: Scientific Nationalism in Modern Japan.* Stanford, CA: Stanford University Press, 2009.

Mizuno, Hiromi, Aaron S. Moore, and John DiMoia, eds. *Engineering Asia: Technology, Colonial Development, and the Cold War Order.* London: Bloomsbury, 2018.

Mody, Cyrus C. M. *Instrumental Community: Probe Microscopy and the Path to Nanotechnology.* Cambridge, MA: MIT Press, 2011.

Moore, Aaron Stephen. *Constructing East Asia: Technology, Ideology, and Empire in Japan's Wartime Era, 1931–1945.* Stanford, CA: Stanford University Press, 2013.

Morange, Michel. *The Black Box of Biology: A History of the Molecular Revolution.* Translated by Matthew Cobb. Cambridge, MA: Harvard University Press, 2020 [1998].

Morris-Suzuki, Tessa. "The Great Translation: Traditional and Modern Science in Japan's Industrialisation." *Historia Scientiarum* 5, no. 2 (1995): 103–16.

———. *The Technological Transformation of Japan: From the Seventeenth to the Twenty-First Century.* Cambridge: Cambridge University Press, 1994.

Müller-Wille, Staffan. "Hybrids, Pure Cultures, and Pure Lines: From Nineteenth-Century Biology to Twentieth-Century Genetics." *Studies in History and Philosophy of Biological and Biomedical Sciences* 38 (2007): 796–806.

Murakami, Hideya. "Classification of the Kōji Mold." *Journal of General and Applied Microbiology* 17 (1971): 281–309.

Myers, Ramon H., and Saburō Yamada. "Agricultural Development in the Empire." In *The Japanese Colonial Empire, 1895–1945,* edited by Ramon H. Myers and Mark R. Peattie, 420–52. Princeton, NJ: Princeton University Press, 1984.

Nakamura, Ellen Gardner. *Practical Pursuits: Takano Chōei, Takahashi Keisaku, and Western Medicine in Nineteenth-Century Japan.* Cambridge, MA: Harvard University Press, 2005.

Nakamura, Takafusa. *Economic Growth in Prewar Japan.* Translated by Robert A. Feldman. New Haven, CT: Yale University Press, 1983.

Nakaoka, Tetsuro. "The Transfer of Cotton Manufacturing Technology from Britain to Japan." In *International Technology Transfer: Europe, Japan and the USA, 1700–1914,* edited by David J. Jeremy, 181–98. Aldershot, UK: Edward Elgar, 1991.

Nakayama, Shigeru. "The Central Laboratory Boom and the Rise of Corporate R&D." In *A Social History of Science and Technology in Contemporary Japan*, vol. 3, *High Economic Growth Period, 1960–1969*, edited by Shigeru Nakayama and Kunio Gotō, 67–77. Melbourne: Trans Pacific Press, 2006.

————. "The International Exchange of Scientific Information." In *A Social History of Science and Technology in Contemporary Japan*, vol. 1, *The Occupation Period, 1945–1952*, edited by Shigeru Nakayama and Kunio Gotō, 237–48. Melbourne: Trans Pacific Press, 2001.

————. "Overseas Study Leave and Participation in International Conferences." In *A Social History of Science and Technology in Contemporary Japan*, vol. 2, *Road to Self-Reliance, 1952–1959*, edited by Shigeru Nakayama and Kunio Gotō, 271–80. Melbourne: Trans Pacific Press, 2005.

Nakayama, Shigeru, Kunio Gotō, and Hitoshi Yoshioka, eds. *A Social History of Science and Technology in Contemporary Japan*. 4 vols. Melbourne: Trans Pacific Press, 2001–2006.

Nappi, Carla. *The Monkey and the Inkpot: Natural History and Its Transformations in Early Modern China*. Cambridge, MA: Harvard University Press, 2009.

Nature Index. "East & Southeast Asia." *Nature* 515 (2014): S73–S75.

Ndiaye, Pap A. *Nylon and Bombs: DuPont and the March of Modern America*. Translated by Elborg Forster. Baltimore: Johns Hopkins University Press, 2007.

Neswald, Elizabeth, and David F. Smith. Introduction to *Setting Nutritional Standards: Theory, Policies, Practices*, edited by Elizabeth Neswald, David F. Smith, and Ulrike Thoms, 1–28. Rochester, NY: University of Rochester Press, 2017.

Neushul, Peter. "Science, Government, and the Mass Production of Penicillin." *Journal of the History of Medicine and Allied Sciences* 48 (1993): 371–95.

Nishiyama, Takashi. *Engineering War and Peace in Modern Japan, 1868–1964*. Baltimore: Johns Hopkins University Press, 2014.

Nolt, James H. *International Political Economy: The Business of War and Peace*. Abingdon, Oxon, UK: Routledge, 2015.

Nyhart, Lynn K. *Biology Takes Form: Animal Morphology and the German Universities, 1800–1900*. Chicago: University of Chicago Press, 1995.

Oberländer, Christian. "The Rise of Western 'Scientific Medicine' in Japan: Bacteriology and Beriberi." In *Building a Modern Japan: Science, Technology, and Medicine in the Meiji Era and Beyond*, edited by Morris Low, 13–36. New York: Palgrave Macmillan, 2005.

O'Bryan, Scott. *The Growth Idea: Purpose and Prosperity in Postwar Japan.* Honolulu: University of Hawai'i Press, 2009.

Odagiri, Hiroyuki, and Akira Goto. *Technology and Industrial Development in Japan: Building Capabilities by Learning, Innovation, and Public Policy.* Oxford: Clarendon Press, 1996.

Okazaki, Tetsuji, and Masahiro Okuno-Fujiwara, eds. *The Japanese Economic System and Its Historical Origins.* Translated by Susan Herbert. Oxford: Oxford University Press, 1999.

Olby, Robert. *The Path to the Double Helix: The Discovery of DNA.* New York: Dover, 2012 [1974].

Oldenziel, Ruth. *Making Technology Masculine: Men, Women and Modern Machines in America, 1870-1945.* Amsterdam: Amsterdam University Press, 1999.

O'Malley, Maureen A. *Philosophy of Microbiology.* Cambridge: Cambridge University Press, 2014.

O'Malley, Maureen A., and John Dupré. "Size Doesn't Matter: Towards a More Inclusive Philosophy of Biology." *Biology and Philosophy* 22 (2007): 155-91.

Omura, Satoshi. "Philosophy of New Drug Discovery." *Microbiological Reviews* 50 (1986): 259-79.

Onaga, Lisa. "Silkworm, Science and Nation: A Sericultural History of Genetics in Modern Japan." PhD diss., Cornell University, 2012.

———. "Toyama Kametaro and Vernon Kellogg: Silkworm Inheritance Experiments in Japan, Siam, and the United States, 1900-1912." *Journal of the History of Biology* 43 (2010): 215-64.

———. "Tracing the Totsuzen in Tanaka's Silkworms: An Exploration of the Establishment of *Bombyx mori* Mutant Stocks." In *Max Planck Institute for the History of Science Preprint* 393. *Making Mutations: Objects, Practices, Contexts,* edited by Luis Campos and Alexander von Schwerin, 109-17. Berlin: MPIWG, 2010.

Ong, Aihwa, and Nancy N. Chen, eds. *Asian Biotech: Ethics and Communities of Fate.* Durham, NC: Duke University Press, 2010.

Otis, Laura. *Membranes: Metaphors of Invasion in Nineteenth-Century Literature, Science, and Politics.* Baltimore: Johns Hopkins University Press, 1999.

———. *Müller's Lab: The Story of Jakob Henle, Theodor Schwann, Emil du Bois-Reymond, Hermann von Helmholtz, Rudolf Virchow, Robert Remak, Ernst Haeckel, and Their Brilliant, Tormented Advisor.* Oxford: Oxford University Press, 2007.

Otsubo, Sumiko. "Between Two Worlds: Yamanouchi Shigeo and Eugenics in Early Twentieth-Century Japan." *Annals of Science* 62 (2005): 205–31.

Otsuka, Yasuo. "Chinese Traditional Medicine in Japan." In *Asian Medical Systems: A Comparative Study*, edited by Charles M. Leslie, 322–40. Berkeley: University of California Press, 1976.

Otter, Chris. "The British Nutrition Transition and Its Histories." *History Compass* 10, no. 11 (2012): 812–25.

———. "Industrializing Diet, Industrializing Ourselves: Technology, Food, and the Body, Since 1750." In *The Routledge History of Food*, edited by Carol Helstosky, 220–46. New York: Routledge, 2015.

Park, Hyunhee. "The Rise of Soju: The Transfer of Distillation Technology from 'China' to Korea during the Mongol Period (1206–1368)." *Crossroads* 14 (2016): 173–204.

Paxson, Heather. *The Life of Cheese: Crafting Food and Value in America.* Berkeley: University of California Press, 2012.

Peattie, Mark R. Introduction to *The Japanese Colonial Empire, 1895–1945,* edited by Ramon H. Myers and Mark R. Peattie, 3–52. Princeton, NJ: Princeton University Press, 1984.

———. "*Nanshin*: The 'Southward Advance,' 1931–1941, as a Prelude to the Japanese Occupation of Southeast Asia." In *The Japanese Wartime Empire, 1931–1945,* edited by Peter Duus, Ramon H. Myers, and Mark R. Peattie, 189–242. Princeton, NJ: Princeton University Press, 1996.

———. "The Nan'yō: Japan in the South Pacific, 1885–1945." In *The Japanese Colonial Empire, 1895–1945,* edited by Ramon H. Myers and Mark R. Peattie, 172–210. Princeton, NJ: Princeton University Press, 1984.

Peters, Erica J. "Taste, Taxes, and Technologies: Industrializing Rice Alcohol in Northern Vietnam, 1902–1913." *French Historical Studies* 27 (2004): 569–600.

Peterson, Mark. *A Brief History of Korea.* New York: Facts On File, 2010.

Podolsky, Scott H. *The Antibiotic Era: Reform, Resistance, and the Pursuit of a Rational Therapeutics.* Baltimore: Johns Hopkins University Press, 2015.

Pollack, Andrew. "Japan's Biotechnology Effort." *New York Times,* August 28, 1984, D-1.

Pratt, Edward E. *Japan's Protoindustrial Elite: The Economic Foundations of the Gōnō.* Cambridge, MA: Harvard University Press, 1999.

Rader, Karen A. *Making Mice: Standardizing Animals for American Biomedical Research, 1900–1955.* Princeton, NJ: Princeton University Press, 2004.

Raj, Kapil. "Beyond Postcolonialism . . . and Postpositivism: Circulation and the Global History of Science." *Isis* 104 (2013): 337–47.

Ramakrishnan, Venki. "Potential and Risks of Recent Developments in Bio-technology: A Speech by Venki Ramakrishnan, President of the Royal Society, AAAS Annual Meeting, Saturday 18 February 2017." https://royalsociety.org/~/media/news/2017/venki-ramakrishnan-aaas-speech-gene-tech-18-02-17.pdf.

Rasmussen, Nicolas. "The Mid-Century Biophysics Bubble: Hiroshima and the Biological Revolution in America, Revisited." *History of Science* 35 (1997): 245–93.

———. "Of 'Small Men,' Big Science and Bigger Business: The Second World War and Biomedical Research in the United States." *Minerva* 40 (2002): 115–46.

Rath, Eric C. "Afterword: Foods of Japan, Not Japanese Food." In *Devouring Japan: Global Perspectives on Japanese Culinary Identity*, edited by Nancy K. Stalker, 312–27. Oxford: Oxford University Press, 2018.

Rath, Eric C., and Stephanie Assmann, eds. *Japanese Foodways, Past and Present*. Champaign: University of Illinois Press, 2010.

Reinhardt, Carsten. *Shifting and Rearranging: Physical Methods and the Transformation of Modern Chemistry*. Sagamore Beach, MA: Science History Publications, 2006.

Revells, Tristan. "Filched Fungi? Bioprospecting and the Circulation of 'Chinese Yeast,' c. 1892–1933." In *Alcohol Flows across Cultures: Drinking Cultures in Transnational and Comparative Perspective*, edited by Waltraud Ernst, 159–85. Abingdon, Oxon, UK: Routledge, 2020.

Reynolds, Douglas R. "Training Young China Hands: Tōa Dōbun Shoin and Its Precursors, 1886–1945." In *The Japanese Informal Empire in China, 1895–1937*, eds. Peter Duus, Ramon H. Myers, and Mark R. Peattie, 210–71. Princeton, NJ: Princeton University Press, 1989.

Rheinberger, Hans-Jörg. *Toward a History of Epistemic Things: Synthesizing Proteins in the Test Tube*. Stanford, CA: Stanford University Press, 1997.

Riskin, Jessica. *The Restless Clock: A History of the Centuries-Long Argument over What Makes Living Things Tick*. Chicago: University of Chicago Press, 2016.

Roosth, Sophia. *Synthetic: How Life Got Made*. Chicago: University of Chicago Press, 2017.

Rose, Nikolas. "The Human Sciences in a Biological Age." *Theory, Culture & Society* 30 (2013): 3–34.

Russell, Edmund. *Evolutionary History: Uniting History and Biology to Understand Life on Earth*. Cambridge: Cambridge University Press, 2011.

———. *War and Nature: Fighting Humans and Insects with Chemicals from World War I to Silent Spring*. Cambridge: Cambridge University Press, 2001.

Samson, Robert A., Huub A. van der Aa, and G. Sybren de Hoog. "Centraalbureau voor Schimmelcultures: Hundred Years Microbial Resource Centre." *Studies in Mycology* 50 (2004): 1–8.

Sand, Jordan. *House and Home in Modern Japan: Architecture, Domestic Space, and Bourgeois Culture, 1880–1930*. Cambridge, MA: Harvard University Press, 2005.

———. "A Short History of MSG: Good Science, Bad Science, and Taste Cultures." *Gastronomica* 5 (2005): 38–49.

Sangodeyi, Funke Iyabo. "The Making of the Microbial Body, 1900s–2012." PhD diss., Harvard University, 2014.

Sankaran, Neeraja. "When Viruses Were Not in Style: Parallels in the Histories of Chicken Sarcoma Viruses and Bacteriophages." *Studies in History and Philosophy of Biological and Biomedical Sciences* 48 (2014): 189–99.

Santesmases, María Jesús. "Gender in Research and Industry: Women in Antibiotic Factories in 1950s Spain." In *Gendered Drugs and Medicine: Historical and Socio-Cultural Perspectives*, edited by Teresa Ortiz-Gómez and María Jesús Santesmases, 61–84. Farnham, Surrey, UK: Ashgate, 2014.

———. "Screening Antibiotics: Industrial Research by CEPA and Merck in the 1950s." *Dynamis* 31 (2011): 407–28.

Sapp, Jan. "The Bacterium's Place in Nature." In *Microbial Phylogeny and Evolution: Concepts and Controversies*, edited by Jan Sapp. Oxford: Oxford University Press, 2005, 1–52.

———. "The Structure of Microbial Evolutionary Theory." *Studies in History and Philosophy of Biological and Biomedical Sciences* 38 (2007): 780–95.

———. *Where the Truth Lies: Franz Moewus and the Origins of Molecular Biology*. Cambridge: Cambridge University Press, 1990.

Sasges, Gerard. *Imperial Intoxication: Alcohol and the Making of Colonial Indochina*. Honolulu: University of Hawai'i Press, 2017.

Sato, Jin. "Formation of the Resource Concept in Japan: Pre-War and Post-War Efforts in Knowledge Integration." *Sustainability Science* 2 (2007): 151–58.

Sawada, Janine Tasca. *Practical Pursuits: Religion, Politics, and Personal Cultivation in Nineteenth-Century Japan*. Honolulu: University of Hawai'i Press, 2004.

Schäfer, Dagmar. *The Crafting of the 10,000 Things: Knowledge and Tech-*

nology in Seventeenth-Century China. Chicago: University of Chicago Press, 2011.

Scheffler, Robin Wolfe. *A Contagious Cause: The American Hunt for Cancer Viruses and the Rise of Molecular Medicine*. Chicago: University of Chicago Press, 2019.

Schmid, Andre. "Colonialism and the 'Korea Problem' in the Historiography of Modern Japan: A Review Article." *Journal of Asian Studies* 59 (2000): 951–76.

Schneider, Justin Adam. "The Business of Empire: The Taiwan Development Corporation and Japanese Imperialism in Taiwan, 1936–1946." PhD diss., Harvard University, 1998.

Schumpeter, E. B., ed. *The Industrialization of Japan and Manchukuo, 1930–1940: Population, Raw Materials and Industry*. New York: Macmillan, 1940.

Schwartzberg, Louie, dir. *Fantastic Fungi*. Moving Art, 2019.

Schwerin, Alexander von, Heiko Stoff, and Bettina Wahrig, eds. *Biologics: A History of Agents Made from Living Organisms in the Twentieth Century*. London: Pickering & Chatto, 2013.

Secord, James. "Knowledge in Transit." *Isis* 95 (2004): 654–72.

Sellers, Christopher. "From Poison to Carcinogen: Towards a Global History of Concerns about Benzene." *Global Environment* 7 (2014): 38–71.

Setoguchi, Akihisa. "Control of Insect Vectors in the Japanese Empire: Transformation of the Colonial/Metropolitan Environment, 1920–1945." *East Asian Science, Technology and Society* 1 (2007): 167–81.

Shurtleff, William, and Akiko Aoyagi. *History of Koji (300 BCE to 2012): Extensively Annotated Bibliography and Sourcebook*. Lafayette, CA: Soyinfo Center, 2012. https://www.soyinfocenter.com/pdf/154/Koji.pdf.

———. "History of Soy Sauce, Shoyu, and Tamari." Chapter from the unpublished manuscript "History of Soybeans and Soyfoods, 1100 B.C. to the 1980s." Lafayette, CA: Soyfoods Center, 2004. https://www.soyinfocenter.com/HSS/soy_sauce1.php.

———. *Jokichi Takamine (1854–1922) and Caroline Hitch Takamine (1866–1954): Biography and Bibliography*. Lafayette, CA: Soyinfo Center, 2012. https://www.soyinfocenter.com/pdf/155/Taka.pdf.

Sigaut, François. "Technology." In *Companion Encyclopedia of Anthropology*, edited by Tim Ingold, 420–59. London: Routledge, 2002 [1994].

Siniawer, Eiko Maruko. *Ruffians, Yakuza, Nationalists: The Violent Politics of Modern Japan, 1860–1960*. Ithaca, NY: Cornell University Press, 2008.

Sivasundaram, Sujit, ed. "Focus: Global Histories of Science." *Isis* 101 (2010): 95–158.

Sivin, Nathan. "Why the Scientific Revolution Did Not Take Place in China—Or Didn't It?" *Chinese Science* 5 (1982): 45–66.

Slingsby, B. T., interview by Erin Schneider and Brian Hutchinson. "Beyond Borders: New Growth and Direction for Japan's Pharmaceutical Industry." National Bureau of Asian Research, January 25, 2012. https://www.nbr.org/publication/beyond-borders-new-growth-and-direction-for-japans-pharmaceutical-industry/.

Smith, Hilary A. *Forgotten Disease: Illnesses Transformed in Chinese Medicine*. Stanford, CA: Stanford University Press, 2017.

Smith, Jenny Leigh. *Works in Progress: Plans and Realities on Soviet Farms, 1930–1963*. New Haven, CT: Yale University Press, 2014.

Smith, Norman. *Intoxicating Manchuria: Alcohol, Opium, and Culture in China's Northeast*. Vancouver: UBC Press, 2012.

Smith, Pamela H. *The Body of the Artisan: Art and Experience in the Scientific Revolution*. Chicago: University of Chicago Press, 2006.

Smith, Thomas C. *Native Sources of Japanese Industrialization, 1750–1920*. Berkeley: University of California Press, 1988.

Smocovitis, Vassiliki Betty. *Unifying Biology: The Evolutionary Synthesis and Evolutionary Biology*. Princeton, NJ: Princeton University Press, 1996.

Solt, George. *The Untold History of Ramen: How Political Crisis in Japan Spawned a Global Food Craze*. Berkeley: University of California Press, 2014.

Somsen, Geert J. "A History of Universalism: Conceptions of the Internationality of Science from the Enlightenment to the Cold War." *Minerva* 46 (2008): 361–79.

Spackman, Christy C. W. "Formulating Citizenship: The Microbiopolitics of the Malfunctioning Functional Beverage." *BioSocieties* 13 (2018): 41–63.

Spary, E. C. *Feeding France: New Sciences of Food, 1760–1815*. Cambridge: Cambridge University Press, 2014.

Spath, Susan Barbara. "C. B. van Niel and the Culture of Microbiology, 1920–1965." PhD diss., University of California, Berkeley, 1999.

Squier, Susan Merrill. *Epigenetic Landscapes: Drawings as Metaphor*. Durham, NC: Duke University Press, 2017.

Stanier, Roger Y., Michael Doudoroff, and Edward A. Adelberg. *The Microbial World*. Englewood Cliffs, NJ: Prentice-Hall, 1957.

Stolz, Robert. *Bad Water: Nature, Pollution, and Politics in Japan, 1870–1950*. Durham, NC: Duke University Press, 2014.

Stranges, Anthony N. "Synthetic Fuel Production in Prewar and World

War II Japan: A Case Study in Technological Failure." *Annals of Science* 50 (1993): 229–65.

Strasser, Bruno J. *Collecting Experiments: Making Big Data Biology*. Chicago: University of Chicago Press, 2019.

———. *La fabrique d'une nouvelle science: La biologie moléculaire à l'âge atomique (1945–1964)*. Florence: Leo S. Olschki Editore, 2006.

Stross, Randall E. *The Stubborn Earth: American Agriculturalists on Chinese Soil, 1898–1937*. Berkeley: University of California Press, 1986.

Sugihara, Kaoru. "The Development of an Informational Infrastructure in Meiji Japan." In *Information Acumen: The Understanding and Use of Knowledge in Modern Business*, edited by Lisa Bud-Frierman, 75–97. London: Routledge, 1994.

Summers, William C. *The Great Manchurian Plague of 1910–1911: The Geopolitics of an Epidemic Disease*. New Haven, CT: Yale University Press, 2012.

Sumner, James. *Brewing Science, Technology and Print, 1700–1880*. London: Pickering and Chatto, 2013.

Swann, John Patrick. "The Search for Synthetic Penicillin during World War II." *British Journal for the History of Science* 16 (1983): 154–90.

Takahashi, Teizo. "A Preliminary Note on the Varieties of *Aspergillus oryzae*." *Journal of the College of Agriculture, Imperial University of Tokyo* 1 (1909): 137–40.

Takahashi, Teizo, and Kin-ichiro Sakaguchi. *Summaries of Papers*. Tokyo: Committee of Commemorative Meeting of 35 Year's Anniversary of Professor Kin-ichiro Sakaguchi, 1958.

Takahashi, Teizo, and Takeharu Yamamoto. "On the Physiological Differences of the Varieties of *Aspergillus Oryzae* Employed in the Three Main Industries in Japan, Namely Saké-, Shôyu-, and Tamari Manufacture." *Journal of the College of Agriculture, Imperial University of Tokyo* 5 (1913): 151–61.

Tamiya, Hiroshi. "The Koji, an Important Source of Enzymes in Japan." In *Proceedings of the International Symposium on Enzyme Chemistry, Tokyo and Kyoto, 1957*, edited by the Organizing Committee, International Symposium on Enzyme Chemistry, Science Council of Japan, 21–24. Tokyo: Maruzen, 1958.

Tamura, Yoshiya. "The English Essays of Minakata Kumagusu: Centering on his Contributions to 'Nature.'" *Japan Foreign Policy Forum* 16 (2013). https://www.japanpolicyforum.jp/archives/society/pt20131007043828.html.

Tanimoto, Masayuki. "Capital Accumulation and the Local Economy: Brewers and Local Notables." In *The Role of Tradition in Japan's Industrialization: Another Path to Industrialization*, edited by Masayuki Tanimoto, 301-22. Oxford: Oxford University Press, 2006.

———. "From Peasant Economy to Urban Agglomeration: The Transformation of 'Labour-Intensive Industrialization' in Modern Japan." In *Labour-Intensive Industrialization in Global History*, edited by Gareth Austin and Kaoru Sugihara, 144-75. Abingdon, Oxon, UK: Routledge, 2013.

———. "The Role of Tradition in Japan's Industrialization: Another Path to Industrialization." In *The Role of Tradition in Japan's Industrialization: Another Path to Industrialization*, edited by Masayuki Tanimoto, 3-44. Oxford: Oxford University Press, 2006.

Tanimoto, Masayuki, ed. *The Role of Tradition in Japan's Industrialization: Another Path to Industrialization*. Oxford: Oxford University Press, 2006.

Teich, Mikuláš. "Fermentation Theory and Practice: The Beginnings of Pure Yeast Cultivation and English Brewing, 1883-1913." *History of Technology* 8 (1983): 117-33.

Thomas, Julia Adeney. "History and Biology in the Anthropocene: Problems of Scale, Problems of Value." *American Historical Review* 119 (2014): 1587-1607.

———. "Reclaiming Ground: Japan's Great Convergence." *Japanese Studies* 34 (2014): 253-63.

———. *Reconfiguring Modernity: Concepts of Nature in Japanese Political Ideology*. Berkeley: University of California Press, 2001.

Thompson, E. P. "Time, Work-Discipline, and Industrial Capitalism." *Past & Present* 38 (1967): 56-97.

Tierney, Robert Thomas. *Tropics of Savagery: The Culture of Japanese Empire in Comparative Frame*. Berkeley: University of California Press, 2010.

Tilley, Helen. *Africa as a Living Laboratory: Empire, Development, and the Problem of Scientific Knowledge, 1870-1950*. Chicago: University of Chicago Press, 2011.

Tonomura, Hitomi. "Gender Relations in the Age of Violence." In *Japan Emerging: Premodern History to 1850*, edited by Karl F. Friday, 267-77. Boulder, CO: Westview Press, 2012.

Tracy, Sarah E. "Delicious Molecules: Big Food Science, the Chemosenses, and Umami." *The Senses and Society* 13 (2018): 89-107.

———. "Tasty Waste: Industrial Fermentation and the Creative Destruction of MSG." *Food, Culture & Society* 22 (2019): 548-65.

Trambaiolo, Daniel. "Vaccination and the Politics of Medical Knowledge in Nineteenth-Century Japan." *Bulletin of the History of Medicine* 88 (2014): 431–56.

Treitel, Corinna. *Eating Nature in Modern Germany: Food, Agriculture and Environment, c. 1870 to 2000.* Cambridge: Cambridge University Press, 2017.

Trouillot, Michel-Rolph. "Anthropology and the Savage Slot: The Poetics and Politics of Otherness." In *Recapturing Anthropology: Working in the Present*, edited by Richard G. Fox, 17–44. Santa Fe, NM: School of American Research Press, 1991.

Tsing, Anna Lowenhaupt. *The Mushroom at the End of the World: On the Possibility of Life in Capitalist Ruins.* Princeton, NJ: Princeton University Press, 2015.

Tsukahara, Togo, and Jianjun Mei. "Putting Joseph Needham in the East Asian Context: Commentaries on Papers about the Reception of Needham's Works in Korea and Taiwan." *East Asian Science, Technology and Society* 14 (2020): 403–10.

Tsutsui, William M. "Landscapes in the Dark Valley: Toward an Environmental History of Wartime Japan." *Environmental History* 8 (2003): 294–311.

———. *Manufacturing Ideology: Scientific Management in Twentieth-Century Japan.* Princeton, NJ: Princeton University Press, 1998.

Udaka, Shigezo. "The Discovery of *Corynebacterium glutamicum* and Birth of Amino Acid Fermentation Industry in Japan." In *Corynebacteria: Genomics and Molecular Biology*, edited by Andreas Burkovski, 1–6. Wymondham, UK: Caister Academic Press, 2008.

———. "Screening Method for Microorganisms Accumulating Metabolites and Its Use in the Isolation of *Micrococcus glutamicus.*" *Journal of Bacteriology* 79, no. 5 (1960): 754–55.

Udaka, Shigezo, and Shukuo Kinoshita. "Studies on L-Ornithine Fermentation I: The Biosynthetic Pathway of L-Ornithine in *Micrococcus Glutamicus.*" *Journal of General and Applied Microbiology* 4 (1958): 272–82.

Udaka, Shigezo, and Shukuo Kinoshita. "Studies on L-Ornithine Fermentation II: The Change of Fermentation Product by a Feedback Type Mechanism." *Journal of General and Applied Microbiology* 4 (1958): 283–88.

Ui, Jun, ed. *Industrial Pollution in Japan.* Tokyo: United Nations University Press, 1992.

Umbarger, H. Edwin. "Feedback Control by Endproduct Inhibition." *Cold Spring Harbor Symposia on Quantitative Biology* 26 (1962): 301–12.

Umemura, Maki. *The Japanese Pharmaceutical Industry: Its Evolution and Current Challenges*. Abingdon, Oxon, UK: Routledge, 2011.

Upham, Frank K. "Unplaced Persons and Movements for Place." In *Postwar Japan as History*, edited by Andrew Gordon, 325–46. Berkeley: University of California Press, 1993.

Van Rooij, Arjan. *The Company That Changed Itself: R&D and the Transformations of DSM*. Amsterdam: Amsterdam University Press, 2007.

Velmet, Aro. *Pasteur's Empire: Bacteriology and Politics in France, Its Colonies, and the World*. Oxford: Oxford University Press, 2020.

Vernon, Keith. "Pus, Sewage, Beer and Milk: Microbiology in Britain, 1870–1940." *History of Science* 28 (1990): 289–325.

Vogel, Ezra F. *Japan as Number One: Lessons for America*. Cambridge, MA: Harvard University Press, 1979.

Vom Saal, Frederick, and Aly Cohen. "How Toxic Chemicals Contribute to COVID-19 Deaths." *Environmental Health News*, April 17, 2020.

Von Verschuer, Charlotte. *Rice, Agriculture, and the Food Supply in Premodern Japan: Gokoku bunka*. Translated by Wendy Cobcroft. Abingdon, Oxon, UK: Routledge, 2016.

Walker, Brett L. *The Lost Wolves of Japan*. Seattle: University of Washington Press, 2005.

———. *Toxic Archipelago: A History of Industrial Disease in Japan*. Seattle: University of Washington Press, 2010.

Warner, Deborah Jean. *Sweet Stuff: An American History of Sweeteners from Sugar to Sucralose*. Washington: Smithsonian Institution, 2011.

Warren, Wilson J. *Meat Makes People Powerful: A Global History of the Modern Era*. Iowa City: University of Iowa Press, 2018.

Watson, James D. *The Double Helix: A Personal Account of the Discovery of the Structure of DNA*. New York: Simon and Schuster, 1996 [1968].

Westney, D. Eleanor. *Imitation and Innovation: The Transfer of Organizational Patterns to Meiji Japan*. Cambridge, MA: Harvard University Press, 1987.

Westwick, Peter J. *Into the Black: JPL and the American Space Program, 1976–2004*. New Haven, CT: Yale University Press, 2006.

Wigen, Kären. *The Making of a Japanese Periphery, 1750–1920*. Berkeley: University of California Press, 1995.

Wittner, David G. *Technology and the Culture of Progress in Meiji Japan*. Abingdon, Oxon, UK: Routledge, 2008.

Wittner, David G., and Philip C. Brown, eds. *Science, Technology, and Medicine in the Modern Japanese Empire*. New York: Routledge, 2016.

Wolfe, Audra J. *Competing with the Soviets: Science, Technology, and the State in Cold War America*. Baltimore: Johns Hopkins University Press, 2013.

Worboys, Michael. "Manson, Ross, and Colonial Medical Policy: Tropical Medicine in London and Liverpool, 1899–1914." In *Disease, Medicine, and Empire: Perspectives on Western Medicine and the Experience of European Expansion*, edited by Roy MacLeod and Milton Lewis, 21–38. London: Routledge, 1988.

Wylie, Alison. "Feminist Philosophy of Science: Standpoint Matters." Presidential Address delivered to the Pacific Division APA. In *Proceedings and Addresses of the American Philosophical Association* 86, no. 2 (2012): 47–76.

Xue, Katherine S. "What Viral Evolution Can Teach Us about the Coronavirus Pandemic." *New Yorker*, April 19, 2020.

Yamashima, Tetsumori. "Jokichi Takamine (1854–1922), the Samurai Chemist, and His Work on Adrenalin." *Journal of Medical Biography* 11 (2003): 95–102.

Yang, Daqing. "Resurrecting the Empire? Japanese Technicians in Postwar China, 1945–9." In *The Japanese Empire in East Asia and Its Postwar Legacy*, edited by Harald Fuess, 185–205. Munich: Iudicium Verlag, 1998.

———. *Technology of Empire: Telecommunications and Japanese Expansion in Asia, 1883–1945*. Cambridge, MA: Harvard University Press, 2011.

Yang, Timothy M. *A Medicated Empire: The Pharmaceutical Industry and Modern Japan*. Ithaca, NY: Cornell University Press, 2021.

———. "Selling an Imperial Dream: Japanese Pharmaceuticals, National Power, and the Science of Quinine Self-Sufficiency." *East Asian Science, Technology and Society* 6 (2012): 101–25.

Yergin, Daniel. *The Prize: The Epic Quest for Oil, Money & Power*. New York: Simon & Schuster, 2008 [1991].

Yi, Doogab. *The Recombinant University: Genetic Engineering and the Emergence of Stanford Biotechnology*. Chicago: University of Chicago Press, 2015.

Yonekura, Seiichiro. "The Functions of Industrial Associations." In *The Japanese Economic System and Its Historical Origins*, translated by Susan Herbert, and edited by Tetsuji Okazaki and Masahiro Okuno-Fujiwara, 180–207. Oxford: Oxford University Press, 1999, 180–207.

Yong, Ed. *I Contain Multitudes: The Microbes within Us and a Grander View of Life*. New York: Ecco, 2016.

Yongue, Julia. "The Introduction of American Mass Production Technology to Japan during the Occupation: The Case of Penicillin." In *Organizing*

Global Technology Flows: Institutions, Actors, and Processes, edited by Pierre-Yves Donzé and Shigehiro Nishimura, 213–29. New York: Routledge, 2014.

Yoshioka, Hitoshi. Introduction to *A Social History of Science and Technology in Contemporary Japan: Vol. 4, Transformation Period 1970–1979*, edited by Shigeru Nakayama and Hitoshi Yoshioka. Melbourne: Trans Pacific Press, 2006, 1–78.

Index

The letter *f* following a page number denotes a figure.

indigenous and traditional fermentation: and alcohol production, 91, 131–32, 138, 142, 145–46, 244n121; and the improvement of sake and shōyu, 15–31, 44, 64, 78, 81, 86; influence on antibiotics research of, 162, 165, 169, 175; and the Japanese approach to microbes, 1–13, 47–49, 90, 175, 205, 209–10; and Japanese knowledge, 102, 109–10, 113–23, 126, 222nn16–17, 224n123, 269n11

industrial alcohol. *See* alcohol, industrial

industrialization: and alcohol production, 123–46, 213; and antibiotics, 13, 153, 159–62; and Japanese colonies, 9, 70; and the Meiji period, 5–6, 9, 16, 27, 209, 269n11; microbes and, 12, 47, 214; and nutrition science and shortages, 12, 51–55, 61–63, 70, 233n14; post–World War I, 3, 9, 52; post–World War II, 10, 174–76, 187, 196, 204, 258n5; and traditional sake and soy sauce brewing, 17–24, 30, 222n16. *See also* economy, Japanese; modernity

inosinic acid (IMP), 78, 187–88, 197, 200–201, 211. *See also* nucleic acid fermentation

Inoue Junnosuke, 233n6

instant ramen, 2, 173, 188–89, 196, 263n72

Institute for Fermentation, Osaka (IFO; Hakkō kenkyūjo), 177–78, 190, 193, 199–200, 208, 230n100, 260n20, 261n30, 265n105

Institute for Nutrition Research (Eiyō kenkyūjo), 71, 85

Institute of Applied Microbiology (IAM; Ōyō biseibutsu kenkyūjo), 157, 169, 188–90, 193–96

Institute of Physical and Chemical Research. *See* Riken

Instrumental Community (Mody), 258n7

Ishidō Toyota, 129–31

Ishige Naomichi, 103

Ishiwara Kanji, 232n4

Isobe Hajime, 235n35

"Isolation, Contamination" (Gradmann), 232n125

Iwadare Tōru, 147

Iwai Ki'ichirō, 131, 246n25

Japan. *See* economy, Japanese; Meiji period; microbes; occupation of Japan

Japan Antibiotics Research Association (JARA; Nihon kōseibusshitsu gakujutsu kyōgikai). *See* Japan Penicillin Research Association (JPRA; Nihon penishirin gakujutsu kyōgikai)

Japan Penicillin Manufacturing Association (JPMA; Shadan hōjin Nihon penishirin kyōkai), 147, 151–53, 156–63, 256n107

Japan Penicillin Research Association (JPRA; Nihon penishirin gakujutsu kyōgikai), 148–53, 156–70, 191, 251n14, 257n117

Jimmu (emperor), 114

Johnson, Ann, 201–2, 258n7

Johnson, Chalmers, 149

Jōkai (journal), 29–30, 33

Journal of Penicillin, 167, 251n14

Jōzōkai (journal), 32–34

JPMA, 147, 151–53, 156–63, 256n107

JPRA, 148–53, 156–70, 191, 251n14, 257n117

Kaji Masanori, 101–2

Kamiya Denbei, 127–30, 133

Kamiya Shun'ichi, 137–38

Kamiya Shuzō, 127–30, 133

kamutachi, 105, 114–18

Katō Benzaburō, 128, 139, 179, 182, 250n80

Katō Hyakuichi, 116

Katsumata Minoru, 152

katsuobushi, 77–78, 187–88. *See also* IMP

Kawamata Shōyu, 34, 82–83

Kay, Lily, 186

kekabi mold, 35, 40–42, 103–4, 109–10, 136–37

Kellner, Oskar, 24–25, 227n36

Kikkawa Hideo, 194

Kikkōman, 34, 83, 229n75

Kimi no na wa (film), 244n123

Kinoshita Shukuo, 179, 182–89, 198, 202, 262n40, 262n56

Kita Gen'itsu, 71

Kitahara Kakuo, 105

Kitasato Institute, 168, 254n52, 256n111

Kitasato Shibasaburō, 254n52

Kodama Shintarō, 187

Kohata Kengorō, 248n60

kōji: and amino and nucleic acid fermentation, 187–88; and flavor technology, 188, 197–98; and the improvement of traditional industry, 12, 15–26, 30–31, 34–40, 44–48, 177, 210, 229n78, 238n116; and industrial alcohol production, 125–27, 131–32, 136–45; and Japanese fermentation knowledge, 1–3, 95, 103–10, 113–16, 119, 122, 194, 197, 229n84; and Japanese knowledge and identity, 3, 90–92, 183, 239n7; nutrition science and sake and, 58, 62, 72, 79, 238n116. *See also* molds; *tanekōji*

Kojiki, 114

Komagata Kazuo, 93–95, 195–96, 255n91

Konno Kenji, 35